Prof. Bose

Thank you for
everything I
learned from you.

Thank you for
showing the
best example of
academic and
professional success
that you are for
me!

Marcelo Godoy

Modeling and Analysis with Induction Generators

THIRD EDITION

POWER ELECTRONICS AND APPLICATIONS SERIES

Muhammad H. Rashid, Series Editor
University of West Florida

PUBLISHED TITLES

Advanced DC/DC Converters
Fang Lin Luo and Hong Ye

Alternative Energy Systems: Design and Analysis with Induction
Generators, Second Edition
M. Godoy Simões and Felix A. Farret

Complex Behavior of Switching Power Converters
Chi Kong Tse

DSP-Based Electromechanical Motion Control
Hamid A. Toliyat and Steven Campbell

Electric Energy: An Introduction, Third Edition
Mohamed A. El-Sharkawi

Electrical Machine Analysis Using Finite Elements
Nicola Bianchi

Modern Electric, Hybrid Electric, and Fuel Cell Vehicles:
Fundamentals, Theory, and Design
Mehrdad Eshani, Yimin Gao, Sebastien E. Gay, and Ali Emadi

Modeling and Analysis with Induction Generators, Third Edition
M. Godoy Simões and Felix A. Farret

Uninterruptible Power Supplies and Active Filters
Ali Emadi, Abdolhosein Nasiri, and Stoyan B. Bekiarov

Modeling and Analysis with Induction Generators

THIRD EDITION

M. Godoy Simões
Colorado School of Mines, Golden, USA

Felix A. Farret
Federal University of Santa Maria, Brazil

CRC Press
Taylor & Francis Group
Boca Raton   London   New York

CRC Press is an imprint of the
Taylor & Francis Group, an informa business

MATLAB® and Simulink® are trademarks of The MathWorks, Inc. and are used with permission. The MathWorks does not warrant the accuracy of the text or exercises in this book. This book's use or discussion of MATLAB® and Simulink® software or related products does not constitute endorsement or sponsorship by The MathWorks of a particular pedagogical approach or particular use of the MATLAB® and Simulink® software.

CRC Press
Taylor & Francis Group
6000 Broken Sound Parkway NW, Suite 300
Boca Raton, FL 33487-2742

© 2015 by Taylor & Francis Group, LLC
CRC Press is an imprint of Taylor & Francis Group, an Informa business

No claim to original U.S. Government works

Printed on acid-free paper
Version Date: 20141007

International Standard Book Number-13: 978-1-4822-4467-0 (Hardback)

Contents

Foreword

The book *Modeling and Analysis with Induction Generators* is an important work about a technology—induction generators—that has the potential to advance applications of renewable energy systems. Several countries—Denmark, Australia, Germany, the United Kingdom, and others—have begun the transition from fossil fuel and nuclear energy to hydro and wind power plants, as a response to increasing concerns about fuel supplies, global security, and climate changes.

Induction generators were used since the beginning of the twentieth century until they were abandoned in the 1960s. With the present high energy costs, rational use and conservation implemented by many processes of heat recovery became important goals. The general consciousness of finite and limited sources of energy on earth and international disputes over the environment and global safety of quality of life have created an opportunity of new, more efficient, less polluting power plants with advanced technologies of control, robustness, and modularity. Induction generators, with their lower maintenance demands and simplified controls, appear to be a good solution for such applications. With their simplicity, robustness, and small size per kilowatt generated, they are favored for small hydro and wind power plants. With the widespread use of power electronics, computers, and electronic microcontrollers, these generators have become easier to administer and are guaranteed for use in a vast majority of applications where they are more efficient up to 500 kVA.

In their 14 chapters, the authors present topics like numerical analysis of stand-alone and multiple induction generators, process of self-excitation, requirements for optimized laboratory experimentation, application of modern vector control, optimization of power transference, use of doubly fed induction generators, computer-based simulations, and social and economic impacts of induction generators.

I am very pleased that the intention of the authors in this new version of the book has been to move from a research-oriented approach toward a more educational approach. I am sure this work will contribute to our better understanding of these electrical machines that in general are not part of academic curricula. I strongly recommend this textbook to a wide audience.

Dr. Francesco Profumo*
Politecnico di Torino
Torino, Italy

---

* Dr. Francesco Profumo was Italy's Minister of Education from 2011 to 2013. He was president of the Italy National Research Council (CNR) and has previously served as chancellor of the Politecnico di Torino from 2005 to 2011.

Preface

During the fall of 2003, the authors decided to bring together their interests and start working on what would be the first edition of this book, published in 2004. The reasoning was that although so many books have been written on induction machines, drives, and motors in general, none existed at that time that would cover specifically how to understand, model, analyze, and simulate induction generators, particularly in the applications of renewable or alternative energy systems. In the second edition, we shortened a few sections and added new ones, trying to make clear some concepts. We have also provided better coverage of doubly fed induction generators and applications of induction generators. Over the years, we noticed how important induction generators became both for stand-alone and grid-connected applications. The number of installations of small- and medium-sized wind energy power plants based on this very easy, cost-effective, and reliable generating machine is remarkable, to the point of making us even more enthusiastic about this subject.

Now, more than a decade after we first started this project, we are very proud to present this third edition, with a new title that focuses on our objectives, that is, to present the fundamentals and advances in modeling and analysis of induction generators. Topics like understanding the process of self-excitation, numerical analysis of stand-alone and multiple induction generators, requirements for optimized laboratory experimentation, application of modern vector control, optimization of power transference, use of doubly fed induction generators, computer-based simulations, and social and economic impacts are presented in order to take the academic realm of the subject to the desks of practicing engineers and undergraduate and graduate students. Our intention in this new edition of the book has been to move from a research-oriented approach toward a more educational approach. Therefore, we have provided several solved problems and further suggested problems at the end of each chapter. We would really love to receive feedback regarding how instructors are using and adapting this textbook in their courses.

Part of our intent is to give ideas and suggest directions for further development in this field; the reader is also referred to other sources for details regarding development. As teachers and researchers, we realize the importance of feedback and appreciate any comments and suggestions for improvements that might add value to the material we have presented.

This book is organized into 14 chapters. Chapter 1 presents definitions; characteristics of primary sources and industrial, commercial, residential, and remote sites; and rural loads, highlighting the selection of suitable electric generators. Chapter 2 presents steady-state modeling of induction generators, with classical steady-state representation, parameter measurements, and peculiarities of the interconnection to the distribution grid. Chapter 3 expounds on transient modeling of induction generators, with a novel numerical representation of state-space modeling that permits generalization of association of generators in parallel, an important subject for wind farms. Chapter 4 introduces in detail the performance of self-excited induction generators, voltage regulation, and the mathematical description of self-excitation. Chapter 5 presents some general characteristics of induction generators in respect to torque vs. speed, power vs. output current, and relationship of air-gap voltage and magnetizing current. Chapter 6 discusses the construction features of induction machines, particularly generator sizing, design, and manufacturing aspects. Chapter 7 presents power electronic devices, requirements for injection of power to the grid, and interfacing with renewable energy systems—ac–dc, dc–dc, dc–ac, and ac–ac conversion topologies—as it applies to the control of induction machines used for motoring and generation purposes. Chapter 8 describes the fundamental principles of scalar control for induction motors/generators and how to control voltage magnitude and frequency to achieve suitable torque and speed. Chapter 9 presents vector control techniques in order to calculate stator current components for decoupling of torque and flux for fast-transient closed-loop response. Chapter 10 presents methods of optimized control for induction generators, with techniques for peak power tracking control of induction generators, with an emphasis on hill-climbing control and fuzzy control. Chapter 11 covers doubly fed induction generators as applicable for high-power renewable energy systems with important pumped-hydro and grid-tied applications, with a detailed transient model suitable for computer simulations. Chapter 12 describes simulation approaches for transient response of self-excited induction generators in several environments, steady-state analysis, and vector control–based induction motoring/generating systems with several examples of modeling in different computer programs. Chapter 13 presents the applications of induction generators based on their variable speed features, which enable the control of a wide set of variables such as frequency, speed, output power, slip factor, voltage tolerance, and reactive power. The most simple application of an induction generator is by directly connecting it to a grid, which is sufficient to guarantee electrical rotation above the synchronous speed to control a scheduled power flow. Stand-alone applications are also common in cases of low levels of power generation. Chapter 14 provides some discussions on economic and social impact issues related to alternative sources of energy with appraisal and consideration of investments for the decision-maker. The appendices provide extra information to support the chapters.

M. Godoy Simões
Denver, Colorado

Felix A. Farret
Santa Maria, Rio Grande do Sul, Brazil

MATLAB® is a registered trademark of The MathWorks, Inc. For product information, please contact:

The MathWorks, Inc.
3 Apple Hill Drive
Natick, MA 01760-2098 USA
Tel: 508-647-7000
Fax: 508-647-7001
E-mail: info@mathworks.com
Web: www.mathworks.com

PSIM® is a registered trademark of Powersim Inc. For product information, please contact:

Powersim Inc.
2275 Research Blvd, Suite 500
Rockville, MD 20850 USA
E-mail : info@powersimtech.com
Web: www.powersimtech.com

Acknowledgments

This book has been in our minds for the last decade. Several presentations, discussions, meetings, debates, and hundreds of e-mails ushered us toward the goal of writing this book. Many colleagues assisted us in the process. We thank the people who patiently read and commented on the drafts, gave practical suggestions, supported in developing simulations and experimentation, and inspired and made us find veiled relationships in the intricacy of the topics. In particular, we thank Luiz Pedro Barbosa, Ivan Chabu, Sudipta Chakraborty, Lineu Belico dos Reis, Jeferson Marian Correa, Colin Grantham, Cigdem Gurgur, Hua Jin, Enes Gonçalves Marra, Bhaskara Palle, José Antenor Pomilio, Dawit Seyoum, and Robert Wood. Thanks also to Prof. Muhammad H. Rashid, for first approaching us with the suggestion of preparing this book, and to Frede Blaabjerg, Leon Freris, David A. Torrey, and Edson H. Watanabe for the initial review. These colleagues made relevant suggestions at the onset of this project and believed in our competence, motivating us to carry on with our venture. We appreciate the photographs provided by WEG Motores and Equacional Elétrica e Mecânica LTDA. We are also grateful to four wonderful people who diligently worked with us during all the important stages of this project: Nora Konopka, Jessica Vakili, and Kathryn Everett from Taylor & Francis Group and Paul Abraham from SPi Global.

Authors

M. Godoy Simões earned his BSc and MSc from the University of São Paulo, Brazil, and a PhD from the University of Tennessee, Knoxville, Tennessee, in 1985, 1990, and 1995, respectively. He received his DSc (*Livre-Docência*) from the University of São Paulo in 1998. He has been an IEEE senior member since 1998.

Dr. Simões was a faculty member at the University of São Paulo, Brazil, from 1989 to 2000, where he was involved in teaching and research in the Department of Mechatronics. He has been with Colorado School of Mines since 2000, working in intelligent-based control for advancing high-power electronics applications for distributed energy systems and smart-grid technology. He is a recipient of the NSF Faculty Early Career Development (CAREER) Award in 2002, one of the most prestigious awards for new faculty members supporting activities of young teacher-scholars. He was appointed as visiting professor at the University of Technology of Belfort-Montbéliard (UTBM) for several summer contracts and twice at the l'École Normale Supérieure (ENS) de Cachan, both in France. He is director of the Center for Advanced Control of Energy and Power Systems (ACEPS), where he is supervising research in fuzzy logic, neural networks, Bayesian networks, and multiagent systems for power electronics, drives, and machines control. His contemporary work builds the academic framework in advancing and integrating the distribution grid in applications defined as "smart-grid technology." He was awarded a Fulbright Fellowship in 2014 to conduct research and educational activities at the University of Aalborg, Denmark.

Dr. Simões published several books in English covering application of induction generators for renewable energy systems, integration of alternative sources of energy, as well as power electronics for integration of renewable energy systems. He has also published the only book in Portuguese language about fuzzy logic modeling and control.

Dr. Simões has served the IEEE in various capacities. He was IEEE PELS educational chairman, IEEE PELS intersociety liaison, and associate editor for *Energy Conversion*, as well as editor for *Intelligent Systems* of *IEEE Transactions on Aerospace and Electronic Systems* and associate editor for *Power Electronics and Drives* of *IEEE Transactions on Power Electronics*. He was transactions paper review chair for IAS/IACC and vice chair of IEEE IAS/MSDAD. Dr. Simões organized several conferences and technical sessions for the IEEE for the past few years. He was the program chair for IEEE PESC 2005, active in the Steering and Organization Committee of the IEEE International Future Energy Challenge, and also the organizer of the 1st PEEW'05, Power Electronics Education Workshop. Dr. Simões coordinated PELS chapter activities in Denver; he is the founder of the IEEE Denver chapter of the Power Electronics Society and served as their first chair for two consecutive years. He served in all administrative capacities, for about a decade with the IEEE IAS Industry Automation and Control Committee, and was one of the founders and served as secretary and chair for the IEEE IES Smart-Grid Committee.

Felix A. Farret received his bachelor's and master's degrees in electrical engineering from the Federal University of Santa Maria (UFSM), Brazil, in 1972 and 1976, respectively. He specialized in electronic instrumentation at Osaka Prefectural Industrial Research Institute, Japan, in 1975. He earned his MSc from the University of Manchester, England, in 1981, and PhD in electrical engineering from the University of London, England, in 1984, followed by a postdoctoral program in alternative energy sources in the Colorado School of Mines, Golden, Colorado, in 2003. He worked as an engineer for operation and maintenance at the State Electric Power Company (CEEE), Rio Grande do Sul, Brazil, in 1973–1974. He was visiting professor in the Division of Engineering at the Colorado School of Mines in 2002–2003. He published his first book in Portuguese entitled *Utilization of Small Energy Sources* (Publisher UFSM, 1st edition in 1999, 2nd edition in 2010, 3rd edition in 2014). He is also the author of *Renewable Energy Sources: Design and Analysis with Induction Generators* (CRC Press, 2004), *Integration of Alternative Sources of Energy* (John Wiley & Sons, 2006), and *Alternative Energy Sources: Design and Analysis with Induction Generators* (2nd edition, CRC Press, 2008), and coauthor of *Power Electronics for Renewable Energy and Distributed Systems* (Springer International, 2013). Currently, he is professor in the Department of Energy Processing, UFSM, Brazil, since 1974, teaching both undergraduate and postgraduate levels; he is involved in research and development in industrial electronics and alternative energy sources. Dr. Farret has worked in a multidisciplinary educational environment linking industrial electronics, quality of generation in power systems, efficiency of business and residential electrical installations, and integration of alternative energy sources. He is professor of a graduate course in electrical engineering at the UFSM in the discipline of small electrical power plants and industrial electronics and the discipline of applied electronics and instrumentation at the undergraduate level. He was a speaker at various national and international events. He coordinated several scientific and technological projects, including alternative sources of energy that were transferred to Brazilian companies such as AES Sul Distributor of Energy, State Company of Electric Power, Hydroelectric Power Plant Palma Nova, RGE Distributor of Energy, and CCE Engineering Controls Ltd, related to the integration of different primary sources, voltage and speed control for induction generators, development and modeling of energy systems with proton exchange membrane fuel cell stacks for applications in low power, minimization of harmonics in high-voltage direct current transmission, and use of superficial geothermal energy to reduce consumption of electric power. In Brazil, Dr. Farret developed several intelligent systems for industrial applications related to integration, sizing, and location of alternative sources of power for distribution and industrial systems. Such sources include fuel cell stacks, hydro generators, and wind, photovoltaic, and geothermal energy storage applications in batteries and other systems such as ac–ac, dc–ac, and ac–dc–ac.

1 Principles of Alternative Sources of Energy and Electric Generation

1.1 SCOPE OF THIS CHAPTER

Induction generators had been used from the beginning of the twentieth century until they were abandoned and almost disappeared in the 1960s. With the dramatic increase in petroleum prices in the 1970s, the induction generator returned to the scene. With the present high energy costs, rational use and conservation implemented by many processes of heat recovery and other similar forms became important goals. The end of the 1980s was characterized by a wider distribution of population over the planet. There were improved transportation and communication systems enabling people to move away from large urban concentrations. As a consequence, an unprecedent growing in power generation caused concerns with the environment leading many isolated communities to built their own power plants. In the 1990s, ideas such as distributed generation began to be discussed more intensively in the media and in research centers. The general consciousness of finite and limited sources of energy on the earth and international disputes over the environment, global safety, and the quality of life have created an opportunity for new, more efficient, less polluting power plants with advanced technologies of control, robustness, and modularity.

In this new millennium, the induction generator, with its lower maintenance demands and simplified controls, appears to be a good solution for such applications. For its simplicity, robustness, and small size per generated kilowatt, the induction generator is favored for small hydro and wind power plants. More recently, with the widespread use of power electronics, computers, and electronic microcontrollers, it has become easier to administer the use of these generators and to guarantee their use for the vast majority of applications where they are more efficient up to around 500 kVA.

The induction generator is always associated with alternative sources of energy. Particularly for small power plants, it has a great economic appeal. Standing alone, it usually reaches maximum of 15 kW. On the other hand, if an induction generator is connected to the grid or to other sources or storage, it can easily approach 100 kW. Very specialized and custom-made wound-rotor schemes enable even higher power. More recently, power electronics and microcontroller technologies have given a decisive boost to induction generators because they enable very advanced and inexpensive types of control, new techniques of reactive power supplements, and asynchronous injection of power into the grid, among other features. This chapter will begin the discussion of these aspects of induction generator technology. The details will be dealt with in later chapters.

1.2 LEGAL DEFINITIONS

In most countries, a small power plant is considered to generate up to 10,000 kW. For a long time, this classification was applied mostly to hydropower plants, but more recently, the term "small power plant" has come to include three different ranges of power generation: micro, mini, and small, as shown in Table 1.1.[1–3] In this book, the term is applied to plants that use all other forms of energy. Micropower plants (up to 100 kW) are distinguished from the others because in many countries they are much simpler, legally as well as technically, than mini power plants (between 100 and 1,000 kW), small power plants (between 1,000 and 10,000 kW), and large power plants (larger than 10,000 kW). Table 1.1 also lists the heads of waterfalls necessary for the various power ranges as recommended by Eletrobrás, the Brazilian energy agency, OLADE, the Latin-American organization for development and education, and the US Department of Energy (DOE).

Mini, small, and large power plants are usually commercial and supply several consumers. Although power plants of these sizes can be asynchronous or synchronous, isolated or directly or indirectly connected to the network, there is a growing concern with the quality and frequency of the voltage they generate. As mini, small, and large power plants supply energy to a considerable number of people, they tend to be governed by complex legal regulations and technical standards. Although their technology is well established, large units can be responsible for a large portion of the total energy being generated in their area; hence, there is an increased possibility of blackouts.

Micropower plants, on the other hand, are practically free of legal regulations. They have a strong dependence on the load connected across their terminals, which may affect their voltage and frequency controls. Because of their small size, they lend themselves to scheduled investments. In general, they have reduced losses and reduced transmission and distribution costs. Furthermore, aggregation of generators allows for programmed interruptions of loads for maintenance. They can supply both

TABLE 1.1
Classification of Small Hydroelectric Power Plants

Size of the Power Plant	Source	Power (kW)	Water Head (m)		
			Low	Average	High
Micro	OLADE	Up to 50	15	15–50	50
	Eletrobrás	Up to 100	As above	As above	As above
	DOE	100	10	Probl. 1.7	Probl. 1.7
Mini	OLADE	50–500	20	20–100	100
	Eletrobrás	100–1,000	As above	As above	As above
	DOE	100–1,000	Probl. 1.7	Probl. 1.7	Probl. 1.7
Small	OLADE	500–5,000	25	25–130	130
	Eletrobrás	1,000–10,000	As above	As above	As above
	DOE	>1,000	Probl. 1.7	Probl. 1.7	Probl. 1.7

stand-alone and/or interconnected loads. When connected to power systems, they can be considered more as loads than as generating points.

With respect to the amount of generated power, three different situations can be considered: (1) sufficient energy for the consumer's own consumption and for sale to or interchange with the distributor; (2) lower generation than the consumer's own consumption to avoid the sale of excess energy (from the point of view of the public network, the generated power is seen more as an apparent reduction of the effective load); and (3) surplus generation for accumulation of energy (water, battery, fuel, coal) to be used for demand reduction in the peak hours of the public network.

As can be inferred from this discussion, the design of a micropower plant should not follow the same steps as that of a larger plant only on a smaller scale. Mini, small, and large power plants are fundamentally different because they are strictly for commercial or community applications, which require rigid voltage and frequency controls. The necessary investments, technical implications, and operational costs are higher and very difficult for an individual consumer to assume them. The bureaucratic formalities are so complicated as to be almost insurmountable everywhere in the world.[2–6]

Most micropower plants, on the other hand, are of the stand-alone type. They are often the only possible solution for a remote site. When they operate connected to the public network, their purpose is to reduce energy bills. There should be a negotiation with the local energy distributor with regard to cost of a kilowatt-hour generated, safety conditions, voltage tolerances, and frequency, among other matters, but to begin operations, most micropower plants need only a formal communication to the competent authorities.

1.3   PRINCIPLES OF ELECTRICAL CONVERSION

Electrical energy conversion is either direct or indirect. Primary energy is said to be converted directly when it is transformed into electricity without going through any intermediate forms of energy. Examples are photovoltaic arrays and thermoelectric. Direct conversions are usually static, that is, without moving parts. Surprisingly, they are not necessarily the most efficient. Indirect conversions like hydro and wind power use one or more intermediary processes.

Direct energy conversions are the most desirable because they are nonpolluting. They are silent, although they are also costly and not very efficient. An example is thermal energy converted directly into electrical power. Chemical energy is also directly converted into electricity through the use of batteries or fuel cells. Mechanical energy is converted into electricity in conventional ac or dc generators or in fluid dynamic converters such as electro-gas-dynamic (EGD) or magneto-hydro-dynamic (MHD) systems. Thermal conversion of energy is limited by the Carnot cycle, which is expressed as a relationship between sources of high and low temperatures, that is, efficiency $= 1 - T_L/T_H$. The other direct conversions are not limited by this externally reversible heat-engine cycle. The attention of many researchers throughout the world is still focused on them.

The thermoelectric generator produces a very clean type of energy, and it has a number of advantages over the conventional thermodynamic heat-engine system, based on the Seebeck effect, the Peltier effect, and the Thompson effect, which describe the generation of voltage by the application of temperature to pairs of different metals or thermocouples. Thermoelectric generators are rugged, reliable, compact, and without moving parts. These generators are confined to very low-power applications because their efficiency is extremely low, something around 5%–10%.

Batteries and fuel cells convert chemical energy into electrical energy. They are similar in that both normally contain two electrodes separated by an electrolyte solution or matrix. In a fuel cell, the fuel reactant, usually hydrogen or carbon monoxide, is fed into porous electrodes and oxygen or air is fed into other porous electrodes. Some batteries can be reloaded and are used to store energy. Energy can also be stored in fuel cells by storing the fuel used to activate the reactions in the cell. Actually, batteries store energy, and fuel cells generate it, though either are based on chemical reactions.

Solar energy is directly converted into electrical energy in photovoltaic cells. These cells are based on the *p–n* junction of semiconductors, usually silicon, in which there is a transient charging process that establishes an electrical field close to the junction. The concentration of doping impurities in the material of the *n*-type is made so high that when it is combined with the *p*-type material, some of the electrons from the *n*-type material flood the nearby holes of the *p*-type material and vice versa, making the *n* material positively charged and the *p* material negatively charged in the area close to the junction. This charging process continues until the electric field or junction potential inhibits further charges to be moved away, making the electron and hole flow the same in both directions of the semiconductor material. An external voltage generated by photons can alter this balance of electrical charges. Photons react with the valence electrons near the *p–n* junction to produce a forward biasing voltage and, thus, electrical current. The problems of photovoltaic energy are low efficiency (less than 20% in good quality commercial systems) and high cost. Furthermore, any terrestrial solar conversion system needs to be integrated with either an energy storage system or with another type of conversion system to supply energy when solar energy is not available. This feature limits even more the use of photovoltaic cells in large power schemes.

The principle of a rotating electrical generator is based on the relative movement between a coil and a magnetic core. If a coil is rotated between a pair of poles (a field) of an electromagnet or a permanent magnet, the output from the rotor will be either alternating or direct current, depending on whether a slip ring (ac) or segmented commutator (dc) is employed. The field windings of a conventional dc generator require a dc power supply, and they can be excited in several ways. Some of these machines must be excited by an external dc power generator. Others are self-excited (shunt, series, or compound), needing only a small amount of residual magnetic flux present in the core of the machine to start the excitation process. AC current generators or alternators also require a dc power supply for the field winding, usually coming from a small dc generator (exciter) connected to the same alternator shaft. These generators can be synchronous or asynchronous; they are synchronous if their frequency is an integer multiple of the frequency of the grid.

When connecting a generator to an ac power grid, care must be taken that the alternator is in phase with the alternating current of the grid at the instant of connection. If the alternator is out of phase, the grid will cause immediate acceleration or deceleration of the rotor and high peaks of voltage and current. Severe damage to the machine may result. Asynchronous generators do not suffer so drastically from this limitation, but they may cause stabilization problems to the network, since their power frequency is not a fixed value.

1.4 BASIC DEFINITIONS OF ELECTRICAL POWER

Since production of electrical power is the goal, it is important to understand some fundamentals of electric power. Let us consider the variables: V_{dc}, I_{dc}, V_{rms}, and I_{rms}, where V_{dc} and I_{dc} are the average values of voltage and current and V_{rms} and I_{rms} are the effective values of voltage and current in the load, including the harmonic components. The dc load power (lossless rectifier) and ac load power in the source, respectively, are defined as follows:

$$P_{dc} = V_{dc}I_{dc} \qquad P_{ac} = V_{rms}I_{rms}$$

$$\eta = \frac{P_{dc}}{P_{ac}} \qquad V_{ac} = \sqrt{V_{rms}^2 - V_{dc}^2} = \sqrt{\sum_{i=1}^{n} V_i^2}$$

where
  η is the efficiency
  V_i is the effective value of the harmonic component of order i

$$\text{Form factor: } FF = \frac{V_{rms}}{V_{dc}}$$

$$\text{Ripple factor: } RF = \frac{V_{ac}}{V_{dc}} = \sqrt{\left(\frac{V_{rms}}{V_{dc}}\right)^2 - 1} = \sqrt{FF^2 - 1}$$

Circuits using transformers need some additional definitions:

$$\text{Transformer utilization factor: } TUF = \frac{P_{dc}}{V_s I_s}$$

where V_s and I_s are the effective values, respectively, of the voltage and current waveforms of the transformer secondary.

The displacement factor, $\cos\phi_1$, is the cosine of the angle ϕ_1 between the fundamentals of voltage V_{s1} and current I_{s1} (Figure 1.1).

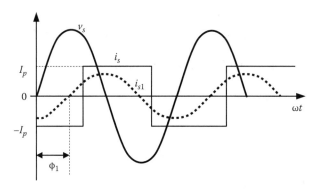

FIGURE 1.1   Input voltage and current waveforms.

The harmonic factors or the total harmonic distortion (*THD*) for voltage and current, respectively, are as follows:

$$THD_{(v)} = \left(\frac{V_{rms}^2 - V_{rms1}^2}{V_{rms1}^2} \right)^{1/2} = \left[\left(\frac{V_{rms}}{V_{rms1}} \right) - 1 \right]^{1/2}$$

$$THD_{(i)} = \left(\frac{I_{rms}^2 - I_{rms1}^2}{I_{rms1}^2} \right)^{1/2} = \left[\left(\frac{I_{rms}}{I_{rms1}} \right)^2 - 1 \right]^{1/2}$$

where V_{rms1} and I_{rms1} are the rms values of the fundamental components of the input voltage V_s and current I_s, respectively.

$$\text{Power factor: } PF = \frac{V_{rms1}I_{rms1}}{V_{rms}I_{rms}} \cos \phi_1 \approx \frac{I_{rms1}}{I_{rms}} \cos \phi_1$$

So the power factor is strongly influenced by the harmonic content of the voltage and current. If the harmonic distortion factor I_{rms1}/I_{rms} is replaced by the inverse of the total harmonic distortion factor and simplified as a whole, we get

$$PF \cong \frac{\cos \phi_1}{\sqrt{\left(THD_{(i)}^2 + 1 \right)}}$$

This approximation applies to normally acceptable cases of the *THD* of voltage that rarely goes beyond 10%.

$$\text{Crest factor: } CF = \frac{I_{s(peak)}}{I_s}$$

$$\text{Short circuit ratio: } SCR = \frac{SCC}{P_{dc}}$$

where *SCC* is the short circuit capacity of the ac source, that is, the ac open circuit voltage (ideal source of voltage) multiplied by the short circuit current of this source.

TABLE 1.2

Ideal Parameters of a Rectifier

$\eta = 100\%$	$V_{ac} = 0$	$RF = 0$	$SCR \to \infty$
$TUF = 1$	$THD_{(v)} = THD_{(i)} = 0$	$PF = DF = 1$	$CF = 1$

TABLE 1.3

Basic Electric Definitions for Sinusoidal Waveforms

Peak voltage	$V_p = \dfrac{V_{max} - V_{min}}{2}$	Active power	$P = \dfrac{1}{T}\displaystyle\int_0^T p(t)\,dt$
rms voltage	$V_{rms} = \dfrac{V_p}{\sqrt{2}}$	Power factor	$\cos\phi = \dfrac{P}{S}$
Peak current	$I_p = \dfrac{I_{max} - I_{min}}{2}$	Reactive power	$Q = S\sin\phi$
rms current	$I_{rms} = \dfrac{I_p}{\sqrt{2}}$	Resistance	$R = Z\cos\phi$
Frequency	$f = \dfrac{1}{T}$	Reactance	$X = Z\sin\phi$
Impedance	$Z = \dfrac{V_{rms}}{I_{rms}}$	Average energy	$W = PT = \displaystyle\int_0^T p(t)\,dt$
Apparent power	$S = V_{rms}I_{rms}$	Total harmonic distortion	$THD = \sqrt{\left(\dfrac{V_{rms}}{V_{rms1}}\right)^2 - 1}$

So the resulting short circuit current is only limited by the source impedance. Any conversion system is usually considered weak when $SCR < 3$ and strong when $SCR > 10$. The ideal parameters for a rectifier are shown in Table 1.2. Other basic electric definitions are provided in Table 1.3.

1.5 CHARACTERISTICS OF PRIMARY SOURCES

Electrical energy systems can be static, rotating, or energy storing. Systems differ with respect to their primary source, whether they are stand-alone or connected to the grid, their degree of interruptibility, and the quality and cost of their output. In general, the electrical power can offer power flux reversion, phase changing, direct current transformation, line-frequency and line-voltage transformation, isolation, and stabilization.

The interruptibility of power sources and dependence on each source are strongly related to the costs of the overall power supply; all aspects of these factors should be taken into consideration at the design stage. Such factors can be listed as allowable outage time, allowable transfer time between normal and standby supplying alternatives, initial costs, operation costs, and maintenance costs.

The vast majority of sources of energy in the world depend on rotating generators: hydro, wind, diesel, general internal combustion engines (ICE), steam turbines, and micro turbines. Solar and fuel cells are static sources and depend on purely electronic systems to convert their electricity.

Energy storage systems may also be classified into static and rotating. Some of these sources are battery, water dam, heat, flywheel, fuel tank, charcoal field, pneumatic container (air or gas), super capacitor, and injection to grid. Superconducting magnetic energy storage (SMES) takes advantage of a new generation of high-temperature superconducting materials. SMES can potentially be used in large and small stationary products. SMESs are currently commercialized for power quality conditioners. It seems that static sources of energy will play a very important role alongside the rotating types since they tend to be cleaner, noiseless, lighter, and less dependent on nonrenewable sources of energy like coal, petroleum, and natural gas.

An inertial flywheel is a heavy wheel coupled to the shaft of a turbine generator group; it works as a regulator of the variations of load in the generating unit, within determined regulation conditions imposed previously, the GD^2 (moment of inertia) of the generator and of the turbine.[5] Manufactured of melted steel or of melted iron, flywheels are considered the best way of storing pure or mechanical energy in terms of simplicity, speed in the delivery and absorption of energy, obtaining of high power, limitless number of cycles (charge and discharge), high reliability, and low maintenance cost. Flywheel technology has been advanced recently by the use of light materials, vacuum operation, and sealed joints capable of withstanding high pressure differences.

Nontechnical factors like politics and economics are also very important and have to be considered. Many nonrenewable sources of energy cause high degrees of pollution and are located in remote places of extreme political and economic instability. So it is very important to find and develop new alternative sources of energy, preferably renewable ones. To do that, it is important to consider the following factors: development time, cost, environmental effects, expectations to meet future demands, robustness, and access routes.

A very simple example of the principles uniting hydro storage of energy and rotating systems of energy generation in the form of an induction generator and wind energy is presented in Section 1.9.

1.6   CHARACTERISTICS OF REMOTE INDUSTRIAL, COMMERCIAL, AND RESIDENTIAL SITES AND RURAL ENERGY

The availability of energy is closely related to the following:

- Supply of electricity, telephone, gas, and water
- Increase and improvements in the networks of transportation and communications
- Construction of schools, hospitals, and clinics
- Access to media (radio, television, newspapers, etc.)

- Increased job opportunities
- Access to agricultural and health services
- Rural exodus

Because of the overall importance of energy and the scarcity of investment capital, it is vital to make maximum use of the available energy resources. Maximum use is not difficult to achieve, but it takes planning and management. Energy planning is a capital investment decision. An important part of energy planning is the sizing of the loads associated with electric power plants.[6,7]

The initial studies leading to sizing plant loads should encompass a sufficiently long time frame. Hydroelectric power plants should not be sized only for the present load or for the immediate future. A power plant capacity has to absorb possible regional economic growth.

Oversizing a project implies idle investments; undersizing leads to repressed demand with a consequent reduction in the quality of the supply and premature aging of the equipment. However, undersizing may be necessary if the capital for a larger plant is not available.

The determination of the power necessary to serve the users of the electric load is made based on the relationship of the hourly electric loads of all their components. In that list of loads should be lighting, appliances, and other electric equipment that consumes energy, characterized by different power and consumption periods. The peak load found defines the necessary power generation for the power plant.

Distributing each component power demand on the installation throughout the operating hours of the day and adding the power demand of those components in each operating period yields the load peak. So an electric load system, working within a schedule and consumption period, is characterized by a coefficient called demand factor, defined as follows:

$$F_d = \frac{D_{av}}{D_{max}}$$

where
D_{av} is the average demand (average load consumed)
D_{max} is the maximum demand (maximum load consumed)

The closer the demand factor F_d is to 1.0, the better the energy consumption distribution of the system. In places where there is intense consumption variation during certain periods, it is necessary to estimate the load distribution in those periods and compare it with the distribution in a normal period of consumption. The highest peak found in the two periods will define the rated power of the power plant to be installed. If there is available potential, the rated power can be increased when necessary as a safety measure, anticipating a change in the load programming, or even to provide for load growth in the future.

Furthermore, to determine the power to be installed, the necessary electric load should be compared with the electric potential of the primary source of energy. One of the following three hypotheses should be considered.

If the available power is larger than the power demanded, one can take advantage of the entire potential, obtaining a surplus of energy. Or in the case of a hydroelectric power plant, the net fall or the water flow can be reduced, for the sake of economy, adjusting the potential to the power demanded. The decrease of water flow reduces the cost of the receiving systems and pipelines, while a decrease in the raw fall reduces the height of the dam.

In the cases of wind or solar energy, such economies have no meaning. Electronic controls that can take maximum advantage of those sources are to be preferred. For biomass micropower plants and other fuels, the production of energy should also be strictly limited to what is required for consumption, due to their high costs and the ease of adding new units when there is an increase in the demand.[7–9]

Where there is a state or private monopoly of electric power distribution, with no possibility of sale of production excess, it is convenient to determine the year-round minimum consumption of energy. Otherwise, the public network can acquire the demand above that limit, drastically reducing project costs. But whenever the public and private demand peaks coincide, they may overload the public network.

If the available potential equals the desired power, the whole energy potential should be used. If the available potential is smaller than the desired power, consumption must be rationalized by selecting loads according to service priorities or by making up the deficit with energy originating from other sources. The interconnection with the public network can be a good way out for such situations. Brownouts can also be implemented. That is, the available rated voltage for the consumer can be slightly reduced. It is known that consumers do not even notice a 5% or 10% reduction in supply voltage, which can obtain a 10%–20% reduction in consumption. The reason for this economy is that the supplied energy is reduced as the square of the voltage reduction.

1.7 SELECTION OF THE ELECTRIC GENERATOR

This book focuses on the induction generator rather than other types of energy converters because of its importance for rotating energy conversion from hydro and wind systems. After the installation site of the power plant is selected, the next step is to select the turbine rating, the generator, and the distribution system. In general, the distribution transformer is sized to the peak capacity of the generator and the capacity of the available distribution network. As a rule, the output characteristics of the turbine power do not follow exactly those of the generator power; they have to be matched in the most reasonable way possible. Based on the maximum speed expected for the turbine and taking into account the cubic relationship between the fluid speed and the generated power, the designer must select the generator and the gearbox so as to match these limits. The most sensitive point here is the correct selection of the rated speed for the generator. If it is too low, the high speed of the primary source fluid, say, wind, will be wasted; if it is too high, the power factor will be harmed. The characteristics of the commercially available turbines and generators must be matched to the requirements of the project with regard to cost, efficiency, and maximum generated power in an iterative design process.

Several types of generators can be coupled to the rotating power turbines: dc and ac types, parallel and compound dc generators, with permanent magnet or variable

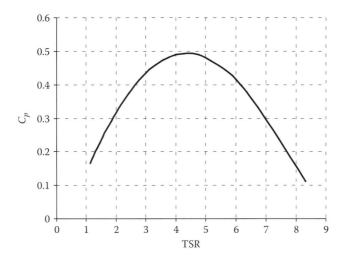

FIGURE 1.2   Value of C_p for wind turbines.

electrical fields, synchronous or asynchronous, and, especially, induction generators. The right choice of the generator depends on a wide range of factors related to the primary source, the type of load, and the speed of the turbine, among others.

In the case of wind energy, there is a control variable known as power coefficient, C_p, which is of utmost relevance. The maximum value of C_p occurs approximately at the same speed as the maximum power in the power distribution curve, and it is defined as the relationship between the tip speed ratio (TSR) of the rotor blade and the wind speed. So the turbine speed must be kept optimally constant at its maximum possible value to capture the maximum wind power. It suggests that to optimize the annual energy capture at a given site, it is necessary for the turbine speed (TSR) to vary to keep C_p maximized as illustrated in Figure 1.2 for a hypothetical turbine. Obviously, the design stress must be kept within the limits of the turbine manufacturer's data since the torque is related to the instant power by $P = T\omega$.

Because of the way it works as a motor or generator, the possibility of variable speed operation, and its low cost compared with other generators, the induction machine offers advantages for rotating power plants in both stand-alone and interconnected applications.

1.8   INTERFACING PRIMARY SOURCE, GENERATOR, AND LOAD

As previously discussed, to think of generation of small amounts of electricity is, as a practical matter, to think of producing energy for one's own individual consumption. Due to the types of loads used today by society in general, only in rare situations can such uses be thought of in commercial or collective sense. A small generator can be a solution for rural areas or for very poor and distant centers of electric power distribution where there is no other option and the loads do not go beyond a few kilowatts. Electric showers, often used in developing countries, air conditioning, electrical stoves, and other modern appliances with high and concentrated consumption of electric power should not be connected to a small generator. So we need to think in terms of a spectrum of power supplies from small (few watts) to large (close to 100 kW).

"Small applications" are, for example:

- Small data transmission equipment or telephony for distant places
- Equipment for scientific experiments or unassisted collection of data in remote areas
- Power supply for electric equipment in places subject to dangerous conditions: fire, explosions, toxic environment, radiation, conflict areas, cliffs, proximity to wildlife, and so on

Whereas "Large applications" are, for example:

- Irrigation
- Facilities in farms with high conditions of comfort
- Reduction in the electric bills of companies or other large consumers
- Storage of energy during off-peak hours of consumption of the public network for reduction of the peak hours of higher demand[6,7]

In all its manifestations, primary energy comes in nature in raw form and has to be captured and adapted to be useful. The process of conversion of energy as nature provides it to us into electric power can be described in three stages: the primary energy, the conversion system, and the electric load (Figure 1.3).[8,9] Among the alternative primary sources of energy, the most common, in order of current importance, are as follows:

- Hydraulic energy represented by the movement of waters
- Wind energy present in the movements of air masses
- Thermal energy from the burning of fuels
- Solar energy (light and heat) found in the sun's rays
- The energy of gases or of biomass generated from the decomposition of plants and animals or from the burning of organic matter like trees and organic garbage
- Electrochemical energy represented by fuel cells and batteries

Energy conversion uses the primary energy in its raw natural state and transforms it into electricity in a useful and efficient form suited to its purpose. The rotating power converter is usually called the primary moving machine because it is the

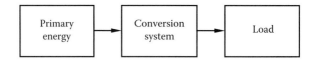

FIGURE 1.3   Basic elements for electric power generation.

one involved directly in the process of transforming energy. Recently, a variety of types of electronic or static power converters have emerged. Power converters are discussed in Chapter 7. The loads of all converters should be adapted to their uses. They can be of direct or alternating current, static or rotating, and electrochemical or mechanical, depending on the electric principle of operation as discussed in Section 1.7.

1.9 EXAMPLE OF A SIMPLE INTEGRATED GENERATING AND ENERGY-STORING SYSTEM

The shaft of a wind power turbine of $P_{wt} = 2$ kW supplies energy for pumping water to a height $h = 10$ m through a hydraulic pump of efficiency $\eta_{hp} = 60\%$ and a mechanical transmission system of efficiency $\eta_{ts} = 80\%$, as depicted in Figure 1.4. What is the storage time of energy under these conditions, and what is the average output power being supplied?

Of all the wind energy in nature, only a small average portion is available during 1 day at a power plant, taking into account periods of high and low wind intensities. This average is represented, in effective terms, by the characteristic factor of the place, F_c, whose typical value is around 0.30. Besides these losses, we must discount

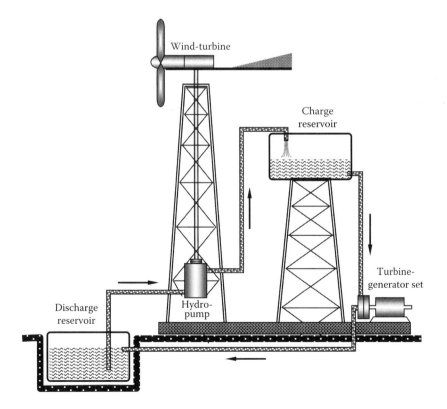

FIGURE 1.4 Regulation of the wind power energy delivery by hydraulic storage.

losses in the wind turbine that are reflected in its efficiency, usually around $\eta_{wt} = 0.35$. So the available wind power at a given place is as follows:

$$P_{wind} = \frac{\rho_{air} \pi R^2 V^3}{2} \tag{1.1}$$

where
ρ_{air} is the air density in kg/m³ (1.2929 kg/m³) at sea level
V is the wind speed in m/s
R is the length of each blade of the turbine from the center of the rotor shaft to the blade tip measured in meters

The net power available on the wind power turbine shaft for the hydraulic pump, for simplicity, is considered, in this example, as being

$$P_{wt} = P_{wind} = 2.0 \text{ kW}$$

The power P_{hp} is supplied from the useful power in the rotor of the wind power turbine, P_{wt}, and expressed as

$$P_{hp} = \eta_{ts} P_{wt}$$

and, then, the average flow of the hydraulic pump will be given by

$$Q_{hp} = \frac{F_c \eta_{hp} P_{wt}}{Hg\rho_{hyd}} \quad \left(\text{in m/s} \right) \tag{1.2}$$

where
P_{hp} is the power of the hydraulic pump
η_{hp} is the efficiency of the hydraulic pump
H is the level difference between the lower and upper reservoirs
ρ_{hyd} is the specific mass of the water (= 1000 kg/m³)

Using the data of this example and Equation 1.2, the flow of the hydraulic pump is as follows:

$$Q_{hp} = \frac{0.30 \times 0.60 \times 0.80 \times 2000 \text{ W}}{\left(10 \times 9.81 \text{ m/s}^2 \times 1000 \text{ kg/m}^3 \right)}$$

$$= 2.936 \; \ell/s = 10{,}570 \; \ell/h$$

To avoid an average accumulation of water in the reservoirs, it is necessary that the average input and output flows be the same, therefore,

$$Q_{hp} = Q_{ht}$$

where Q_{ht} is the flow of the hydraulic turbine of the generating group.

The electric output power of the hydro generator set is given by

$$P = \eta_{ht}\eta_{eg}Q_{eg}Hg\rho_{hyd} \qquad (1.3)$$

where

η_{ht} is the efficiency of the hydraulic turbine
η_{eg} is the efficiency of the electric generator

or

$$P = 0.90 \times 0.95 \times 2.936 \times 10^{-3} \text{ m}^3/\text{s} \times 10 \text{ m} \times 9.81/\text{s}^2 \times 1000 \text{ kg/m}^3 = 247 \text{ W}$$

For those values, if an ordinary water reservoir of 1000 L is used with a calculated flow of 2.936 L/s (i.e., steadily generating 247 W), there will be an autonomy of $1000/2.936 = 5.67$ min (without wind). That means, it can steadily light four ordinary bulbs of 60 W, if the wind is not absent for more than 5 min.

1.10 SOLVED PROBLEMS

1. Classify small power plants according to their power ranges.

   Solution

   Small power plants can be classified as (i) micro—up to 100 kW, for example, micro-hydro, wind power for hospitals, school buildings, and stand-alone systems; (ii) mini—up to 1000 kW, for example, mini-hydro, wind power for large residential complex, villages, and industries; and (iii) small—above 1000 kW, for example, small hydro, wind park for grid interconnection, commercial users, and balance of power flow.

2. Given a potential site for a wind power of 10 kW and a terrain elevation difference of 15 m and using the factors given in the example in Section 1.9, calculate the volume of reservoir water required to allow 30 min of steady energy without wind.

   Solution

   $$P_{tw} = 10 \text{ kW}$$

   $$Q_{bh} = \frac{F_e\eta_{bh}\eta_{ts}P_{tw}}{Mg\gamma} = \frac{0.3 \times 0.6 \times 0.8 \times 1000}{15 \times 9.8 \times 1000} = 9.79 \text{ L/s}$$

   $$Q_{th} = Q_{bh} = 9.79 \text{ L/s}$$

   $$P = \eta_{th}\eta_{ge}Q_{th}Hg\gamma = 1231.7 \text{ W} \quad \text{for } \eta_{th} = 0.9 \text{ and } \eta_{ge} = 0.95$$

   In order to supply 1231.7 W for 30 min, a reservoir with volume $Q_{th} \times 30 \times 60 = 17,662$ L is necessary.

1.11 SUGGESTED PROBLEMS

1.1 Why are micropower plants so different from larger ones? In what cases are they recommended?

1.2 Discuss which of the conventional alternative sources of energy is more suitable for use with induction generators to provide a steady supply of energy with minimum equipment.

1.3 Calculate the power factor of an electronic system subject to a reasonable amount of harmonics given as $THD_{(v)} = 1.0\%$, the displacement factor is 0.92. What would be the calculated error if harmonic components were not taken into consideration? If the effective output voltage was harmonic-free, but if a harmonic content of 10% was allowed in the effective output current, how much power would be measured with no power filters?

1.4 A transformer winding was designed to be fed by an ac voltage of 110 V and to produce an effective current of 10.5 A. By mistake, the transformer was fed from this same voltage but rectified in a full wave converter. Consider a loss-less rectifier and no voltage drops across the connections. Estimate what the apparent power applied to the transformer would be in both cases if its primary resistance was $R = 1.8\ \Omega$.

1.5 Collect information in your local library or on the Internet in order to complete Table 1.1.

REFERENCES

1. Pádua, J.C.U. and Castro, M.M.O., *Small Hydro Power Plants Handbook (Manual de pequenas centrais hidroelétricas)*, Eletrobrás/DNAEE, Brasília, Brazil, 1985.
2. Legislation on small power plants, *Energy Power Magazine (Revista Força Energética)*, 2(5), 7–8, MEPB Publishers, Porto Alegre, Brazil, December 1993.
3. *Diário Official (Official Publication), Section 1, extra-edition.* Brasília, Brazil, No. 129-A, year CXXXIII, July 08, 1995.
4. Krämer, K.G., Schneider, P.W., and Styczynski, Z.A., Use of energy in the power network and options for the Polish power system. In: *Seminar of the Institute of Power Transmission and High Voltage Technique, University of Stuttgart*, February, 1997. Sponsored by the European Community Directorate-General XVII, Energy, Ver. 29, January 1997.
5. Stroev, V.A. and Gremiakov, A.A., Energy storages in electrical power systems. In: *Seminar of the Institute of Power Transmission and High Voltage Technique*, Electrical Power Systems Department, Moscow Power Engineering Institute, University of Stuttgart, Stuttgart, Germany, February, 1997. Sponsored by the European Community Directorate-General XVII, Energy, Ver. 29, January 1997.
6. Inversin, A.R., *Micro-Hydropower Plant Handbook*, National Rural Electrification Cooperative Association (NRECA), Washington, DC, 1986.
7. REA Staff, *Modern Energy Technology*, (Vols. I and II), Research and Education Association (REA), New York, 1975.
8. Culp Jr., A.W., *Principles of Energy Conversion*, McGraw-Hill Book Company, New York, 1979.
9. Decher, R., *Energy Conversion: Systems, Flow Physics and Engineering*, Oxford University Press, New York, 1994.

2 Steady-State Model of Induction Generators

2.1 SCOPE OF THIS CHAPTER

The induction machine offers advantages for hydro- and wind power plants because of its easy operation either as a motor or as a generator, robust construction, natural protection against short circuits, and low cost compared with other generators. Only the case of connection of the induction generator to the infinite bus—the point of the distribution network with no voltage drops across the generator terminals for any load situation (ideal voltage source)—will be considered in this chapter.

Many commercial power plants in developed countries were designed to operate in parallel with large power systems, usually supplying the maximum amount of available primary energy for conversion in their surroundings (wind, solar, or hydro). This solution is very convenient because the public network controls voltage and frequency while static and reactive compensating capacitors can be used for correction of the power factor and harmonic reduction. Aspects related to voltage regulation, stand-alone generation, and output power are further discussed.

2.1.1 INTERCONNECTION AND DISCONNECTION OF THE ELECTRIC DISTRIBUTION NETWORK

When interconnected to the distribution network, the induction machine should have its speed increased until equal to the synchronous speed. The absorbed power of the distribution network in these conditions is necessary to overcome the iron losses. The energy absorbed by the shaft to maintain itself in synchronous rotation is necessary to overcome mechanical friction and air resistance. If the speed is increased, a regenerative action happens, but without supplying energy to the distribution network. This happens when the demagnetizing effect on the current of the rotor is balanced by a stator component capable of supplying core losses. In this situation, the generator is supplying its own iron losses. From this point on, the generator begins to supply power to the load.

The electrical frequency is the number of times per second a rotor pole passes in front of a certain stator pole. If p is the number of poles, the synchronous speed is determined by

$$n_s = \frac{120 f_s}{p} \qquad (2.1)$$

where
n_s is the synchronous mechanical speed in rpm
f_s is the synchronous frequency in Hz

When the induction machine is operating as a generator, the maximum current is reached when the power transferred to the shaft with load varies within a slip factor range given by

$$s = \frac{\left(n_s - n_r\right)}{n_s} \tag{2.2}$$

where
  s is the slip factor
  n_r is the rotor speed in rpm

When considering interconnection to or disconnection from the electric distribution network, some matters affecting efficiency and electric safety should be observed:

- *Efficiency*: There is a rotation interval above the synchronous speed at which efficiency is very low. If the choice of the moment of interconnection is between synchronous rotation and satisfactory efficiency, great precision is not necessary in the sensor that authorizes the connection.
- *Safety*: It is necessary to determine the maximum rotation at which disconnection from the distribution network should happen so that the control system acts to brake the turbine under speed-controlled operation. The disconnection should be planned for the electric safety of the generator in case of a flaw in the control brake.[1-4] Besides, if the plant is interconnected to the distribution network, it is important to switch it off during maintenance, as much for the sake of the local team as for the electric power company team. Although IEEE Standards have been recently released to the public, interconnection guidelines may vary among local utilities and should be checked during the design of the installation.[5-9]

2.1.2 ROBUSTNESS OF INDUCTION GENERATORS

From the preceding discussion, it is clear that the induction generator accepts constant and variable loads, starts either loaded or without load, is capable of continuous or intermittent operation, and has natural protection against short circuit and overcurrents through its terminals. That is, when the load current goes above certain limits, the residual magnetism falls to zero and the machine is de-excited. When this happens, there are four methods available for its remagnetization: (1) maintain a spare capacitor always charged and, when necessary, discharge it across one of the generator's phases, (2) use a charged battery, (3) use a rectifier fed by the distribution network, or (4) keep the induction generator running for about 10 min so the reduced presence of the residual magnetism in the iron can be fully recovered.

An induction generator with a squirrel cage rotor can be specifically designed to work with wind or hydro turbines, in other words, with a larger slip factor, with more convenient deformation in the torque curve, winding sized to support higher saturation current, an increased number of poles, and so on. Such steady-state conditions must be foreseen at the design stage of the power supply.

2.2   CLASSICAL STEADY-STATE REPRESENTATION OF THE ASYNCHRONOUS MACHINE

An equivalent circuit of the induction machine, also known as the per-phase equivalent model,[9–11] is represented in Figure 2.1. In this figure, R_1 and X_1 are the resistance and leakage reactance, respectively, of the stator; R_m and X_m are the loss resistance and the magnetizing reactance; and R_r and X_r are the resistance and the reactance of the rotor. This model is limited to the case of sinusoidal and balanced excitation.

The induction generator does not differ much in construction from the induction motor dealt with in Chapter 6. In its operation, it resembles the transformer except that the transformer secondary is a stationary (nonrotating) part. For this reason, it is common to use the transformer model to represent the induction generator. However, it should be observed that in spite of the fact that the magnetizing curves of both machines are similar in form, in the characteristic BxH (or ΦxNI or VxI) of these machines, the slope and the saturation area of the magnetomotive force (mmf) curve of the induction generator is much less accentuated than that of a good-quality transformer. This is due to the air gap in the induction generator that reduces the coupling between the primary and secondary windings. This means that the high reluctance caused by the air gap increases the magnetizing current required to obtain the same level of magnetic flux in the core, and X_m will be much smaller than it would be in a transformer.

The internal primary voltage of the stator, E_1, and the secondary voltage of the rotor, E_r, are coupled by an ideal transformer with an effective transformation ratio, a_{rms}. This a_{rms} value is easy to determine for a motor with a cylindrical rotor; it is the relationship between the number of stator phase turns and the number of rotor phase turns modified by the differences introduced by the pitch and distribution factors. On the other hand, for the squirrel cage rotor, as in most induction generators, determining the transformation rate becomes extremely difficult because there are no distinct windings. In any situation, the voltage induced in the rotor, E_r, generates a current that circulates through the short-circuited rotor.

The equivalent circuits of the induction generator and of the transformer differ fundamentally, in that in the induction generator, the rotor voltage is subject to a variable frequency, making E_r, R_r, and X_r also variable. Everything depends on the slip factor, that is, on the difference between the rotor speed and the speed of the rotating magnetic field of the stator. As previously mentioned, with

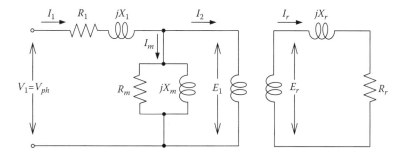

FIGURE 2.1   A rotating transformer-like model of the induction generator.

a rotation below the synchronous speed, the machine works as motor; above the synchronous speed ω_s, it works as generator.

To better understand the induction phenomenon in the asynchronous generator, it should first be understood what happens to the voltage across stator windings when the machine rotation goes from zero (or blocked rotor) to a rotation above the synchronous speed, ω_s. This can be summarized by saying that the larger the difference of speed or the slip factor between the magnetic fields of the rotor, ω_r, and the stator, ω_s, the larger the induced voltage on the rotor. The smallest of these voltages happens when there is no relative rotation between the rotor and the stator, $\omega_r = \omega_s$. The induced voltage on the rotor is directly proportional to the slip factor, s, and from this definition, the induced voltage on the rotor for any speed, E_r, can be established with respect to the blocked rotor, E_{r0}, that is[11-15]:

$$E_r = sE_{r0} \tag{2.3}$$

where s is the slip factor given by

$$s = \frac{f_s - \dfrac{p}{2}\dfrac{n_r}{60}}{f_s} = \frac{n_s - n_r}{n_s} \tag{2.4}$$

where
  n_r is the rotor speed (rpm)
  $f_s = pn_s/120$

The rotor frequency, f_r, is also related to the electric frequency, $f_e = f_s$, through the following expression:

$$f_r = \frac{p}{120}(n_s - n_r)\frac{n_s}{n_s} = sf_s \tag{2.5}$$

Notice that f_r is the relative speed between the stator and rotor magnetic fields. The circulating current in the rotor will depend on its impedance, the resistance, and the inductance, which alter slightly due to the skin effect for the usual values. However, only the inductance is affected in a more complex way by the slip factor, depending on the self-inductance of the rotor and on the frequency of the induced voltage and current. The reactance is usually presented as the following function of the slip factor:

$$X_r = 2\pi f_r L_{r0} = 2\pi sf_s L_{r0} = sX_{r0} \tag{2.6}$$

where X_{r0} is the blocked rotor reactance.

Figure 2.2 displays the equivalent circuit of the rotor impedance $Z_{r0} = R_r + j\omega_s L_{r0} = R_r + jX_{r0}$ as a function of the slip factor given by

$$I_r = \frac{\overrightarrow{E_r}}{Z_r} = \frac{sE_{r0}}{R_r + jsX_{r0}} = \frac{E_{r0}}{\dfrac{R_r}{s} + jX_{r0}}$$

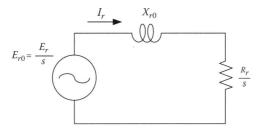

FIGURE 2.2 Equivalent circuit of the rotor.

or

$$Z_r = \frac{\overrightarrow{E_r}}{\overrightarrow{I_r}} = \frac{R_r}{s} + jX_{r0} \tag{2.7}$$

For small slip factors ($s \to 0$), the rotor impedance becomes predominantly resistive, and the rotor current can be considered to vary linearly with s. With very large slip factors, the rotor impedance plus skin effect will impose a nonlinear relationship.

Similarly to what is done in the representation of transformers, the model presented in Figure 2.1 can be converted to the model shown in Figure 2.3, converting the secondary parameters per phase to primary parameters per phase by using the transformation ratio a_{rms}:

$$E_1 = a_{rms} E_r \tag{2.8}$$

$$I_2 = \frac{I_r}{a_{rms}} \tag{2.9}$$

$$Z_2 = a_{rms}^2 \left(\frac{R_r}{s} + jX_{r0} \right) \tag{2.10}$$

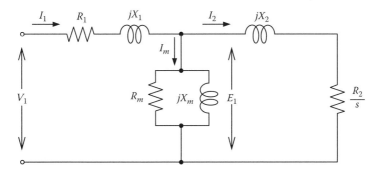

FIGURE 2.3 Equivalent circuit per phase of the induction generator.

or yet

$$Z_2 = \frac{R_2}{s} + jX_2 \tag{2.11}$$

with Z_2, R_2, and X_2 being the rotor values referred to the stator.

2.3 GENERATED POWER

The power balance in an induction machine or any machine can be expressed as follows:

$$P_{out} = P_{in} - P_{losses} \tag{2.12}$$

The output power, P_{out}, can be given in the form of three balanced voltages and three balanced currents, each one shifted by 120° and expressed in line values as follows:

$$P_{out} = \sqrt{3}V_\ell I_\ell \cos\phi \tag{2.13}$$

Like any other electric machine, an induction generator has inherent losses that can be combined in the following expression:

$$P_{losses} = P_{statorCu} + P_{iron} + P_{rotorCu} + P_{frict+air} + P_{stray} \tag{2.14}$$

The copper losses in the stator are obtained by $P_{statorCu} = I_1^2 R_1$. The iron losses are due to the hysteresis current (magnetizing) and Foucault current (current induced in the iron). It should be said at this point that the stator and rotor iron losses usually appear mixed, although in the latter they are much smaller and more difficult to distinguish. So the loss resistance, R_m, in the equivalent circuit of the induction generator practically represents these losses with the other mechanical losses. Subtracting these losses from the input power, the air gap losses remain dissipated in the power transfer from the stator to the rotor. The rotor copper losses, $P_{rotorCu} = I_2^2 R_2$, remaining due to the transformation of mechanical into electrical power should be subtracted from the power received by the rotor. In these losses are included the friction losses and the ones resulting from the rotor movement against the air and spurious losses. Spurious losses can be combined with other losses such as high frequency losses from stator and rotor dents and parasite currents produced by fast flux pulsations when dents and grooves move from their relative positions.[11–15] Finally, the output power is the net power in the shaft, also called shaft useful power.

As rotation increases, friction losses increase in ventilation (against the air) and spurious losses. In compensation, losses in the core decrease up to synchronous speed; these are called rotating losses. Such losses are considered approximately constant because some of their components increase with rotation and others decrease.

When voltage appears across the terminals of the induction generator for a given constant rotation, represented by the equivalent circuit of Figure 2.3, there is an average equivalent impedance per phase given by

$$Z = R_1 + jX_1 + \cfrac{1}{\cfrac{1}{jX_m} + \cfrac{1}{R_m} + \cfrac{1}{\cfrac{R_2}{s} + jX_2}} \tag{2.15}$$

Quantitatively, a current I_1 circulates through this impedance per phase that produces losses in the stator winding of the three phases whose total is given by

$$P_{statorCu} = 3I_1^2 R_1 \tag{2.16}$$

The iron core losses come from the stator and rotor circuits and are lumped together in a shunt resistor R_m. Sometimes, these core losses are lumped with the mechanical losses and subtracted from them to obtain the net mechanical power. So they may be given as follows:

$$P_{iron} = \frac{3E_1^2}{R_m} \tag{2.17}$$

The three-phase power transferred from the rotor to the stator through the air gap can be obtained from Equations 2.13, 2.16, and 2.17:

$$P_{airgap} = \sqrt{3} V_\ell I_\ell \cos\phi + 3I_1^2 R_1 + \frac{3E_1^2}{R_m} \tag{2.18}$$

On the other hand, from Figure 2.3, the only possible dissipation of the total power for the three phases corresponding to the secondary part of the circuit is through the rotor resistance as follows:

$$P_{airgap} = 3I_2^2 \frac{R_2}{s} \tag{2.19}$$

However, from the rotor equivalent circuit shown in Figure 2.2, the resistive losses are given by

$$P_{rotorCu} = 3I_r^2 R_r \tag{2.20}$$

In an ideal transformer, there are no losses; the rotor power remains unaltered when referred to the stator and, therefore, from Equation 2.19,

$$P_{rotorCu} = 3I_2^2 R_2 = sP_{airgap} \tag{2.21}$$

The mechanical power converted into electricity, for a negative slip s, is measured by the difference of the air gap power minus the dissipated power in the rotor, indicated by Equation 2.22, or

$$P_{converted} = 3I_2^2 \frac{R_2}{s} - 3I_2^2 R_2 \qquad (2.22)$$

which when simplified becomes

$$P_{converted} = P_{mec} = 3I_2^2 R_2 \frac{1-s}{s} = (1-s)P_{airgap} \qquad (2.23)$$

2.4 INDUCED TORQUE

Converted torque is instantaneous torque defined in terms of power in the shaft. It results from a force, F, which is concentrated on the shaft with surface of radius R exerting a torque on it with respect to the center of its cross section. In Figure 2.4, we see the work, dW, that this force exerts to move any point on the shaft to a distance $d\ell$:

$$dW = Fd\ell$$

where $d\ell = Rd\theta$ and, as a consequence, $dW = FRd\theta$.

By definition, the torque is $T = FR$. Therefore, $dW = Td\theta$. Deriving both sides of this equality with respect to time, we get

$$P = T\omega \qquad (2.24)$$

This induced torque in the machine is defined as the torque generated by the conversion of the mechanical power into internal electric power, which differs from the available torque T in the shaft because of the negative torques of friction and movement against the air. In such case, this torque is given by

$$T_{converted} = \frac{P_{converted}}{\omega_r} \qquad (2.25)$$

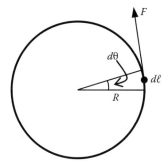

FIGURE 2.4 Torque on the shaft of a rotating machine.

Recalling from Equation 2.4 that $\omega_r = (1 - s)\omega_s$, from Equations 2.23 and 2.25, we can deduce a relationship between power and torque that does not change except for ω_s and see that it is therefore more directly useful to estimate the power transferred through the air gap as follows:

$$P_{airgap} = \omega_s T_{converted} \tag{2.26}$$

Two components of electrical torque are induced in the rotor: one due to the variable magnetic field of the stator and the other due to the rotor movement. Let us suppose that a sinusoidal magnetic field, Φ, is applied on the squirrel cage rotor, as shown by the flux lines represented in Figure 2.5. This sinusoidal field is characterized by an alternating elevation and reduction of the magnetic flux whose variations induce an electromotive force (emf) on the conductive segments of the rotor cage according to Faraday's induction law:

$$e = \frac{d\phi}{dt}$$

The induced emf on the rotor provokes a current obeying the right-hand rule and, in turn, a magnetic intensity around each conductor as represented in Figure 2.5a. The interaction between the magnetic intensities of stator and rotor generates, on each conductor segment, a force F_1 whose direction goes toward the center of the shaft, and the net torque is null:

$$F_1 = I\ell B$$

On the other hand, for a constant magnetic density, B (in Wb/m²), in the stator, when the rotor is driven by an external force F_2 at a tangent linear speed v (in m/s),

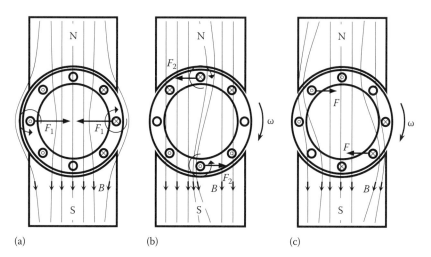

(a) (b) (c)

FIGURE 2.5 Magnetic torque of the induction generator. (a) Torque due to an increasing B, (b) torque due to rotor movement and static B, and (c) resulting torque.

according to Faraday's law of electromagnetic induction, it will induce an emf on each conductor segment of length ℓ of the rotor (in meters) given by

$$e = B\ell v$$

This induced emf, in turn, produces a current through the rotor segments trying to oppose the movement that has generated it, according to Lenz's law, as shown in Figure 2.5b. Again, as one can see, the net torque opposes the torque that has generated it.

Dynamically, that is, when the field is variable and the rotation is different from zero at the same time, both torques that previously had the same magnitude and opposed directions now are unbalanced with the highest torque in the direction of the initial rotor movement, according to Figure 2.5c. The magnetic field resulting from the stator is distorted although not represented in Figure 2.5.

Let us suppose now that the machine is rotating without load. Let us designate by B the resulting flux that will only exist due to the magnetizing current I_m, which, in turn, is proportional to the stator effective voltage, $E_1 \cong V_{ph}$. These values will be constant for a constant V_{ph}. Actually, under load, there is a small voltage drop across the stator winding due to load variations (current I_1) through the stator impedance Z_1, which is very small, as a rule. Therefore, without load, the magnetic flux can be represented as in Figure 2.6a where δ is also the angle between I_r and I_m. The power factor angle, θ_r, of the rotor circuit is given by

$$\theta_r = \tan^{-1}\left(\frac{2\pi sf_s L_2}{R_2}\right) \tag{2.27}$$

The torque induced by the rotor movement alone is given by

$$T = kB_r B \sin \delta \tag{2.28}$$

where
    k is a proportionality factor
    $\delta = \theta_r + 90°$

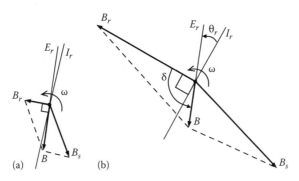

FIGURE 2.6 Magnetic fields of the induction generator. (a) Light loads and (b) heavy loads.

The torque may be determined from the rotor current, the total magnetic field, and the power factor. These vectors are represented in Figure 2.6.

As the magnetic field is quite reduced in the rotor, the induced torque will also consume power only for the losses. However, if there is current through the stator winding, there will be an increase in the slip factor and a larger relative movement between rotor and stator. The larger the slip factor, the larger will be the induced power on the rotor through E_r and, therefore, the larger will be I_r and B_r. The angle between the rotor current and its magnetic field increases. The frequency of the rotor increases according to $f_r = sf_s$, therefore, it increases the rotor reactance, causing a larger delay of I_r and of the consequent magnetic field regarding V_r, according to the diagram of Figure 2.6b. The magnetic field B_r increases as much as δ ($> 90°$), almost having a torque compensation effect according to Equation 2.28 with predominance of the first on the second.

With the increase of the exerted torque on the rotor, there will be a point, known as the disrupting point or point of maximum torque, where the term "$\sin \delta$" in Equation 2.28 begins to decrease in larger proportions than B_r can increase, tending to a reduction in the power supplied until total collapse occurs. Further discussions and characteristics about this subject are dealt with in Section 5.2.

2.5   REPRESENTATION OF INDUCTION GENERATOR LOSSES

The stator power transferred from the rotor through the air gap is consumed in driving the rotor and in copper losses whose values can be represented independently in the machine equivalent model. This value can be deduced directly from Equation 2.23, being defined as an equivalent resistor per phase to represent the input power:

$$R_{2eq} = R_2 \frac{(1-s)}{s} \tag{2.29}$$

So the equivalent circuit of Figure 2.3 can be transformed as shown in Figure 2.7 with the copper losses separated from the converted power and neglecting losses, R_m.

To estimate the generator efficiency, we use

$$\eta = \frac{P_{out}}{P_{in}} = 100\% \tag{2.30}$$

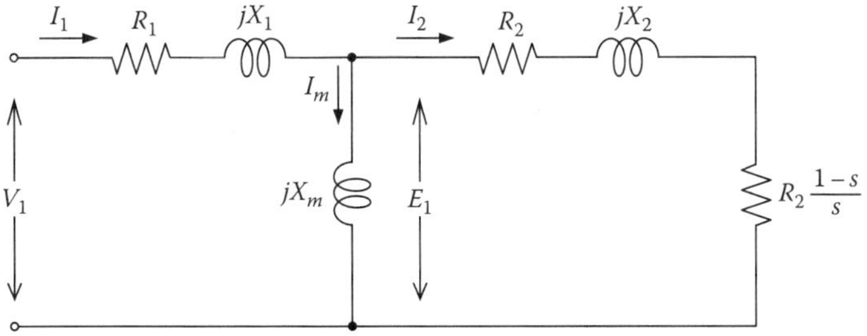

FIGURE 2.7   Separate representation per phase of the copper losses and the power converted in the rotor.

or using Equations 2.13 and 2.14, we get

$$\eta = \frac{P_{in} - P_{losses}}{P_{in}} 100\% = \frac{-3I_2^2 R_2 \dfrac{1-s}{s} - P_{losses}}{-3I_2^2 R_2 \dfrac{1-s}{s}} \qquad (2.31)$$

The efficiency of the induction generator is not lower than that of the synchronous generator, although some manufacturers claim otherwise (see Section 5.8).[2] Even so, as we will see later on, induction generators have been used in practice only for low power, more commonly in hydro and wind electric generation.

As has already been said, from the active power per phase supplied to the machine, P, we should discount the stator copper losses, $I_1^2 R_1$, the hysteresis losses, parasites currents, friction, and air opposition, represented by E_1^2/R_m. The amount of these losses varies with the frequency, temperature, and operating voltage. However, this model is sufficiently approximate for the rated values for most applications, which, as a rule, do not stray far from the rated values. Thus, we may have the following:

$$P_h = I_1^2 R_1 + \frac{E_1^2}{R_m} \qquad (2.32)$$

Recall that s is negative for a generating operation. The net output power, P_{out}, of the machine can be obtained from Equations 2.13, 2.21, and 2.31 as

$$P_{out} = P_{mec} - P_{losses} = -3I_2^2 R_2 \frac{1-s}{s} - 3I_1^2 R_1 - \frac{3E_1^2}{R_m} - 3I_2^2 R_2 - P_{frict+air} - P_{stray}$$

or, simplified, as

$$P_{out} = P_{airgap} - P_{st-losses} = -3I_2^2 R_2 \frac{1}{s} - 3I_1^2 R_1 - \frac{3E_1^2}{R_m} - P_{frict+air} - P_{stray} \qquad (2.33)$$

2.6   MEASUREMENT OF INDUCTION GENERATOR PARAMETERS

Evidently, measurement of induction generator parameters follows the same methods as for induction motors found in the classic texts and standard test procedures about electrical machines (for example, IEEE Std. 112). They can be useful in models and control algorithms under variable load. Similar to the methods used in transformers, these parameters should be precisely established in laboratory tests of short circuit and open circuit conditions. The losses are combined in Table 2.1.

For the conventional short circuit and no-load tests discussed in any basic text on alternate machines and defined by IEEE Std. 112, we need two wattmeters, three ampere meters, and a voltmeter, under the usual hypothesis that the three-phase

TABLE 2.1

Standard Loss Distribution in an Induction Machine

Type of Loss	Description
Friction and windage	Mechanical loss due to bearing (and brush) friction and windage
Core	Loss in iron at no load
Stator ($I_1^2 R_1$)	$I_1^2 R_1$ loss in stator windings
Rotor ($I_2^2 R_2$)	$I_2^2 R_2$ loss in rotor windings (and brush-contact loss of wound rotor machines)
Stray load	Stray loss in iron and eddy-current losses in conductors

Source: IEEE Std. 112-1991, *IEEE Standard Test Procedure for Polyphase Induction Motors and Generators.*

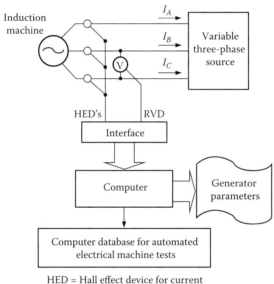

FIGURE 2.8 Automatic test to obtain induction machine parameters.

source is well designed for amplitude and phase unbalances. If it is available, a computerized bank of tests can be used with sensors for the instantaneous values of voltage and current, as sketched in Figure 2.8. With these values, direct measure of the average value of power can be skipped; it can be obtained by processing only two variables (instantaneous voltage and current). Other alternatives also exist.

To test the machine without load (open circuit), we measure the rotating losses and the magnetizing current. The blocked rotor test measures the circuit parameters. In other words, if $s = 1$ in Figure 2.3 (short circuit), R_2/s becomes equal to R_2 or almost zero, as much as X_2.

2.6.1 BLOCKED ROTOR TEST ($s = 1$)

The blocked rotor test is the equivalent of the preceding test for a short-circuited secondary transformer. The voltage, usually very low, is slowly increased across the terminals of the motor with the rotor locked until the current reaches the rated value being measured, then, voltage, current, and power. The equivalent circuit used for $s = 1$ is shown in Figure 2.9 where R_m and X_m have been subtracted from the complete equivalent circuit of Figure 2.3. Notice that the magnetizing parameters, R_m and X_m, are a lot larger, respectively, than R_2 and X_2, and, for simplicity, they can be neglected.

An important aspect to be considered in the blocked rotor test would be that the rotor assumes the frequency of the test source when, in reality, this is around 1–3 Hz for slip factors of 2% and 4%. The slip factor can be larger in motors designed for low noise, as in air conditioning fan motors. To compensate for this phenomenon, the use of 15 Hz is recommended (25% of rated frequency and at rated current according to IEEE Std. 112).[9]

Once V_ℓ, I_ℓ, and P_{in} have been measured in the blocked rotor test, Equation 2.13 enables us to obtain the power factor of the motor. The modules of the total impedance are, respectively,

$$\left|Z_{blocked}\right| = \frac{V_{ph}}{I_\ell} = \frac{V_\ell}{\sqrt{3}I_\ell} \tag{2.34a}$$

$$\cos\phi = \frac{P_{in}}{\sqrt{3}V_\ell I_\ell} \tag{2.34b}$$

From Equation 2.34, we obtain the values of summations of resistances and reactances (at the test frequency) of the stator and rotor defined as follows:

$$R_1 + R_2 = \left|Z_{blocked}\right|\cos\phi \tag{2.35}$$

$$X_1 + X_2 = \left|Z_{blocked}\right|\sin\phi \tag{2.36}$$

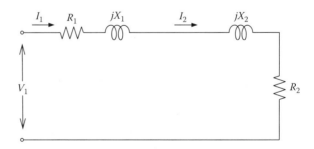

FIGURE 2.9 Approximate circuit of the induction generator with blocked rotor.

To obtain R_1, usually, an approximate measure is used in dc current from ordinary multitesters across the stator terminals. As conventional multitesters use dc current for measurements of resistances, of course, they will not include in the measure any reactance or reaction of the rotor and only R_1 will oppose the passage of current. Here, the effects of heating, frequency, and, most importantly, skin effects on the motor are not included. In the three-phase windings, the measure across the motor terminals always will measure the resistance of two windings in series. As a consequence,

$$R_1 \approx \frac{V_{dc}}{2I_{dc}} \qquad (2.37)$$

With this value of the stator winding resistance, R_2 can be determined from Equation 2.35.

As the sum of reactances in Equation 2.36 is proportional to the measured frequency, if the rated frequency is different from the test frequency, then the results should be multiplied by $F = f_{rated}/f_{test}$.

In practice, commercial motors make individual contributions of the rotor and the stator to the total reactance of the machine. The reactance X_1 contributes something between 30% and 50% and X_2 between 50% and 70%.[10,11,15]

2.6.2 No-Load Test ($s = 0$)

The no-load test begins by making the machine work as an induction motor, freely rotating under the rated voltage without any load on the shaft. In this way, the power dissipation just overcomes friction and air opposition; these are mechanical losses. The slip factor will be minimal (on the order of 0.001 or less). With this slip factor, and from Figure 2.7, we can see that the resistance corresponding to the converted power is much larger than the rotor losses in the copper or in the rotor reactance. The losses are concentrated in friction, core iron, and air windage opposition, $R_{frict+iron+air}$. The equivalent circuit will be transformed, then, as displayed in Figure 2.10.

Notice that $R_{frict+iron+air}$ is parallel to X_m for $s \approx 0$, but much larger, making $I_2 \approx 0$; therefore, the copper losses in the rotor are negligible. As a result, it can be established that

$$P_{in} = 3I_1^2 R_1 + P_{iron} + P_{frict+air} + P_{stray} \qquad (2.38)$$

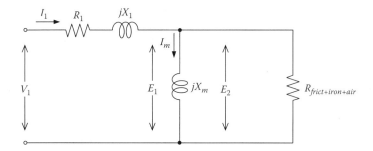

FIGURE 2.10 Equivalent circuit of the induction generator at no load.

Since the current necessary to establish the magnetic field in induction motors is quite large due to the high reluctance of the air gap, the resistances due to iron losses, R_{iron}, and that of the converted power, $R_2(1 - s)/s$, both in parallel with X_m, the magnetizing reactance becomes much smaller than that total parallel resistance. The power factor will be very small. This circuit will be predominantly inductive and expressed as follows:

$$\frac{V_{ph}}{I_{10}} \cong \left(X_1 + X_m \right) \tag{2.39}$$

where I_{10} is the stator current with the motor without load.

2.7 FEATURES OF INDUCTION MACHINES WORKING AS GENERATORS INTERCONNECTED TO THE DISTRIBUTION NETWORK

As stated earlier (Equation 2.22), when $n_r > n_s$, the developed mechanical power becomes negative. In other words, when $n_r > n_s$, the asynchronous machine begins to consume mechanical power working as an electric power generator. In electro-magnetic terms, for rotations above the synchronous, the relative speed between the rotating flux and the rotor itself changes sign for speeds below the synchronous. Therefore, the rotor voltage and current change direction (change sign) and, there-fore, V_1 and E_1 do the same. Equations transformed by the sign change and describing this phenomenon from Figure 2.3 can be combined as follows[11–13]:

$$E_1 = V_1 + I_1 R_1 + j I_1 X_1 \tag{2.40}$$

$$E_2 = I_2 \frac{R_2}{s} + j I_2 X_2 \tag{2.41}$$

$$E_1 = E_2 = -I_m Z_m = \left(I_1 - I_2 \right) Z_m \tag{2.42}$$

In Equation 2.41, the slip factor s is negative, and the representation of these equations is shown in Figure 2.11.[1,10,11] As s is negative in the generator, the secondary emf sE_2 is leading by 90° with respect to the magnetic field B (and opposed to the direction of the secondary voltage if it is a motor). For the same reason, the voltage drop $j I_2 s X_2$ is lagging regarding the current I_2 in 90°, and I_2 is leading with respect to sE_2. The stator voltage E_1 is leading V_{ph}.

In the induction generator, the magnetizing current produces the magnetizing field B. For this, synchronous generators or other electrical sources are used to feed external circuits in association with induction generators. Some interesting conclusions can be drawn from the earlier observations. As the current I_m reaches values around 20%–40% of I_r and supplies the generator at the circuit voltage, the excitation power (in kVA) will also reach 20%–40% of the generator's rated values.

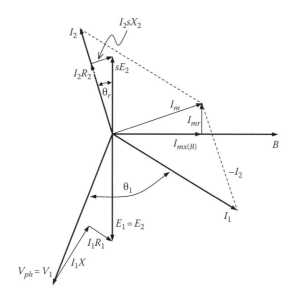

FIGURE 2.11 Vector diagram of the induction generator.

Therefore, if two to four induction generators of the same output are installed in a distribution network, their excitation needs will pull all the capacity of a synchronous generator of similar capacity, recalling that the excitation power of synchronous generators is smaller than 1%. Such differences in the need for excitation power are the main argument against the induction generators. Besides that, the current I_m is lagging behind the current I_2 by approximately 90°, and, consequently, the parallel operation of induction generators with synchronous generators reduces the power factor even when the external load is purely active. We see that the synchronous generator should supply the lagging current as much for the usual case of inductive loads as for the induction generator. For this reason, the induction generator has a double disadvantage with respect to the synchronous: it is not capable of supplying lagging current for the load, and, on top of that, it will draw lagging current from the distribution network.

Connecting induction generators to the distribution network is a quite simple process, as long as interconnection and protection guidelines are followed with the local utilities. The rotor is made to turn in the same direction in which the magnetic field is rotating, as closely as possible to synchronous speed to avoid unnecessary speed clashes. A phenomenon similar to the connection of motors or transformers in the distribution network will happen.

The active power supplied by an induction generator to the circuit, similarly to what happens with the synchronous generators, can be controlled by speed variation, in other words, controlled by the mechanical primary power.

In the case of the stand-alone operation of induction generators, the magnetizing current can be obtained from the self-excitation process. The generator can supply a capacitive current, because, as seen in Figure 2.11, the current I_1 is leading V_{ph}. This is due to the fact that the mechanical energy of rotation can only influence the active component of the current; it does not affect the reactive component. Therefore, the rotor does not supply reactive current. As a component of the magnetizing current in the main field direction, $I_{mx(B)}$, is needed to maintain this flux, another reactive current is also needed to maintain the dispersion fluxes. Such currents should be supplied by a leading current with respect to the stator current.

As the induction generator depends on the parallel synchronous machine to get its excitation, the short circuit current that it can supply depends on the voltage drop produced across the terminals of the synchronous generator. A very intense short circuit transient current arises, however, of extremely short duration. If the voltage across the terminals goes to zero, the steady-state short circuit current is zero. A small current is supplied in the case of a partial short circuit since the maximum power the induction generator can supply with fixed slip factor and frequency is proportional to the square of the voltage across its terminals (see Equation 5.1). The incapacity of sustaining short circuits greatly reduces possible damage caused by electrical and mechanical stresses. As a consequence, it allows the use of short circuit power of smaller capacity and cost when compared with the case when only synchronous generators provide all the capacity of the facilities.

If the induction generator is self-excited on a stand-alone basis with respect to the distribution network, a very high load current or a short circuit across its terminals alters the value of the effective exciting capacitance or it may remove it altogether. In that case, there is no excitation possibility; voltage and current collapse immediately to zero.

The induction generator has no problems with oscillations, as long as it does not have to work at the synchronous speed. All the load variation is accompanied by a speed variation and a small phase displacement, much the same as with the synchronous generator. The mechanical speed variations of the primary machine driving the generator are so small that they only produce minor variations of load.

In short, the advantages of the induction generator are as follows: (1) a robust and solid rotor, (2) the machine cannot feed large short circuit currents and is free of oscillations, (3) the rotor construction is suitable for high speeds, (4) it does not need special care with regard to synchronization, and (5) voltage and frequency are regulated automatically by the voltage and frequency of the distribution network in parallel operation.

Its disadvantages are as follows: (1) the power factor is determined by the slip factor and has very little to do with the power factor of the load when working in parallel with synchronous machines, and (2) synchronous machines need to supply both the load and the induction generator with lagging reactive power so they have to operate at a power factor higher than that of the load.

2.8   HIGH-EFFICIENCY INDUCTION GENERATOR

A high-efficiency induction generator is commercially available as a high-efficiency induction motor, except for some peculiarities. Therefore, the same care must be taken in design, materials selection, and manufacturing processes for building a high-efficiency generator.

The main advantages of the high-efficiency induction generator compared with the conventional induction generator are better voltage regulation, less loss of efficiency with smaller loads, less oversizing when generators of lower power cannot be used, reduced internal losses, and, therefore, lower temperatures, less internal electric and mechanical stress, and, thus, increased useful life. The constraints are the need for larger capacitors for self-excitation. High-efficiency induction generators should not be used for self-excited applications.

The efficiency of the high-efficiency generator compared with the standard ones differs by more than about 10% for small power ratings (up to 50 kW) and about 2% for higher powers (above 100 kW). It is therefore highly recommended for micro-power plants. Rated efficiencies are normalized, and they should have guaranteed minimum values stated by the manufacturer on the plate of the machine for each combination of power versus synchronous speed.

The investment return time (payback period) for these high-efficiency generators will be smaller in an inversely proportional way to the local tariff values in the cases of injection into the public distribution network. This investment will be increased by the number of operating hours, increased by the increased difference in the efficiency with respect to the conventional ones, and smaller for small load power with respect to the rated power of the machine. The investment return time can go from 0.9 to 3 or 4 years for micropower plants.

High-efficiency generators are better suited to stand the harmful effects of the harmonic generated by nonlinear loads (power converters) because they have higher thermal margin and smaller losses.

2.9 DOUBLY FED INDUCTION GENERATOR

The doubly fed induction generator, also known as the Scherbius variable speed driver, is a wound rotor machine with slip rings to allow control of the rotor winding current. The rotor circuit is connected to an external variable frequency source via slip rings, and the stator is connected to the grid network as illustrated in Figure 2.12. The speed of the doubly fed induction generator, which is usually limited to a 2:1 range, is controlled by adjusting the frequency of the external rotor source of current.

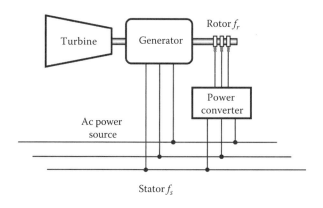

FIGURE 2.12 Doubly fed induction generator.

These machines have not been very popular due to the easy wearing out of the slip rings. More recently, with the development of new materials and power electronics, the doubly fed induction generator has made a comeback for up to several hundreds of kilowatt ratings. Power converters whose cost for large units becomes irrelevant usually make up the need for a variable frequency source for the rotor. More details are given in Chapter 11.

2.10 SOLVED PROBLEMS

1. Determine the necessary primary mechanical torque of a turbine for driving an induction generator of 1.0 kW, 2.3 kV, 60 Hz, 12 poles, and $s = -0.03$.

   Solution

$$P_{conv} = 1000\,\text{W} \quad \text{so,} \quad P_{airgap} = \frac{P_{conv}}{1-s} = \frac{1000}{1-0.03} = 1030.9\,\text{W}$$

Then

$$n_s = \frac{120 f_s}{p} = \frac{120 \times 60}{12} = 600\,\text{rpm, so}$$

$$\omega_s = 2\pi \frac{600}{60} = 20\pi\,\text{rad/s}$$

$$T_{airgap} = \frac{P_{airgap}}{\omega_s} = \frac{1030.9}{20\pi} = 16.41\,\text{N}\,\text{m} \quad \Rightarrow T_{mech} = T_{airgap} = 16.41\,\text{N}\,\text{m}$$

2. What is the slip factor of a 100 hp, 6 pole, 60 Hz, three-phase induction generator to achieve its rated power when the rotor current is 10 A and its rotor resistance is 0.407 Ω? What would be the turbine torque under these conditions?

   Solution

$$P = 100\,\text{hp} = 74{,}600\,\text{W}$$

$$P_{conv} = 3 \times I_r^2 \times R_r \times \frac{1-s}{s}$$

$$\Rightarrow 74{,}600 = 3 \times 10^2 \times 0.407 \times \frac{1-s}{s}$$

$$\frac{1-s}{s} = 610.97$$

$$1 - s = 610.97s \Rightarrow 1 = 611.97s \Rightarrow s \approx 0.002$$

$$P_{airgap} = \frac{74{,}600}{1-s} = 74{,}722.1\,\text{W}$$

$$\therefore T_{mech} = \frac{P_{airgap}}{\omega_s} = 594.62\,\text{N m}$$

2.11 SUGGESTED PROBLEMS

2.1 A 10 hp, Y-connected, 60 Hz, 380 V, three-phase induction generator is used to drive a small micropower plant. The generator's load is 500 W for a 4 A line current at 1740 rpm. The rotor resistance is 4.56 Ω at 75°C. The losses are equal to 28 W. A no-load test of the machine gave the following results: $P = 215$ W, $I = 3.0$ A, and $V = 380$ V. Calculate the output power, the efficiency, and the power factor for the given load. If the generator was Δ-connected with the same data as mentioned earlier, what would be the difference in your calculations?

2.2 Explain the difference between rotor speed and rotor frequency. For $s = -0.01$ and a rotor speed of 1740 rpm, what is the rotor frequency for a 60 Hz induction generator?

2.3 Explain why the efficiency of an induction generator can be smaller when working as an asynchronous motor under certain conditions.

2.4 Calculate the efficiency of a 110 V, 60 Hz induction generator connected directly to the grid with the following parameters: $R_m = 36.2$ Ω; losses equal to 0.8% of the rated power; $s = -0.03$; $R_1 = 0.2$ Ω; $R_2 = 0.15$ Ω; $X_1 = 0.42$ Ω; $X_2 = 0.43$ Ω.

REFERENCES

1. Lawrence, R.L., *Principles of Alternating Current Machinery*, McGraw-Hill Book Co., New York, 1953, p. 640.
2. Kostenko, M. and Piotrovsky, L., *Electrical Machines*, Vol. II, Mir Publishers, Moscow, Russia, 1969, p. 775.
3. Smith, I.R. and Sriharan, S., Transients in induction motors with terminal capacitors, *Proc. IEEE*, 115, 519–527, 1968.
4. Murthy, S.S., Bhim Singh, M., and Tandon, A.K., Dynamic models for the transient analysis of induction machines with asymmetrical winding connections, *Electr. Mach. Electromech.*, 6(6), 479–492, November/December 1981.
5. IEEE Std. 1547, *Standard for Interconnecting Distributed Resources with Electrical Power Systems*.
6. IEEE Std. (Draft) P1547.1, *Standard Conformance Test Procedures for Interconnecting Distributed Energy Resources with Electric Power Systems*.
7. IEEE Std. (Draft) P1547.2, *Application Guide for IEEE Standard 1547: Interconnecting Distributed Resources with Electric Power Systems*.

8. IEEE Std. (Draft) P1547.3, *Guide for Monitoring, Information Exchange, and Control of Distributed Resources Interconnected with Electric Power Systems.*

9. IEEE Std. 112–1991, *IEEE Standard Test Procedure for Polyphase Induction Motors and Generators.*

10. Liwschitz-Gärik, M. and Whipple, C.C., *Alternating Current Machines (Máquinas de Corriente Alterna)*, Companhia Editorial Continental, authorized by D. Van Nostrand Company, Princeton, NJ, 1970, p. 768.

11. Chapman, S.J., *Electric Machinery Fundamentals*, 3rd edn., McGraw-Hill International Edition, New York, 1999.

12. Barbi, I., *Fundamental Theory of the Induction Motor (Teoria Fundamental do Motor de Indução)*, Ed. da UFSC, Eletrobrás, Florianópolis, Brazil, 1985.

13. Langsdorf, A.S., *Theory of Alternating Current Machines (Teoría de Máquinas de Corriente Alterna)*, McGraw-Hill Book Co., New York, 1977, p. 701.

14. Hancock, N.N., *Matrix Analysis of Electric Machinery*, Pergamon, Oxford, U.K., 1964, p. 55.

15. Grantham, C., Determination of induction motor parameter variations from a frequency stand still test, *Electric Mach. Power Syst.*, 10, 239–248, 1985.

3 Transient Model of Induction Generators

3.1 SCOPE OF THIS CHAPTER

This chapter will explain how steady-state terminal voltage builds up during the self-excitation process and during the recovery of voltage during perturbations across the terminal voltage and stator current due to load changes. A set of state equations are derived to obtain the instantaneous output voltage and current in the self-excitation process. In the example in this chapter, the equations describe an induction machine with the following rated data: 220/380 V, 26/15 A, 7.5 kW, and 1765 rpm. We also show that the equations in this chapter can be easily used to calculate output voltage, including the variation of mutual inductance with the magnetizing current discussed further in Chapter 4. Furthermore, we will present a general matrix equation for simulation of the parallel aggregation of induction generators (IG). In any case, for every machine at a given speed, we show that there is a minimum capacitance value causing self-excitation in agreement with the steady state case shown in Chapter 2. We will discuss rotor parameter variation and demonstrate that it has little effect on the accuracy of these calculations and that comparable results can be achieved using the standard open circuit and locked rotor parameters. Saturation effects are also considered in this chapter by taking into account the nonlinear relationship between magnetizing reactance and the magnetizing current of a machine; using this approach, the mutual inductance, M, will be shown to vary continually.

3.2 INDUCTION MACHINE IN TRANSIENT STATE

During self-excitation, asynchronous or IGs exhibit transient phenomena that are very difficult to model from an operational point of view; the impact of self-excitation is more pronounced in generators with heavier loads.[1-3] A crucial problem to be avoided is the demagnetization of the IG. To analyze an IG in a transient state, we will apply the Park transformation (also known as the Blondel or Blondel-Park transformation), which is associated with the general theory of rotating machines used in this chapter.

The Park transformation specifies a two-phase primitive machine with fixed stator windings and rotating rotor windings to represent fixed stator windings (direct axis) and pseudo-stationary rotor windings (quadrature axis) (see Figure 3.1). Any machine can be equivalent to a primitive machine with an appropriate number of coils on each axis.[4-6]

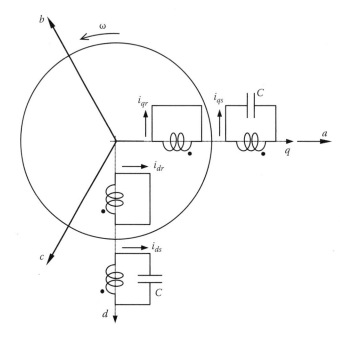

FIGURE 3.1   Representation of an unloaded SEIG in d–q axis.

Saturation effects are also considered in this chapter by taking into account a non-linear relationship between the magnetizing reactance and the magnetizing current of the machine. Under this approach, the mutual inductance M varies continuously.

3.3   STATE SPACE–BASED INDUCTION GENERATOR MODELING

Figure 3.1 represents a primitive machine through a Park transformation as applied to the induction generator. Currents i_{ds} and i_{qs} refer to the stator currents, and I_{dr} and I_{qr} to the rotor currents, in the direct and quadrature axis, respectively. The angular speed $\omega = d\theta/dt$ is the mechanical rotor speed.

As can be seen in Figure 3.1, no external voltage is applied across the rotor or the stator windings. This is the standard form of stationary reference axis used in the machine theory texts.[7–10] However, there is an additional component: the self-excitation capacitor C.

3.3.1   No-Load Induction Generator

A self-excited induction generator (SEIG) with a capacitor is initially considered to be operating at no load.[1,2] The relationships between the resulting voltages and currents, direct and quadrature, can be obtained from Figures 3.1 and 3.2 and Equation 3.1. That expression represents the ordinary symmetrical three-phase machine connected to a three-phase bank of identical parallel capacitors. The reference position is put on the stator under every normal operating condition,

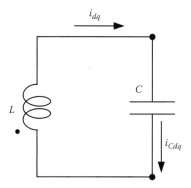

FIGURE 3.2   Equivalent stator d–q circuit of an unloaded SEIG.

including transient ones. The subscript s refers to the stator, while r refers to the rotor. The voltage drop across the capacitor (direct axis, v_{cd} and quadrature axis, v_{cq}) is included in the matrix in the expression $1/pC$. The transformer ratio from the stator to the rotor is assumed to be unity, and transformations should be introduced in the rotor parameters when there is a need to refer them to the stator:

$$
\begin{bmatrix} v_{ds} \\ v_{qs} \\ v_{dr} \\ v_{qr} \end{bmatrix} =
\begin{bmatrix}
R_1 + L_1 p + \dfrac{1}{pC} & 0 & Mp & 0 \\
0 & R_1 + L_1 p + \dfrac{1}{pC} & 0 & Mp \\
Mp & \omega M & R_2 + L_2 p & \omega L_2 \\
-\omega M & Mp & -\omega L_2 & R_2 + L_2 p
\end{bmatrix}
\cdot
\begin{bmatrix} i_{ds} \\ i_{qs} \\ i_{dr} \\ i_{qr} \end{bmatrix}
\tag{3.1}
$$

where
  v_{ds} and v_{qs} are, respectively, the direct and quadrature stator voltages
  v_{dr} and v_{qr} are, respectively, the direct and quadrature rotor voltages

The mutual inductance between stator and rotor (M) varies with the current through the windings and the relative positions between stator and rotor. So the d–q coupled voltage drops are due to the relative position of the coils as well as the current variation through them. Therefore, an additional term ipM is included in the analytical solution; that is, for the mutual flux given by $\phi_{mi} = Mi$, an additional voltage drop is established as follows:

$$
v = \frac{d\phi_m}{dt} = M\frac{di}{dt} + i\frac{dM}{dt} = Mpi + ipM
\tag{3.2}
$$

The first term of Equation 3.2 represents current variation due to generator stator current. The second term, due to variations of the rotor speed, takes into account the short period of the calculation step until self-excitation reaches its steady state. It can be considered either constant and equal to ω or variable and recalculated at each step according the desired variation law from zero to the steady state value.[9–12]

3.3.2 State Equations of SEIG with Resistive Load, R

A resistive load, R, is added in parallel with the self-excitation capacitor. There is a voltage across this load that will be the same as that of the self-excitation capacitor, as illustrated in Figure 3.3. So taking i_{Ld} as the direct axis current through the load, we have

$$v_{Ld} = V_{Cd} = Ri_{Ld}$$ (3.3)

Therefore,

$$i_{Cd} = Cpv_{Cd} = CpRi_{Ld}$$

As $i_{ds} = i_{Cd} + i_{Ld}$, then $i_{ds} = RCpi_{Ld} + i_{Ld}$ that, rearranged as a function of i_{Ld}, results in

$$i_{Ld} = \frac{i_{ds}}{RCp+1}$$ (3.4)

Replacing Equation 3.4 in 3.3 yields

$$v_{Ld} = \frac{R}{RCp+1}i_{ds}$$ (3.5)

Similarly, the quadrature load voltage can be obtained as

$$v_{Lq} = \frac{R}{RCp+1}i_{qs}$$ (3.6)

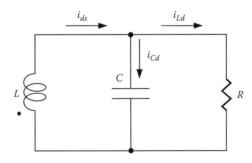

FIGURE 3.3   Stator direct axis component with an *RLC* load.

So Equation 3.1 assumes the following form to take into account the R load:

$$\begin{bmatrix} v_{ds} \\ v_{qs} \\ v_{dr} \\ v_{qr} \end{bmatrix} = \begin{bmatrix} R_1 + L_1 p + \left(\dfrac{R}{RCp+1} \right) & 0 & Mp & 0 \\ 0 & R_1 + L_1 p + \left(\dfrac{R}{RCp+1} \right) & 0 & Mp \\ Mp & \omega M & R_2 + L_2 p & \omega L_2 \\ -\omega M & Mp & -\omega L_2 & R_2 + L_2 p \end{bmatrix} \cdot \begin{bmatrix} i_{ds} \\ i_{qs} \\ i_{dr} \\ i_{qr} \end{bmatrix}$$

(3.7)

3.3.3 STATE EQUATIONS OF SEIG WITH *RLC* LOAD

We will now consider an *RLC* load, which is supposed to be an *RL* load in parallel with the self-excitation capacitor. Therefore, an overall value C will be considered in the following discussion. As was explained in the previous section, the voltage across the self-excitation capacitor is the same as that across the *RL* load. Therefore, with $v_{Ld} = Ri_{Ld} + Lpi_{Ld}$, for the current through the capacitor, we have

$$i_{Cd} = Cpv_{Ld} = \left(RCp + LCp^2 \right) i_{Ld}$$

In the same way, as $i_{ds} = i_{Cd} + i_{Ld}$, we have $i_{ds} = (RCp + LCp^2)i_{Ld} + i_{Ld}$, or

$$i_{Ld} = \frac{i_{ds}}{RCp + LCp^2 + 1}$$

(3.8)

and therefore, $v_{Ld} = (R + Lp)i_{Ld}$ or

$$v_{Ld} = \frac{R + Lp}{RCp + LCp^2 + 1} i_{ds}$$

(3.9)

Similarly,

$$v_{Lq} = \frac{R + Lp}{RCp + LCp^2 + 1} i_{qs}$$

(3.10)

Including the load inductance effects, Equation 3.1 becomes

$$
\begin{bmatrix} v_{ds} \\ v_{qs} \\ v_{dr} \\ v_{qr} \end{bmatrix} =
\begin{bmatrix}
R_1 + L_1 p \\ \quad + \left(\dfrac{R+LP}{RCp+LCp^2+1} \right) & 0 & Mp & 0 \\[3ex]
0 & \begin{array}{c} R_1 + L_1 p \\ + \left(\dfrac{R+Lp}{RCp+LCp^2+1} \right) \end{array} & 0 & Mp \\[3ex]
Mp & \omega M & R_2 + L_2 p & \omega L_2 \\
-\omega M & Mp & -\omega L_2 & R_2 + L_2 p
\end{bmatrix}
\cdot
\begin{bmatrix} i_{ds} \\ i_{qs} \\ i_{dr} \\ i_{qr} \end{bmatrix}
$$

$$(3.11)$$

The impedance matrix obtained earlier is similar to that in the classic literature for induction machines, except for the expression of the voltage drop across the self-excitation capacitor (Equations 3.9 and 3.10), the voltage drop across the generator terminals, and the additional terms corresponding to each load type. Notice that if the load includes a capacitor, it can be added in parallel with the self-excitation capacitor in the matrix representation or in another form linked with the load. Therefore, independently of the load type, the following system of equations holds $[v] = [Z][i]$. For the determination of the instantaneous currents, it is enough to obtain $[v] = [Z]^{-1}[i]$.

To find the voltage across the self-excitation capacitor, it is enough to take the case of an SEIG feeding an RL load. Knowing that i_{Ld} and i_{Lq} correspond to the currents through the load in the direct and quadrature axis, respectively, we have

$$pv_{Ld} = \frac{i_{ds}}{C} - \frac{i_{Ld}}{C} \tag{3.12}$$

$$pv_{Lq} = \frac{i_{qs}}{C} - \frac{i_{Lq}}{C} \tag{3.13}$$

The solution to the system of equations given earlier describes the behavior of currents and voltages during the self-excitation process and in steady operation, besides allowing the verification of load variations. In the solution of Equation 3.11, the alterations of the magnetizing reactance as a function of the magnetizing current must take into account through corrections to satisfy the magnetic saturation data, as discussed in Section 3.3.1.

For a more realistic representation of IGs, the parameters of the machine are obtained through conventional short circuit and open circuit tests, as discussed in Chapter 4. These tests supply resistance values and rotor and stator reactances for a given frequency. The conventional numeric methods of solving simultaneous equation systems seem to be sufficiently appropriate.[11–13] If greater precision is needed, the resistance and reactance variations of the rotor as a function of magnetism and of frequency should be considered.

It is important to observe that for the self-excitation process to initiate the machine must possess a residual magnetism. If this magnetism does not exist, it must be generated by connecting a battery or a net-connected rectifier, by using capacitors previously charged, or by keeping the rotor rotating for some time with the machine in a no-load condition. The effect of the residual magnetism is described in the equations deduced previously in the terms v_{dr}, v_{qr}, v_{ds}, and v_{qs}. Such voltages vary according to the hysteresis curve of the motor. Therefore, they cannot have a constant value because this would generate incorrect results.

3.4 PARTITION OF SEIG STATE MATRIX WITH *RLC* LOAD

Our objective in this section is to find a model to separate machine parameters from load and self-excitation parameters. Equation 3.11 represents the transient electric behavior of a self-excited stand-alone IG, so describing the aggregation of more than one IG is easy.

Consider the state equations that describe the voltage across the terminals of the generator to the current through any *RLC* load. From Figure 3.3, we get

$$i_{Cd} = i_{ds} - i_{Ld} \tag{3.14}$$

or

$$pv_{Ld} = \frac{i_{Cd}}{C} = \frac{i_{ds}}{C} - \frac{i_{Ld}}{C} \tag{3.15}$$

Similarly, for the quadrature current, we have

$$pv_{Lq} = \frac{i_{qs}}{C} - \frac{i_{Lq}}{C} \tag{3.16}$$

From the load voltage $v_{Ld} = Ri_{Ld} + Lpi_{Ld}$, in Figure 3.3, we get the state values for the direct current through the load:

$$pi_{Ld} = \frac{V_{Ld}}{L} - \frac{R}{L}i_{Ld} \tag{3.17}$$

Similarly,

$$pi_{Lq} = \frac{V_{Lq}}{L} - \frac{R}{L}i_{Lq} \tag{3.18}$$

We can now establish the voltage and current relationships between the IG and the self-excitation capacitor. From matrix Equation 3.11 and Equations 3.9 and 3.10, we have

$$v_{ds} = R_1 i_{ds} + L_1 p i_{ds} + v_{Ld} + M p i_{dr} \tag{3.19}$$

$$v_{qs} = R_1 i_{qs} + L_1 p i_{qs} + v_{Lq} + M p i_{qr} \tag{3.20}$$

$$v_{dr} = M p i_{ds} + \omega M i_{qs} + \omega L_2 i_{qr} + R_2 i_{dr} + L_2 p i_{dr} \tag{3.21}$$

$$v_{qr} = -\omega M i_{ds} + M p i_{qs} + R_2 i_{qr} + L_2 p i_{qr} - \omega L_2 i_{dr} \tag{3.22}$$

Isolating the differential terms of the direct and quadrature currents of the rotor in Equations 3.19 and 3.20, we get

$$p i_{dr} = \frac{1}{M}\left(v_{ds} - R_1 i_{ds} - L_1 p i_{ds} - v_{Ld}\right) \tag{3.23}$$

$$p i_{qr} = \frac{1}{M}\left(v_{qs} - R_1 i_{qs} - L_1 p i_{qs} - v_{Lq}\right) \tag{3.24}$$

Substituting Equation 3.24 in 3.22, and isolating $p i_{qs}$, we get

$$p i_{qs} = \frac{M}{M^2 - L_1 L_2}\left(v_{qr} + \omega M i_{ds} - R_2 i_{qr} + \omega L_2 i_{dr}\right) - \frac{L_2}{M^2 - L_1 L_2}\left(v_{qs} - R_1 i_{qs} - v_{Lq}\right) \tag{3.25}$$

Assuming the following leakage coefficient,

$$K = \frac{1}{M^2 - L_1 L_2}$$

Then

$$p i_{qs} = MK\left(v_{qr} + \omega M i_{ds} - R_2 i_{qr} + \omega L_2 i_{dr}\right) - L_2 K\left(v_{qs} - R_1 i_{qs} - v_{Lq}\right) \tag{3.26}$$

Similarly, for the direct axis,

$$p i_{ds} = MK\left(v_{dr} - \omega M i_{qs} - R_2 i_{dr} - \omega L_2 i_{qr}\right) - L_2 K\left(v_{ds} - R_1 i_{ds} - v_{Ld}\right) \tag{3.27}$$

Substituting Equation 3.26 in 3.24, rearranging and combining like terms, we get

$$pi_{qr} = \frac{(1+L_1L_2K)}{M}\left(v_{qs} - R_1i_{qs} - v_{Lq}\right) - L_1K\left(v_{qr} + \omega Mi_{ds} - R_2i_{qr} + \omega L_2i_{dr}\right)$$

Notice that $MK = \dfrac{(1+L_1L_2K)}{M}$ and, therefore,

$$pi_{qr} = MK\left(v_{qs} - R_1i_{qs} - v_{Lq}\right) - L_1K\left(v_{qr} + \omega Mi_{ds} - R_2i_{qr} + \omega L_2i_{dr}\right) \qquad (3.28)$$

Similarly, for the direct axis,

$$pi_{dr} = MK\left(v_{ds} - R_1i_{ds} - v_{Ld}\right) - L_1K\left(v_{dr} - \omega Mi_{qs} - R_2i_{dr} + \omega L_2i_{qr}\right) \qquad (3.29)$$

Grouping the state Equations 3.15, 3.16, and 3.26 through 3.29 in a matrix form for $i_{Ld} = 0$ and $i_{Lq} = 0$, we get the reduced matrix equation describing the self-excitation process at no load:

$$p\begin{bmatrix} i_{ds} \\ i_{qs} \\ i_{dr} \\ i_{qr} \\ v_{Ld} \\ v_{Lq} \end{bmatrix} = K\left\{ \begin{bmatrix} R_1L_2 & -\omega M^2 & -R_2M & -\omega ML_2 & L_2 & 0 \\ \omega M^2 & R_1L_2 & \omega ML_2 & -R_2M & 0 & L_2 \\ -R_1M & \omega ML_1 & R_2L_1 & \omega L_1L_2 & -M & 0 \\ -\omega ML_1 & -R_1M & -\omega L_1L_2 & R_2L_1 & 0 & -M \\ 1/CK & 0 & 0 & 0 & 0 & 0 \\ 0 & 1/CK & 0 & 0 & 0 & 0 \end{bmatrix} \begin{bmatrix} i_{ds} \\ i_{qs} \\ i_{dr} \\ i_{qr} \\ v_{Ld} \\ v_{Lq} \end{bmatrix} \right.$$

$$\left. + \begin{bmatrix} -L_2 & 0 & M & 0 \\ 0 & -L_2 & 0 & M \\ M & 0 & -L_1 & 0 \\ 0 & M & 0 & -L_1 \\ 0 & 0 & 0 & 0 \\ 0 & 0 & 0 & 0 \end{bmatrix} \begin{bmatrix} v_{ds} \\ v_{qs} \\ v_{dr} \\ v_{qr} \end{bmatrix} \right\} \qquad (3.30)$$

Grouping the state Equations 3.15 through 3.18 and 3.26 through 3.29, we get the full matrix equation, including the representation of the load connection:

$$
p\begin{bmatrix} i_{ds} \\ i_{qs} \\ i_{dr} \\ i_{qr} \\ v_{Ld} \\ v_{Lq} \\ i_{Ld} \\ i_{Lq} \end{bmatrix} = K \left\{ \begin{bmatrix} R_1L_2 & -\omega M^2 & -R_2M & -\omega ML_2 & L_2 & 0 & 0 & 0 \\ \omega M^2 & R_1L_2 & \omega ML_2 & -R_2M & 0 & L_2 & 0 & 0 \\ -R_1M & \omega ML_1 & R_2L_1 & \omega L_1L_2 & -M & 0 & 0 & 0 \\ -\omega ML_1 & -R_1M & -\omega L_1L_2 & R_2L_1 & 0 & -M & 0 & 0 \\ 1/CK & 0 & 0 & 0 & 0 & 0 & -1/CK & 0 \\ 0 & 1/CK & 0 & 0 & 0 & 0 & 0 & -1/CK \\ 0 & 0 & 0 & 0 & 1/LK & 0 & -R/LK & 0 \\ 0 & 0 & 0 & 0 & 0 & 1/LK & 0 & -R/LK \end{bmatrix} \right.
$$

$$
\left. \times \begin{bmatrix} i_{ds} \\ i_{qs} \\ i_{dr} \\ i_{qr} \\ v_{Ld} \\ v_{Lq} \\ i_{Ld} \\ i_{Lq} \end{bmatrix} + \begin{bmatrix} -L_2 & 0 & M & 0 \\ 0 & -L_2 & 0 & M \\ M & 0 & -L_1 & 0 \\ 0 & M & 0 & -L_1 \\ 0 & 0 & 0 & 0 \\ 0 & 0 & 0 & 0 \\ 0 & 0 & 0 & 0 \\ 0 & 0 & 0 & 0 \end{bmatrix} \cdot \begin{bmatrix} v_{ds} \\ v_{qs} \\ v_{dr} \\ v_{qr} \end{bmatrix} \right\} \tag{3.31}
$$

The induction motor might also be represented by a matrix equation that does not include the self-excitation capacitor derived from the same equation set used for SEIG. This equation is

$$
p\begin{bmatrix} i_{ds} \\ i_{qs} \\ i_{dr} \\ i_{qr} \end{bmatrix} = K \left\{ \begin{bmatrix} R_1L_2 & -\omega M^2 & -R_2M & -\omega ML_2 \\ \omega M^2 & R_1L_2 & \omega ML_2 & -R_2M \\ -R_1M & \omega ML_1 & R_2L_1 & \omega L_1L_2 \\ -\omega ML_1 & -R_1M & -\omega L_1L_2 & R_2L_1 \end{bmatrix} \begin{bmatrix} i_{ds} \\ i_{qs} \\ i_{dr} \\ i_{qr} \end{bmatrix} \right.
$$

$$
\left. + \begin{bmatrix} -L_2 & 0 & M & 0 \\ 0 & -L_2 & 0 & M \\ M & 0 & -L_1 & 0 \\ 0 & M & 0 & -L_1 \end{bmatrix} \begin{bmatrix} v_{ds} \\ v_{qs} \\ v_{dr} \\ v_{qr} \end{bmatrix} \right\} \tag{3.32}
$$

Equations 3.30 through 3.32 come in the classical form of the state equations as

$$p[x] = [A][x] + [B][u] \qquad (3.33)$$

that is,

$$p\begin{bmatrix} i_G \\ v_C \\ i_L \end{bmatrix} = \begin{bmatrix} G \\ C \\ L \end{bmatrix}\begin{bmatrix} i_G \\ v_C \\ i_L \end{bmatrix} + [B]\begin{bmatrix} v_G \end{bmatrix}$$

where the indexes G, C, and L refer, respectively, to the partitioned matrix of the IG, the self-excitation capacitor bank, and the load. The $[x]$ vector is the matrix, $[i_G \, v_C \, i_L]^T$, and the submatrices $[G]$, $[C]$, and $[L]$ are defined as

$$[G] = K\begin{bmatrix} R_1L_2 & -\omega M^2 & -R_2M & \omega ML_2 & L_2 & 0 & 0 & 0 \\ \omega M^2 & R_1L_2 & -\omega ML_2 & -R_2M & 0 & L_2 & 0 & 0 \\ -R_1M & \omega ML_1 & R_2L_1 & \omega L_1L_2 & -M & 0 & 0 & 0 \\ -\omega ML_1 & -R_1M & -\omega L_1L_2 & R_2L_1 & 0 & -M & 0 & 0 \end{bmatrix}$$

$$[C] = \begin{bmatrix} C_G \vdots C_T \end{bmatrix}\begin{bmatrix} 1/C & 0 & 0 & 0 & 0 & 0 & -1/C & 0 \\ 0 & 1/C & 0 & 0 & 0 & 0 & 0 & -1/C \end{bmatrix}$$

$$[L] = \begin{bmatrix} 0 & 0 & 0 & 0 & 1/L & 0 & -R/L & 0 \\ 0 & 0 & 0 & 0 & 0 & 1/L & 0 & -R/L \end{bmatrix}$$

The excitation vector $[u] = [v_G]$ of Equation 3.33 is premultiplied by the matrix of parameters of the excitation source, $[B]$, and it defines the voltages corresponding to the residual magnetism:

$$[B] = \begin{bmatrix} -L_2 & 0 & M & 0 \\ 0 & -L_2 & 0 & M \\ M & 0 & -L_1 & 0 \\ 0 & M & 0 & -L_1 \\ \hline 0 & 0 & 0 & 0 \\ 0 & 0 & 0 & 0 \\ \hline 0 & 0 & 0 & 0 \\ 0 & 0 & 0 & 0 \end{bmatrix}$$

A more compact way of representing an aggregation of SEIGs would be to consider the self-excitation capacitor as part of the load across the machine terminals. In this case, the matrix of Equation 3.33 would be transformed as it proceeds:

$$p\begin{bmatrix} i_G \\ x_G \end{bmatrix} = \begin{bmatrix} G \\ E \end{bmatrix} \begin{bmatrix} i_G \\ x_G \end{bmatrix} + [B][v_G]$$

where

$$[x_G] = \begin{bmatrix} v_C \\ i_L \end{bmatrix}$$

$$[E] = \begin{bmatrix} 1/C & 0 & 0 & 0 & 0 & 0 & -1/C & 0 \\ 0 & 1/C & 0 & 0 & 0 & 0 & 0 & -1/C \\ 0 & 0 & 0 & 0 & 1/L & 0 & -R/L & 0 \\ 0 & 0 & 0 & 0 & 0 & 1/L & 0 & -R/L \end{bmatrix}$$

Figure 3.4 shows the Excel plots of the transient self-excitation voltage process with a sudden load switching at 7.6 s for a stand-alone SEIG. The IG operated at 1780 rpm with a dc motor as prime mover. A capacitor bank of 160 μF in a star configuration is supplying reactive power for the machine, and a 120 Ω–22.5 mH star load was connected after the generator was completely excited. Remnant magnetism in the machine core is also taken into account. We see that the self-excitation follows a variation law similar to the process of magnetic saturation of the core and that a level of stable output voltage is only reached when the machine core is saturated.

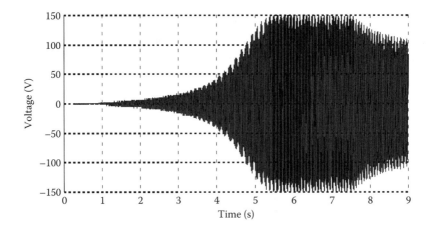

FIGURE 3.4 Voltage growth during self-excitation at no-load and at-load switching.

3.5 GENERALIZATION OF THE ASSOCIATION OF SELF-EXCITED GENERATORS

The method used for a stand-alone machine can be extended to multiple n generators operating in parallel. Each machine needs a self-excitation capacitor because each one has its own curve at no load and, as a consequence, a different self-excitation curve. However, for the numeric solution, there will only be a capacitor and a common load for all the generators. Equation 3.31 can be generalized even for an infinite bus bar, by the following equation:

$$
p\begin{bmatrix} i_{G1} \\ i_{G2} \\ \vdots \\ i_{Gn} \\ -- \\ v_L \\ -- \\ i_L \end{bmatrix} = \begin{bmatrix} G_1 & & & & & L_1 \\ & G_2 & & & & L_2 \\ & & \ddots & & & \vdots \\ & & & G_n & \vdots & L_n \\ \hline C_G & C_G & \cdots & C_G & C_T \\ 0 & 0 & \cdots & 0 & L_T \end{bmatrix} \cdot \begin{bmatrix} i_{G1} \\ i_{G2} \\ \vdots \\ i_{Gn} \\ -- \\ v_L \\ -- \\ i_L \end{bmatrix} + \begin{bmatrix} B_1 & & & & \\ & B_2 & & & \\ & & \ddots & & \\ & & & B_n & \\ \hline 0 & 0 & \cdots & 0 \end{bmatrix} \cdot \begin{bmatrix} v_{G1} \\ v_{G2} \\ \cdots \\ v_{Gn} \end{bmatrix}
$$

$$(3.34)$$

where

$$
i_{Gi} = i\begin{bmatrix} i_{dsi} \\ i_{qsi} \\ i_{dri} \\ i_{qri} \end{bmatrix} \quad v_{Gi} = \begin{bmatrix} v_{dsi} \\ v_{qsi} \\ v_{dri} \\ v_{qri} \end{bmatrix} \quad v_L \begin{bmatrix} v_{Ld} \\ v_{Lq} \end{bmatrix} \quad \text{and} \quad i_L = \begin{bmatrix} i_{Ld} \\ i_{Lq} \end{bmatrix}
$$

$$
G_i = K\begin{bmatrix} R_{1i}L_{2i} & -\omega M_i^2 & -R_{2i}M_i & -\omega M_i L_{2i} \\ \omega M_i^2 & R_{1i}L_{2i} & \omega M_i L_{2i} & -R_2 M_i \\ -R_{1i}M_i & \omega M_i L_{1i} & R_{2i}L_{1i} & \omega L_{1i}L_{2i} \\ -\omega M_i L_{1i} & -R_{1i}M_i & -\omega L_{1i}L_{2i} & R_{2i}L_{1i} \end{bmatrix} \quad \text{and} \quad L_i = \begin{bmatrix} L_{2i} & 0 & 0 & 0 \\ 0 & L_{2i} & 0 & 0 \\ M_i & 0 & 0 & 0 \\ 0 & M_i & 0 & 0 \end{bmatrix}
$$

$$
[C_G] = \begin{bmatrix} 1/C & 0 & 0 & 0 \\ 0 & 1/C & 0 & 0 \end{bmatrix} \quad \text{and} \quad [C_T] = \begin{bmatrix} 0 & 0 & -1/C & 0 \\ 0 & 0 & 0 & -1/C \end{bmatrix}
$$

where $C = \Sigma C_k$ ($k = 1, 2,..., n$) (summation of the individual self-excitation capacitances of the n generators already connected in parallel).

That is, during simulation, the capacitance C is the summation of the self-excitation capacitances of the generators already existing in the parallel aggregation.

$$L_T = \begin{bmatrix} 1/L & 0 & -R/L & 0 \\ 0 & 1/L & 0 & -R/L \end{bmatrix}$$

$$B_i = K \begin{bmatrix} -L_{2i} & 0 & M & 0 \\ 0 & -L_{2i} & 0 & M_i \\ M_i & 0 & -L_{1i} & 0 \\ 0 & M_i & 0 & -L_{1i} \end{bmatrix}$$

The computer simulation models the behavior of currents and voltages of the IG group during the self-excitation process and in steady state, besides allowing the verification of alterations in the output voltage and variations in the load current that occur. Observe that the representation of Equation 3.31 is in the classical form of state space, and, therefore, it is also adapted for the analysis of eigenvalues and eigenvectors.

The dimensions of the submatrices given by Equation 3.34 are shown in Table 3.1.

Matrix Equation 3.34 considers that pv_{Ld} is the result of the current through the capacitor; that is, the sum of each one of the currents of the associated generators is subtracted from the current through the load. So, $pv_{Ld} = [(\Sigma i_{Gk}) - i_{Ld}]/C$. For instance, in the case of three IGs aggregated in parallel, as shown in Figure 3.5, the total current through the capacitor would be $i_C = (i_{G1} + i_{G2} + i_{G3}) - i_{Ld}$.

Notice that the direct and quadrature axis voltages across the load terminals or across the excitation capacitor soon after the parallel connection of every new generator are given, respectively, by

$$V_{Ld} = \frac{C_n V_{Ld(n)} + C_{n+1} V_{Ld(n+1)}}{C_n + C_{n+1}} \qquad V_{Lq} = \frac{C_n V_{Lq(n)} + C_{n+1} V_{Lq(n+1)}}{C_n + C_{n+1}}$$

where
   n refers to the number of associated generators
   $n + 1$ refers to the new generator connected to the former set of parallel generators

TABLE 3.1

Dimensions of the Submatrices in Equation 3.33

Submatrix	Dimension	Submatrix	Dimension
G_i	4×4	G	$4n \times 4n$
C_G	2×4	L_G	$4n \times 4$
C	$2 \times 4n$	B_i	$4n \times 4n$
L_T	2×4	C_T	2×4
0_A	$2 \times 4n$	0_B	$4 \times 4n$
x	$(4n + 4) \times 1$	U	$4n \times 1$

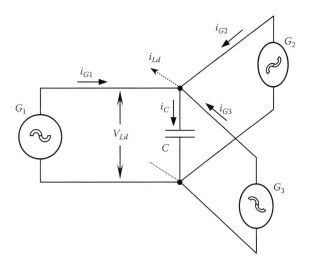

FIGURE 3.5 Current contributions through capacitor *C*.

Brown and Grantham demonstrated that merely taking into account the variation of the magnetization reactance is not enough if the other machine parameters are supposed to be constants.[11,12] Even for steady-state analysis, unknown parameter variations of the rotor introduce considerable errors in the analysis of machine performance with rotor displacement current. In the same way, for machines not specifically designed for displacement current, better results can be obtained in the behavior analysis of these machines if the variations in the machine parameters are considered.

3.6 RELATIONSHIP BETWEEN TORQUE AND SHAFT OSCILLATION

It is important to define the torque of each machine in order to observe its oscillation with respect to the others. Torque base in p.u. is then defined as follows:

$$T_B = \frac{S_B}{\omega_s} \tag{3.35}$$

where S_B is the rated apparent power base in kVA.

Analysis of the SEIG using the state variable representation starting from the well-known theory of the induction machines produces instantaneous values for direct and quadrature current axis to analyze the increases in voltages and currents during the self-excitation process and those due to disturbances caused by load variations. If necessary, it is also possible to calculate the net transient electric torque. This torque is discussed in the literature[11,13–15] as

$$T = \frac{\text{Power}}{\omega_r} = -\left(\frac{3}{2}\right)\left(\frac{p}{2}\right) Mi_{dr}i_{qs} + \left(\frac{3}{2}\right)\left(\frac{p}{2}\right) Mi_{qr}i_{ds} - 2H\frac{p}{\omega_b}\omega_r \qquad (3.36)$$

where

H is the rotor inertia constant
p is the number of stator poles
ω_r is the angular speed of the rotor
ω_b is the base speed, commonly specified as the one obtained from the parameter tests

3.6.1 Oscillation Equation

To determine the oscillation or the angular displacement among the aggregated machines in a group during transient outcomes, it is necessary to solve the differential equations describing the rotor movements of the machines. The electric torque acting on the rotor of only one machine, following the mechanical laws related to rotating masses, is

$$T = \frac{M_j}{g}\alpha \qquad (3.37)$$

where

T is the algebraic sum of all applied torques to the axis in N m
M_j is the moment of inertia in N m²
g is the acceleration of the gravity (= 9.81 m/s²)
α is the angular mechanical acceleration in rd/s²

The electrical angle θ_e is related to the mechanical angle, θ_m by $\theta_e = (p/2)\theta_m$, or, using the relationship $f = pn/120$ (being n in rpm), we get

$$\theta_e = \frac{60 f \theta_m}{n} \qquad (3.38)$$

The angular displacement δ of the rotor with respect to the rotating synchronous reference axis is

$$\delta = \theta_e - \omega_s t$$

The angular variation with respect to the reference axis is established by

$$\frac{d\delta}{dt} = \frac{d\theta_e}{dt} - \omega_s = \omega - \omega_s \qquad (3.39)$$

and the angular acceleration is

$$\frac{d^2\delta}{dt^2} = \frac{d^2\theta_e}{dt^2}$$

Combining this equality with the second derivative of Equation 3.38 with respect to time, we get

$$\frac{d^2\delta}{dt^2} = \frac{60f}{n}\frac{d^2\theta_m}{dt^2} \tag{3.40}$$

where $\frac{d^2\theta_m}{dt^2} = \alpha$ is the mechanical acceleration.

Therefore, substituting Equation 3.40 in 3.37, we get the net torque:

$$T = \frac{M_j}{g}\frac{n}{60f}\frac{d^2\delta}{dt^2} \tag{3.41}$$

It is convenient to express the torque in p.u. The torque base T_B is defined as the necessary torque to supply the rated power at the rated speed.

Therefore, a convenient form of the torque in p.u. is

$$T(\text{p.u.}) = \frac{T}{T_B} = \frac{\dfrac{M_j}{g}\dfrac{p\pi}{f}\left(\dfrac{n}{60}\right)^2}{S_B}\frac{d^2\delta}{dt^2} \tag{3.42}$$

The inertia constant H of a machine is defined as the kinetic energy at the rated speed in kW–s/kVA, this last one given by

$$E_c = \frac{1}{2}\frac{M_j}{g}\omega_s^2 = HS_B$$

where $\omega_s = \dfrac{p\pi n_s}{60}$ at the nominal speed.

As a result, we have

$$H = \frac{\dfrac{1}{2}\dfrac{M_j}{g}(p\pi)^2\left(\dfrac{n}{60}\right)^2}{S_B}$$

which, when substituted in Equation 3.42, gives

$$T(\text{p.u.}) = \frac{2H}{p\pi f}\frac{d^2\delta}{dt^2} \tag{3.43}$$

The sum of all torques acting on the rotor of a generator includes the input mechanical torque of the primary machine, the rotating torque losses (friction, air opposition, and copper plus core losses), electrical output torques, and damping torques from the primary machine, the generator, and the interconnected electric system. The electric and mechanical torques acting on the rotor of a motor are of opposite polarities, and they are the result of the electrical and mechanical loads. Despite the rotating and damping losses, the accelerating torque T_a is given by the difference between the mechanical and electrical torques: $T_a = T_m - T_e$. The electrical torque is the net torque given by Equation 3.43. Therefore,

$$T_m - T_e = T(\text{p.u.}) = \frac{2H}{p\pi f}\frac{d^2\delta}{dt^2} \tag{3.44}$$

As small variations of speed, torque, and power (mechanical and electrical) remain constant in p.u., Equation 3.44 becomes

$$\frac{d^2\delta}{dt^2} = \frac{p\pi f}{2H}\left(P_m - P_e\right) \tag{3.45}$$

For numeric solutions, it is interesting to separate this equation in two simultaneous equations of first order (Equations 3.39 and 3.45):

$$\frac{d\omega}{dt} = \frac{d^2\delta}{dt^2} = \frac{p\pi f}{2H}\left(P_m - P_e\right) \tag{3.46}$$

$$\frac{d\delta}{dt} = \omega - 2\pi f \tag{3.47}$$

If the transient values of torque and power are known during the numeric solution process, they can be used to correct the variations of angular speed using the relationship $P = T\omega$.

3.7   TRANSIENT SIMULATION OF INDUCTION GENERATORS

Nowadays, there are several computer packages used in the simulation studies of electric machines and particularly in the transient modeling of IGs as presented in this chapter. MATLAB® was used to implement differential matrix Equation 3.31 as discussed in Chapter 12. Such an equation describes the behavior of currents and voltages of an IG during the self-excitation process and in steady operation. It also allows verification of the alterations that happen in the output voltage and variations in the load current.

We must emphasize, at this point, that to begin the process of self-excitation of an IG, there must be a small amount of residual magnetism present. This needs to be taken into account in the computer simulation of the self-excitation process; otherwise, it is not possible to start the numerical method of integration. At the beginning of the integration process of Equation 3.31, a pulse function has to be present whose value fades away as soon as the first iterative step has started. This observation is very important in terms of the dynamic understanding of the self-excitation phenomenon, because it could be used during the occurrence of fortuitous core demagnetizing of the machine, recommending a small source of applied voltage to the real machine for recovery of its active state.

The beginning of the self-excitation process can be slow or fast, depending on the construction of the machine, the way the self-excitation process is done, and the level of residual magnetism. It will be fast if the machine is made to rotate at its rated rotation speed and the self-excitation capacitors are then connected. The transient process in this case takes, typically, about 1 s. For small machines, this process can be shorter. If the machine is started with the self-excitation capacitor already connected, the process will take, typically, about 10 s.[13,15]

Once obtained, the state variables should be returned to the phase sequence *a-b-c* in order to obtain the results for a three-phase generator and not a two-phase as given by the Park transformation. For this, the voltages across each phase of the stator are established as follows[8-11]:

$$v_A = v_{qs}$$

$$v_B = -\frac{1}{2} v_{qs} - \frac{\sqrt{3}}{2} v_{ds} \tag{3.48}$$

$$v_C = -\frac{1}{2} v_{qs} + \frac{\sqrt{3}}{2} v_{ds}$$

Similarly, the voltages across the rotor winding are as follows:

$$v_a = v_{qr}$$

$$v_b = -\frac{1}{2} v_{qr} - \frac{\sqrt{3}}{2} v_{dr} \tag{3.49}$$

$$v_c = -\frac{1}{2} v_{qr} + \frac{\sqrt{3}}{2} v_{dr}$$

When connecting several SEIGs in parallel, the common transient voltage established across their terminals could be seen as the voltage of a parallel connection between one generator and a group of several generators whose power could be a lot larger than the one just connected. In these cases, a program for the simulation of only two generators of distinct characteristics could be a good approximation. In the more general case, the solution presented in this chapter is quite reasonable for most of the practical problems.

TABLE 3.2

Rated Parameters of Sample Induction Generators G_1 and G_2

Generator G_1		Generator G_2	
$P = 1$ HP	$R_s = 0.32\ \Omega$	$P = 1$ HP	$R_s = 0.39\ \Omega$
1750 rpm, 60 Hz	$X_{ls} = 0.8\ \Omega$	1750 rpm, 60 Hz	$X_{ls} = 0.93\ \Omega$
$V = 120/220$ V	$R_r = 0.41\ \Omega$	$V = 120/220$ V	$R_r = 0.44\ \Omega$
$I = 11.5/5.5$ A	$X_{lr} = 0.8\ \Omega$	$I = 11.5/5.5$ A	$X_{lr} = 0.93\ \Omega$

Source: Colorado School of Mines Department of Electrical Engineering and Computer Science, Golden, CO.

3.7.1 EXAMPLE OF TRANSIENT MODEL OF AN INDUCTION GENERATOR

In order to validate the self-excitation process, an induction machine G_1 was used with the nameplate data shown in Table 3.2. The iterative solution of Equation 3.11 was used to simulate the IG from the beginning of the self-excitation process at 0.0 ms up to 2.0 ms, using the Colorado School of Mines power laboratory values provided earlier for the generator.

3.7.2 EFFECT OF *RLC* LOAD CONNECTION

The *RLC* load may be seen as a combination of R and L in parallel with the self-excitation capacitance. Figure 3.4 displays the output voltage of the generator for the beginning of the self-excitation process and switching of the load impedance at 7.6 s. It is possible to observe the voltage drop caused at the instant of the connection traduced by a decrease of the load impedance, in other words, a power increase.

As a consequence of Ohm's law, there is a reduction in the load current, as shown in Figure 3.4. No decaying component appears except at the very beginning of the self-excitation process, due to the residual magnetism, as discussed before.

3.7.3 LOSS OF EXCITATION

Figure 3.6 displays the loss of excitation of the generator caused by an excessive increase of the load current without the corresponding voltage increase at $t = 2.6$ s. Generator G_1 was self-excited with a 160 µF capacitance bank, and a 45 Ω–45.5 mH star load was applied after the generator was fully excited. It took about 2.5 s for the voltage to collapse completely. The current increase could be compensated for by an increase in reactive power or an increase in the rotor speed, which did not happen in this example. However, it is known that the voltage tends to increase, even inside of the saturation band of the iron, as the terminal capacitance increases. Therefore, to avoid core demagnetization of the generator, there should be a settled control strategy adopted to allow variation of either the excitation capacitance or the rotor speed according to changes in the load current.

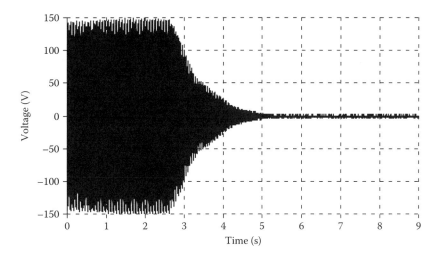

FIGURE 3.6   Terminal voltage collapse due to an excessive increase of the load current.

3.7.4 PARALLEL CONNECTION OF INDUCTION GENERATORS[14,15]

Table 3.2 displays the experimental data of the two IGs G_1 and G_2 used in this example. Figure 3.7 shows the transient process of load connection and generator parallel switching surges of two SEIGs. Generator G_1 was self-excited with a 160 μF capacitance at 1800 rpm. A 130 Ω–22.5 mH star connected load was applied at 2 s from the beginning of the simulation time. Generator G_2 was previously self-excited with 160 μF at 1800 rpm and connected in parallel at $t = 3.75$ s. The sudden and brief collapse in voltage is due to differences in phase voltage between the two generators

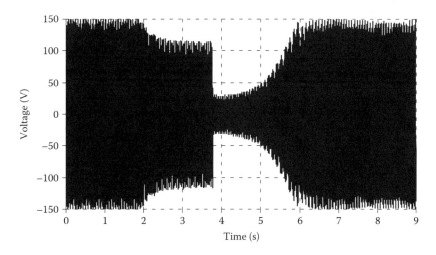

FIGURE 3.7   Transient switching voltage for a parallel connection of SEIGs operating at distinct voltage levels.

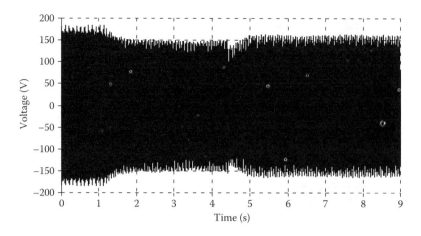

FIGURE 3.8 Transient switching voltage for a parallel connection of SEIGs operating at identical voltage levels.

at the paralleling instant. Full common voltage was recovered at about $t = 6.0$ s. The rotor speed variation of generators G_1 and G_2 during parallel operation were carefully observed in the laboratory and incorporated in the simulation illustrated in Figure 3.7. At exactly the instant of parallel connection of the two generators, a heavy dip in the overall speed of the machines may happen.

Three points in time in those graphs should be observed. The first is the voltage reduction across generator G_1 terminals when the R–L load was switched on. The second is the partial voltage collapse at $t = 3.75$ s, and its recovery up to the steady-state parallel common voltage level at about 6.0 s. The third is the recovered voltage of generators G_1 and G_2 with respect to the no-load voltage level of G_2 alone, representing an appreciable voltage difference at the instant of parallel connecting. These three points can be clearly and closely observed in both theoretical and experimental setups.

Figure 3.8 shows the simulated plot of the transient switching process of the generator when parallel connection was made between SEIGs operating at identical voltage levels. Generator G_1 was self-excited with 180 μF capacitance, and a 130 Ω–22.5 mH star load was applied at $t = 1.2$ s. Generator G_2 was already self-excited with 160 μF and was connected in parallel at $t = 4.4$ s. Full common voltage was recovered at $t = 5.2$ s. As the voltages were similar, there was no collapse of the kind observed in the previous case, and the voltage and speed dips were not very pronounced.[7]

3.8   CONCEPTS COVERED IN THIS CHAPTER TO HELP PRACTICAL DESIGN

This chapter developed a transient modeling framework. We saw the growth of voltage and current during the self-excitation process and its stabilization in steady state. The models enable the visualization of the transient disturbances that the stator output voltage and current suffer when a load is connected to the terminals. The connection of parallel generators was considered as a load connection across each of the IG terminals.

An IG simulation takes a very short time. It must be emphasized that the machine needs a residual magnetism so that the self-excitation process can be started; it cannot be zero at the beginning of the process. During calculations, the parameter variation should follow the curve of the machine hysteresis.

The value of the mutual inductance M should vary continually, following a nonlinear relationship between the gap voltage and the terminal magnetizing current in such a way that correct results are obtained. For even more precise results, it is advisable that the variation of frequency be included during the simulation. A more detailed analysis, including the variation of frequency, is presented in Chapter 5.

This description of the behavior of IGs provides the basis for simulations, taking into consideration the core demagnetization of the generator and transient phenomena. This description in more detail is of fundamental importance so that asynchronous micropower plants can operate with higher safety margins.

3.9 SOLVED PROBLEMS

1. Why is the equivalent p.u. circuit model of the IG, whose stator and rotor frequencies are denoted by F and v, respectively, inadequate to represent the transient state of the capacitor SEIG?

Solution

The equivalent per-phase circuit model is inadequate for the transient state representation of an SEIG because the per-phase model has the slip as a variable, so it is necessary to calculate the output as a function of the instantaneous slip. When the machine is running in steady state with a fixed slip, this model can be useful for calculating steady-state torque, power, voltages, and currents. But for the SEIG, there are variable operating conditions in the speed, slip, and saturation of the magnetizing inductance that makes steady-state conditions not to be held. Therefore, the per-phase model cannot be used for transient analysis. In addition, for capacitive excited SEIG, the phases of the IG are correlated during transients, which should also be taken into account while modeling the machine.

2. Develop a procedure to represent the variation of the mutual inductance M during the process of self-excitation of the IG, establishing its limitations and the necessary equations to include them in a computer program for this purpose.

Solution

The variation of magnetizing inductance is the main factor in the process of voltage buildup and stabilization in operating SEIGs. Since the magnetizing inductance at a rated voltage is given by the relationship between V_g and I_m, the result is a highly nonlinear function. For better results in terms of representation, a table of experimental values usually gives the magnetizing inductance used in laboratory setups as a polynomial series or an analytical adjusted expression. In Section 4.3, it is possible to see that the polynomial representation of the magnetizing inductance taken from experimental results can be reasonably given for practical purposes by a fourth-order curve fits, such as

$$I_m = a_0 + a_1 V_{ph} + a_2 V_{ph}^2 + a_3 V_{ph}^3 + a_4 V_{ph}^4$$

Another way of representing the magnetizing curve is a linear piecewise association within certain ranges of the magnetizing current. The magnetizing reactance
is given as $X_m = \omega L_m$, and the reactance can be defined in segments, such as

$$X_m = \begin{vmatrix} a_0 & \text{for } 0 \ll I_m \ll I_1 \\[2mm] \dfrac{a_1}{(I_m + b_1)} & \text{for } I_1 \ll I_m \ll I_2 \\[3mm] \dfrac{a_2}{(I_m + b_2)} & \text{for } I_2 \ll I_m \ll I_3 \\[3mm] \dfrac{a_3}{(I_m + b_3)} & \text{for } I_3 \ll I_m \ll I_4 \\[3mm] \dfrac{a_4}{(I_m + b_4)} & \text{for } I_4 \ll I_m \ll I_5 \end{vmatrix}$$

The technical literature has reported several possible ways to relate the air-gap
voltage to the magnetizing current, showing that the relationship of V_g and I_m can
be established through the following nonlinear equation:

$$V_g = FI_m \left(K_1 e^{K_2 I_m^2} + K_3 \right)$$

where
     K_1, K_2, and K_3 are constants to be determined
     V_g is the air-gap voltage across the magnetizing reactance (without external
       access)
     F is the frequency in p.u., defined as $F = \dfrac{f}{f_{base}}$, where f is the rotor frequency
     f_{base} is the reference frequency used in the tests to obtain the excitation
       curve (usually it is 60 or 50 Hz from the utility grid)

In this case, the magnetizing reactance can be obtained directly by

$$X_m = \omega L_m = \frac{V_g}{I_m} = F \left(K_1 e^{K_2 I_m^2} + K_3 \right)$$

3.10   SUGGESTED PROBLEMS

3.1 Given the instantaneous stator phase voltages $v_A = V_m \sin(\omega t)$, $v_B = V_m \sin(\omega t - 120°)$, and $v_C = V_m \sin(\omega t + 120°)$ and the rotor instantaneous phase voltages $v_a = V_m \sin(\omega t)$, $v_b = V_m \sin(\omega t - 120°)$, and $v_c = V_m \sin(\omega t + 120°)$, of an induction machine, for $V_m = 110\sqrt{2}\ V\ e\ f = 60$ Hz, calculate $[v_{ds} v_{qs} v_{dr} v_{qr}]$.

3.2 Based on the data in Table 3.2, calculate the values of matrix $[v_{ds} v_{qs} v_{dr} v_{qr}]$ assuming a squirrel cage rotor.

3.3 Calculate the eigenvalues and eigenvectors of Equation 3.11 to establish the oscillating modes of the operation of the IG. Hint: Find the roots of the denominator of the transfer function V/I and evaluate the modes with positive real part.

3.4 Using MATLAB/Simulink®, write a computer program to represent the SEIG in its transient state.

3.5 Explain why an IG cannot self-excite under the following conditions: (1) no residual magnetism, (2) after too much inductive load, and (3) after overcurrent. What you have to do in order to allow the IG to self-excite again?

3.6 Discuss the four best-known practical methods of generating residual magnetism.

3.7 Discuss the advantages of a matrix partition to represent the transient state of an aggregation of SEIGs.

3.8 Using the state variable theory for the generalization of an aggregation of SEIGs, write a matrix system for an aggregation of three IGs with distinct machine parameters. What are the dimensions of the necessary submatrices for this?

3.9 Describe an induction motor using the parameters of the partitioned matrix of Equation 3.30 in the classical state equation form.

REFERENCES

1. Smith, I.R. and Sriharan, S., Transients in induction motors with terminal capacitors, *Proc. IEEE*, 115, 519–527, 1968.
2. Murthy, S.S., Bhim Singh, M., and Tandon, A.K., Dynamic models for the transient analysis of induction machines with asymmetrical winding connections, *Electr. Mach. Electromech.*, 6(6), 479–492, 1981.
3. Murthy, S.S., Malik, O.P., and Tandon, A.K., Analysis of self-excited induction generators, *Proc. IEEE*, 129(6), 260–265, 1982.
4. Watson, D.B. and Milner, I.P., Autonomous and parallel operation of self-excited induction generators. *Int. J. Elect. Eng. Education*, 22, 365–374, 1985.
5. Murthy, S.S., Nagaraj, H.S., and Kuriyan, A., Design-based computational procedure for performance prediction and analysis of self-excited induction generators using motor design packages, *Proc. IEEE*, 135(1), 8–16, 1988.
6. Grantham, C., Sutanto, D., and Mismail, B., Steady state and transient analysis of self-excited induction generators, *Proc. IEEE*, 136(2), 61–68, 1989.
7. Barbi, I., *Fundamental Theory of the Induction Motor (Teoria Fundamental do Motor de Indução)*, UFSC Press, Eletrobrás, Florianópolis, Brazil, 1985.
8. Langsdorf, A.S., *Theory of the Alternating Current Machines (Teoría de Máquinas de Corriente Alterna)*, McGraw-Hill Book Co., New York, 1977, p. 701.
9. Singh, S.P., Bhim Singh, M., and Jain, P., Performance characteristics and optimum utilization of a cage machine as capacitor excited induction generator, *IEEE Trans. Energy Convers.*, 5(4), 679–684, 1990.
10. Hallenius, K.E., Vas, P., and Brown, J.E., The analysis of a saturated self-excited asynchronous generator, *IEEE Trans. Energy Convers.*, 6(2), 336–345, 1991.
11. Brown, J.E. and Grantham, C., Determination of induction motor parameter and parameter variations of a 3-phase induction motor having a current-displacement rotor, *Proc. IEEE*, 122(9), 919–921, 1975.
12. Grantham, C., Determination of induction motor parameter variations from a variable frequency standstill test, *Electr. Mach. Power Systems, United States,* 10, 239–248, 1985.

13. Wang, L. and Lee, C.-H., A novel analysis on the performance of an isolated self-excited induction generator, *IEEE Trans. Energy Convers.*, 12(2), 109–117, 1997.

14. Farret, F.A., Schramm, D.S., and Neves, L.S., Modeling of the self-excited induction generator: Transient and steady state (Modelagem de geradores de indução autoexcitados: estados transitório e permanente), in *Tecnologia Magazine, UFSM Press, Santa Maria, RS,* 16(1–2), 49–59, 1995.

15. Hancock, N.N., *Matrix Analysis of Electrical Machinery*, 2nd edn., Pergamon Press, Oxford, U.K., 1974.

4 Self-Excited Induction Generators

4.1 SCOPE OF THIS CHAPTER

As previously discussed in Chapter 2, the induction generator has a serious limitation: an inherent need for reactive power. It consumes reactive power when connected to the distribution network, and, actually, it needs an external reactive source permanently connected to its stator windings to provide output voltage control. This source of reactive power keeps the current going through the machine windings needed to generate the variable magnetic field. The movement of the rotor conductors causes the voltage across the induction generator terminals. Another way of inducing this voltage is to use the residual magnetism associated with an external capacitor that generates current by rotor movement inside this magnetic field, inducing voltage as described in the following.

The advantages of the induction generator are enhanced under optimized conditions of performance, as discussed in Chapter 10.

4.2 PERFORMANCE OF SELF-EXCITED INDUCTION GENERATORS

Interaction among the operating states of the primary source of energy, the induction generator, the self-excitation process, and the load state will define the global performance of the power plant.[1,2] Performance is greatly affected by the random character of some of the variables involved related to the availability of primary energy and to the way consumers use the load. Therefore, the performance of induction generators depends on appropriate power plant specifications at the design stage, in particular, the following items:

- Parameters of the induction machine
  - Operating voltage
  - Rated power
  - Rated frequency used in the parameter measurements
  - Power factor of the machine
  - Rotor speed
  - Capacity for acceleration
  - Isolation class
  - Operating temperature

- Carcass type
- Ventilation system
- Service factor
- Noise
- Load parameters
  - Power factor
  - Starting torque and current
  - Maximum torque and current
  - Generated harmonics
  - Form of connection to the load: directly to the distribution network or through converters
  - Load type: resistive, inductive or capacitive, constant or variable, and passive or active
  - Evolution of the load over time
- Self-exciting process
  - Degree of iron saturation of the generator caused by the choice of capacitor
  - Fixed or controlled self-excitation capacitor
  - Speed control
- Type of primary source: hydro, wind, biomass, or combinations

Nonlinear loads like electronic power converters generate many harmonics, and they can have a variable power factor (PF) if some control techniques are not adopted. The harmonics are minimized with the installation of filters or the utilization of supplementary signals. In general, the cost of passive filters is relatively small with respect to the cost of the speed control in the conventional power plants exerted by the electronic variation of frequency. The self-excitation capacitor in stand-alone power plants or with electric or electronic control of the load contributes favorably in these cases. Supplementary signals for harmonic minimization, in spite of being a relatively recent technology and not very well established yet, seem to be more promising techniques for such situations. The IEEE Std. 519 establishes the limits of 2% of harmonic content for single- and three-phase induction motors (except category N, that is, conventional) and 3% for high efficiency (H and D). The other standard is IEC1000-2-2.

In the case of stand-alone operation of power plants, the connection of a capacitor bank across the terminals of the induction generator is necessary, as displayed in Figure 4.1, to supply its need for reactive power. Notice that in practical schemes it is advisable to connect each excitation capacitor across each motor winding phase, in either Δ–Δ or Y–Y. The necessary capacitance of the bank will depend on the primary energy and on the instantaneous load as discussed in the following sections.

For a SEIG, the equivalent circuit in per unit values (p.u.) is shown in Figure 4.2. This circuit can be used to represent a more generic form of power plant.[2] The frequency effect on the reactance should be considered if it is used at different frequencies from the base frequency in Hertz f_b at which the parameters of the

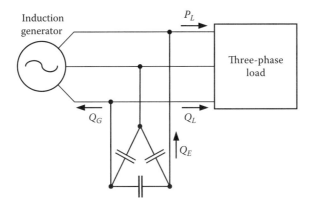

FIGURE 4.1 Capacitor self-excited induction generator. P_L, active power to the load; Q_L, reactive power to the load; Q_G, reactive power to the generator; Q_E, reactive power to the excitation.

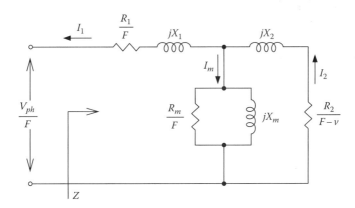

FIGURE 4.2 Generic model per phase of the induction generator.

machine were measured. For this purpose, if F is the p.u. frequency, a relationship can be defined between the self-excitation frequency f_{exc} and the base frequency f_b (usually 60 Hz)[3–5]:

$$F = \frac{f_{exc}}{f_b} = \frac{\omega_{exc}}{\omega_b}$$

In a more generic way, the inductive reactance parameters can be defined for the base frequency as $X = F\omega L$. Figure 4.2 displays the generic equivalent circuit in steady-state per phase of the SEIG with all circuit parameters divided by F, making the source voltage equal to V_{ph}/F. From the definition of the secondary resistance (rotor resistance) shown in Figure 2.3 and from Equation 2.11, the following modification can be used to correct R_2/s to take into account changes in the stator and rotor p.u. frequencies:

$$\frac{R_2}{F_s} = \frac{R_2}{F\left(1 - \dfrac{n_r}{n_s}\right)} = \frac{R_2}{F - v}$$

where v is the rotor speed in p.u. referred to the test speed used for the rotor.

Although the variation of the magnetizing reactance due to the magnetic saturation is taken into account in transient studies of induction generators, the other parameters are not considered constant.[4,5] The use of uncorrected parameters may lead to rough mistakes in the representation of the machine rotor with current changes. In Chapter 3, the general characteristics of the induction generator under the transient state are discussed in more detail.

4.3   MAGNETIZING CURVES AND SELF-EXCITATION

The magnetizing curve, also known as the saturation or excitation curve, is related directly to the quality of the iron, core dimensions, overall geometry, and coil windings. In other words, the induction generator characteristics determine the terminal voltage for a given magnetizing current through the windings (see Figure 4.3). As discussed in Section 5.5, the magnetizing curve is customarily represented either by a polynomial or by a nonlinear expression, like the following:

$$L_m = a_0 + a_1 V_{ph} + a_2 V_{ph}^2 + a_3 V_{ph}^3 + a_4 V_{ph}^4$$

$$X_m = F\omega_s L_m = \frac{V_g}{I_m} = F\left(K_1 e^{k_2 I_m^2} + K_3\right)$$

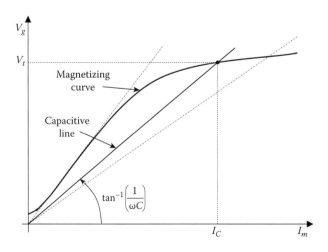

FIGURE 4.3   Magnetizing curve of the induction generator.

The operation of the induction generator as a function of the terminal voltage (at a given frequency) needs the magnetizing current I_m to describe the operation. That is determined by feeding the machine as an induction motor without load and measuring the current as a function of the terminal voltage variation. This curve starts at the value of the residual magnetism (zero current) existing before the beginning of the test in the iron hysteresis curve of the machine.

It is interesting at this point to compare the induction generator and the dc generator; the exciting capacitor of the induction generator takes the place of the field resistor of the dc generator. When the shunt dc generator starts, the residual magnetism in the field generates an initial voltage, which produces a field current that in turn produces more field voltage and so on. The process goes until the iron is saturated. This is a self-excitation process very similar to what happens with the induction generator, as we will see further on. In this case, the residual magnetism in the iron produces a small voltage that produces a capacitive current (a delayed current). This current produces an increased voltage that provokes a higher increase of the capacitive current, and so on, until the iron saturation of the magnetic field. Without residual magnetism, neither the dc generator nor the induction generator can produce any voltage.

It is important to emphasize that it is very difficult to lose residual magnetism completely; a minimum value always remains. In the case of complete loss, however, there are four commonly used techniques for recovering it: (1) Make the machine rotate at no load and a high speed until the residual magnetism is recomposed; in extreme cases, this is not possible. (2) Use a battery to cause a current surge in one of the machine windings. (3) Maintain a charged high-capacity capacitor—it can be one of the electrolytic types—to cause a current surge as in the previous method. (4) Use a rectifier fed from the network to substitute for the battery in the second method.

A way of obtaining a more coherent magnetizing curve with the frequency-dependent nature of the induction motor parameters is to use a secondary driving machine coupled to its shaft. With this, a constant rotation is always guaranteed during the tests, once a good terminal voltage control may produce a wide speed and frequency variation of the rotor.[6–8]

For operation as a stand-alone generator, the SEIG should be connected to a three-phase bank of capacitors. Like the excitation curve, the capacitive reactance (Figure 4.3) will be a straight line passing through zero whose slope is

$$X_C = \frac{1}{\omega C}$$

As has already been said, the magnetizing current lags behind the terminal voltage by about 90°, depending on the losses of the motor in a no-load condition while the current through the capacitor is approximately 90° ahead.

The value of C can be chosen for a given rotation in such a way that the straight line of the capacitive reactance intercepts the magnetizing curve at the point of the desired rated voltage (V_{rated}). This means that the intersection of these two lines is the point at which the necessary reactive power of the generator is supplied only by capacitors (the resonant point). Therefore, to have the excitation process, this value should be between the straight-line slope passing through the origin and the tangent to the most sloped part of the excitation curve (the air gap line), also passing through the origin. The current corresponding to the interception point should not be much above the rated current of the machine (see Figure 4.3). In the first case, the output voltage will be unstable, once there are infinite common points between the two lines. The second extreme end is justified by the maximum current capacity that the motor windings can support.

Notice in Figure 4.3 that the magnetizing curve does not pass through zero, which is very typical of the laboratory tests as the core is rarely fully demagnetized in such tests. This displacement from the zero is due to the residual magnetism that should be corrected to compensate for the distortion.

In addition, the mutual inductance is essentially nonlinear. For numerical computation, it should be determined for each value of the instantaneous magnetizing current I_m. To solve this problem, it is suggested that the no-load curve of the machine be initially obtained and then the terminal voltage found is divided by the product of the magnetizing current times $2\pi f_{base}$.

4.4 MATHEMATICAL DESCRIPTION OF THE SELF-EXCITATION PROCESS

The mathematical description of the self-excitation process can be based on the Doxey model, which comes from the classical steady-state model of the induction machine presented in Figure 2.7, adapted for the induction generator as discussed in this section. As the excitation impedance is much larger than any one of the winding impedances, it is common for simplicity to represent the induction machine with separate resistances of the losses and including the load as in Figure 4.4.

In the specific case of the SEIG, the circuit of Figure 4.4 combines machine resistances and reactances separate from the parallel load circuit as in Figure 4.5.[2]

The shunt resistance R_m represents core losses; it can be determined by making the machine rotate experimentally at no load and measuring the active power per phase P_0, the average voltage per phase V_0, and the average current per phase I_0. As the machine is rotating at no load, the only loads to be gained by the primary machine are the losses by hysteresis, parasite currents in the core, friction, air

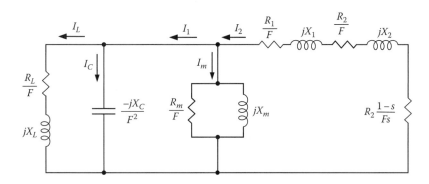

FIGURE 4.4 Simplified classic model of the SEIG.

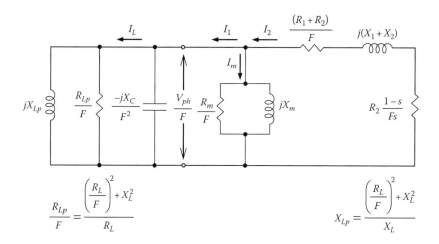

FIGURE 4.5   Compact equivalent circuit of the parallel loaded induction generator.

resistance, and other spurious losses. So, with the help of another external machine, the induction machine is made to turn at the synchronous rotation ($s = 0$), to allow separation of the mechanical losses from the total losses. Then the losses by hysteresis and parasite currents in the core can be represented by[9–11]

$$R_m = \frac{V_0^2}{P_0 - I_0^2 R_1} \tag{4.1}$$

Such losses are dependent on the temperature, voltage levels used, and, above all, on the frequency. However, for a rough estimate like the one we are making here, the value of R_m as given by Equation 4.1 is acceptable.[6–10]

If it is taken into consideration that any load impedance magnitude Z can be transformed from a series to a parallel configuration based on the classical concepts of circuit,

$$Z = R_s + jX_s = \frac{jX_p R_p}{R_p + jX_p} = \frac{R_p X_p^2 + jX_p R_p^2}{R_p^2 + X_p^2} \tag{4.2}$$

where

$$R_p = \frac{R_s^2 + X_s^2}{R_s} = \frac{|Z_s|^2}{R_s} \quad \text{and} \quad X_p = \frac{R_s^2 + X_s^2}{X_s} = \frac{|Z_s|^2}{X_s}$$

So, for Z_L/F, we have

$$\frac{Z_L}{F} = \frac{R_L}{F} + jX_L = \frac{jX_{Lp} \dfrac{R_{Lp}}{F}}{\dfrac{R_{Lp}}{F} + jX_{Lp}} \tag{4.3}$$

Equations 4.2 and 4.3 enable us to determine the load power angle as

$$\theta_L = \tan^{-1}\left(\frac{FX_L}{R_L}\right) = \tan^{-1}\left(\frac{R_{Lp}}{FX_{Lp}}\right) \tag{4.4}$$

In addition, from Equations 4.2 and 4.3, it can be easily shown that

$$R_{Lp} = \frac{|Z_L|^2}{FR_L} \quad X_{Lp} = \frac{|Z_L|^2}{F^2 X_L} \quad P_L = \frac{V_{ph}^2}{FR_{Lp}} = \frac{V_{ph}^2}{|Z_L|^2}R_L \tag{4.5}$$

An even simpler model of the SEIG can be obtained making the circuit of Figure 4.6 the equivalent of that of Figure 4.5.

Notice in Figure 4.6 that the influence of an inductive load on the self-excitation capacitance is to decrease its effective value. In a quantitative form, it can bring down this value from the load parallel inductance X_{Lp} in parallel with X_c to obtain the effective self-exciting capacitance:

$$C_{eff} = C - \frac{L}{R_L^2 + (\omega_s L)^2} \tag{4.6}$$

From Equation 4.6 it is possible to conclude that small values of L do not have much influence on C_{eff}. However, for high values of L, this inductance can have a decisive effect on the output voltage because it approximates the straight line of the capacitive reactance slope to the linear portion of the excitation curve slope (the air gap line), reducing drastically the terminal voltage until total collapse. The positive side of this matter is that when the load resistance is too small, it will help to discharge the self-excitation capacitor more quickly, taking the generator to the de-excitation process. This is a natural protection against high currents and

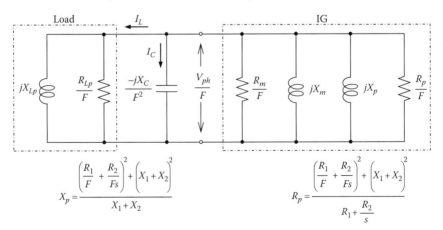

FIGURE 4.6   Doxey simplified model for the induction generator.

short-circuits. On the other hand, the increase of the self-excitation capacitance is limited by the heavy iron saturation when the intersection point on the straight line of the capacitive reactance enters into the saturation portion of the machine characteristic for values beyond the rated current value of the generator. The winding heats up quickly, which can cause permanent damage in the isolations and in the magnetic properties of the iron.[12–17]

To evaluate the performance of the induction generator, it should be taken into account that the only power entering in the circuit is from the primary machine in the form of active power given by Equation 2.23. The reactive power should also be balanced. For calculation effects, the active power balance can be obtained from Figures 4.5 and 4.6 for the active and reactive powers, respectively, by

$$\sum P = I_2^2 R_2 \frac{1-s}{s} + I_2^2 (R_1 + R_2) + \frac{V_{ph}^2}{R_m} + \frac{V_{ph}^2}{R_{Lp}} = 0 \tag{4.7}$$

$$\sum Q = \frac{V_{ph}^2}{X_p} + \frac{V_{ph}^2}{X_m} - \frac{V_{ph}^2}{X_c} + \frac{V_{ph}^2}{X_{Lp}} = 0 \tag{4.8}$$

where

$$X_m = \omega_s L_m$$

The first term of Equation 4.7 is the mechanical energy being supplied to the generator. The sign of this term is inverted when $s > 1$ or $s < 0$; in other words, when the induction machine is working as a brake or as a generator.

Dividing Equation 4.7 by I_2^2 and simplifying, we get

$$\frac{R_2}{s} + R_1 + \frac{V_{ph}^2}{I_2^2} \frac{1}{R_{mL}} = 0 \tag{4.9}$$

where

$$\frac{1}{R_{mL}} = \frac{1}{R_m} + \frac{1}{R_{Lp}}$$

From Figure 4.5, we get

$$\frac{V_{ph}^2}{F^2 I_2^2} = \left(\frac{R_2}{F_s} + \frac{R_1}{F} \right)^2 + (X_1 + X_2)^2 \tag{4.10}$$

and, with that, Equation 4.9 becomes

$$\frac{R_2}{s} + R_1 + \left[\left(\frac{R_2}{s} + R_1 \right)^2 + F^2 (X_1 + X_2)^2 \right] \frac{1}{R_{mL}} = 0$$

or

$$\left(\frac{R_2}{s} + R_1\right) R_{mL} + \left(\frac{R_2}{s} + R_1\right)^2 + F^2 \left(X_1 + X_2\right)^2 = 0 \tag{4.11}$$

Equation 4.11 is an equation of the second order, and if the other parameters of the machine are constant, it can be used to determine the corresponding value of s for the conditions of primary power and load given by

$$s = \frac{2R_2}{-2R_1 - R_{mL} \pm \sqrt{R_{mL}^2 - 4F^2 \left(X_1 + X_2\right)^2}} \tag{4.12}$$

The radical in the denominator of Equation 4.12 will always be negative to satisfy the practical condition that if $X_1 + X_2$ were very small, the load R_{mL} would have almost no influence on s, which cannot be the case. Therefore, it is convenient sometimes to approximate Equation 4.12 by the following expression:

$$s \cong \frac{R_2}{R_1 + R_{mL}} = -\frac{R_2 \left(R_m + R_{Lp}\right)}{R_1 R_m + R_1 R_{Lp} + R_m R_{Lp}} \tag{4.13}$$

A numerical process may be necessary when using Equation 4.12 instead of Equation 4.13 because the calculation of F depends on ω_s, which depends on $\omega_{s'}$ which, in turn, depends on F.

The voltage regulation also depends on the variation of X_m, which can be determined by solving Equation 4.8 as follows:

$$X_m = \frac{1}{F^2 \omega_s C - \dfrac{1}{X_p} - \dfrac{1}{X_{Lp}}} \tag{4.14}$$

The efficiency can be estimated by

$$\eta = \frac{P_{out}}{P_{in}} = \frac{P_{in} - P_{losses}}{P_{in}} \tag{4.15}$$

The input power is the mechanical power, which should be subtracted from the copper losses ($R_1 + R_2$) and the losses represented by the resistance R_m. Using Equation 4.7 once again for the three-phase case without the portion corresponding to the load power ($R_{Lp} \rightarrow \infty$) together with Equation 4.15, and remembering what is said about s in Equation 4.7, we get

$$P_{out} = P_{in} - P_{losses} = 3I_2^2 R_2 \frac{1-s}{s} + 3I_2^2 (R_1 + R_2) + \frac{3V_{ph}^2}{R_m} = -\frac{3V_{ph}^2}{R_{Lp}}$$

where

$$P_{in} = 3I_2^2 R_2 \frac{1-s}{s}$$

Therefore, using these values in Equation 4.15 and simplifying, we get

$$\eta = \frac{I_2^2 R_2 \frac{1-s}{s} + I_2^2 (R_1 + R_2) + \frac{V_{ph}^2}{R_m}}{I_2^2 R_2 \frac{1-s}{s}} \qquad (4.16)$$

Dividing the numerator and the denominator of Equation 4.16 by I_2^2 and using Equation 4.10 again, we get

$$\eta = \frac{\left(\frac{R_2}{s} + R_1\right) + \frac{1}{R_m}\left[\left(\frac{R_2}{s} + R_1\right)^2 + F^2 (X_1 + X_2)^2\right]}{R_2 \frac{1-s}{s}} \qquad (4.17)$$

A plot of Equation 4.17 showing the output power and IG low/high efficiency is depicted in Figure 4.7. It is noticeable the heavy influence of the machine losses for light loads affecting the overall IG efficiency.

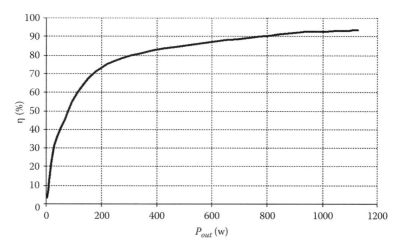

FIGURE 4.7 Influence of the output power on the efficiency of the SEIG.

Besides the generator parameters being studied, the complete solution for the performance of the induction generator depends on the self-excitation capacitance C, on the angular mechanic frequency of the rotor ω_r, and on the load impedance Z_L. The block diagram of Figure 4.8 illustrates the sequence of calculations required to obtain the load current, the output power, and the overall efficiency of the generator.

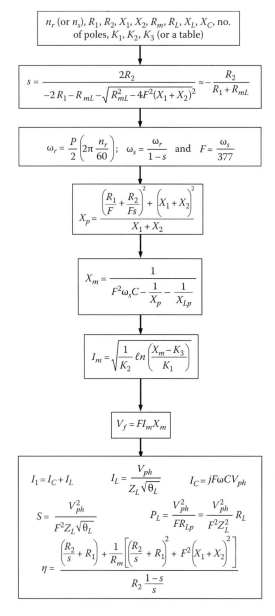

FIGURE 4.8 Calculation flowchart of the performance of the SEIGs.

Furthermore, the magnetizing reactance can be determined point by point in the laboratory instead of using a nonlinear equation adjusted for losses that can enter into the calculations under a table form.

4.5 SERIES CAPACITORS AND COMPOSED EXCITATION OF INDUCTION GENERATORS

Another analogy between the dc generator and the induction generator can be seen with the connection of capacitors in series across the generator terminals rather than in parallel in a form of composed excitation (Figure 4.9).[11,18–20]

The voltage–current characteristic is plotted for the loaded generator with a constant lagging PF. The capacitive reactive power increases with the load increase, partially compensating the reactive power demanded by the load as sketched in Figure 4.10.

4.6 THREE-PHASE GENERATORS OPERATING IN SINGLE-PHASE MODE

A three-phase generator can be converted into a single-phase generator, which is usually derated to approximately 80% of their nominal machine rating. Figure 4.11 shows the connection of two capacitors. The impedance Z_L represents the consumer load connected in parallel across the ballast load R_B. The ballast maintains the total generator load at a constant value. In fact, the ballast load is a variable load where its resistance is controlled to maintain the consumer load plus the ballast load at a constant value.[21]

Of course, such a single-phase connection suffers of poor voltage regulation, and the system is stable only for a very limited range of resistive load. This is normally true for very small hydro schemes, where the voltage and frequency variation are usually well tolerated, such as to power lighting, heating, cooking, and maybe ironing clothes.

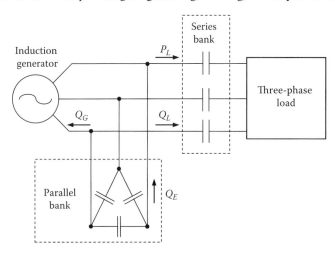

FIGURE 4.9 Composed series capacitors to compensate the SEIG's PF.

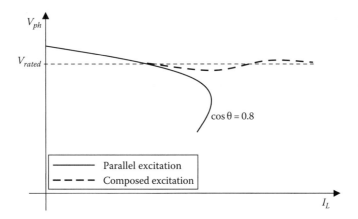

FIGURE 4.10   Influence of the self-excitation capacitance on the terminal voltage.

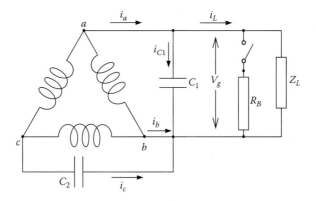

FIGURE 4.11   Single-phase output from a three-phase induction generator.

Usually, a very small hydro solution is used for isolated and remote areas. Using Equations 4.18 and 4.19, and assuming that the machine is operating as a balanced three-phase system, the phasor diagram is constructed as shown in Figure 4.12. As capacitor C_2 is connected across phases b and c, current i_c is perpendicular to the voltage vector V_{bc}. In order to obtain a balanced operation, the following two conditions must be satisfied:

$$\overline{i_a} = \overline{i_L} + \overline{i_{C1}} \tag{4.18}$$

$$\overline{i_b} = -\left(\overline{i_a} + \overline{i_c}\right) \tag{4.19}$$

$$\theta = 60° \quad \text{and} \quad \left|\overline{i_c}\right| = \left|\overline{i_a}\right| \tag{4.20}$$

Following this analysis, it is observed that in order to obtain a balanced operation of the three-phase motor, capacitor C_1 should be selected such that $\left|\overline{i_L}\right| = \sqrt{3}\left|\overline{i_{C1}}\right|$ and

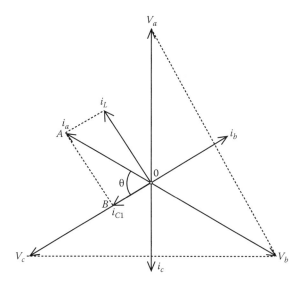

FIGURE 4.12 Phasor diagram for the single-phase connection.

also $\left|\overline{i_c}\right| = 2\left|\overline{i_{C1}}\right|$, which leads to the result that capacitor C_2 should be equal to $2C_1$. When an induction generator is used in this way, particular care must be taken over the connection of the capacitor C_2. If capacitor C_2 is connected across phase a–c instead of c–b, then the resulting phasor diagram is shown in Figure 4.12. In this case, the generator will run as an unbalanced system. It is noticeable that the current through one of the windings of the induction generator becomes twice that of the other winding currents. Under this condition, the winding generator will overheat. So, it is very important to have the correct connection of capacitor C_2.

In typical induction generator–based small hydro schemes, the turbines are run-of-the-river type, where the water input and the mechanical power provided to the generator are not usually controlled. Traditionally a wood or metal barrier sliding in grooves is used to control water levels and flow. But, for example, if a water-wheel is used as a turbine, or whenever the generator operates under manual control of the sluice gate, and the consumer load changes, the generated voltage and frequency will also change. If the load is light, the generator speed increases, leading to a runaway condition. The control technique used to maintain the generated voltage and frequency at their rated values is to keep the total load connected to the machine at a near constant value, using the ballast load as shown in Figure 4.13. Since the terminal voltage under this condition is a constant value, a voltage sensor is used to control the ballast load.

The ballast load can be implemented in several ways.[20–22] One way of obtaining a variable load is to use a resistor with two antiparallel thyristors operating in a phase control mode as shown in Figure 4.14. By changing the firing angle α, the fundamental value of the current through the resistor–thyristor circuit can be controlled. When $\alpha = 0°$, there is a full current impressed through the resistor–thyristor circuit, that is, maximum load. When $\alpha = 180°$, the current through the resistor–thyristor circuit is zero, and a programmable ballast load is possible for values of α in between 0° and 180°.

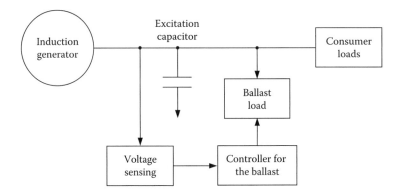

FIGURE 4.13 Typical small hydro scheme with a ballast load control.

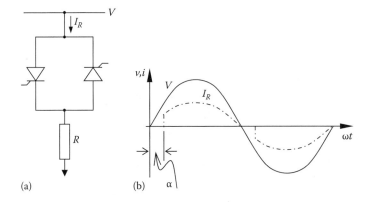

FIGURE 4.14 Thyristor phase controlled ballast load. (a) Phase control with antiparallel thyristors and (b) α-phase control mode.

Of course, as α is increased, the displacement factor (as well as the distortion factor) of the resistor–thyristor circuit increases, thus requiring more reactive power. This circuit will draw a reactive current from the excitation capacitors, reducing the effective capacitance available to supply a magnetization current to the induction generator, which causes a slight reduction in the output generator voltage. Another possible circuit, which does not absorb reactive power, uses a number of resistors with a switched thyristor scheme. The circuit is shown in Figure 4.15 where the back-to-back thyristor pair operates either as a closed or open switch (on-off). Therefore, the load is varied by controlling the number of parallel resistor–thyristor circuits, which are *on*. The load has a stepwise variation, so a smooth control is not possible. Therefore, the thyristor-based circuits have some drawbacks.

In order to get a better resolution from the circuit depicted in Figure 4.15, resistors are selected in a binary weighted configuration. If $R_1 = R$, $R_2 = 2R$, and $R_3 = 4R$, then the load can be varied from 0 to $7R$, in steps of R.

A circuit that is very smooth in respect to a varying ballast load can be implemented with a full-wave rectifier followed by a transistorized chopper operating at

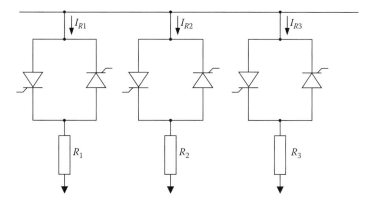

FIGURE 4.15   Thyristor phase stepwise controlled ballast load.

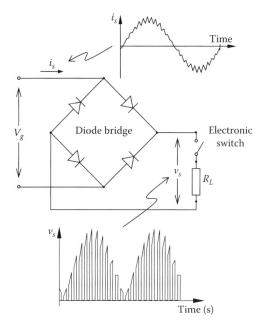

FIGURE 4.16   High-frequency ballast showing the voltage across the load and the supply-side current for a 50% loading in the machine.

a high-frequency, as indicated in Figure 4.16. If such frequency is at least 10 times higher than the fundamental induction generator frequency, the system will be nearly at unit PF for any pulse-width modulation (duty-ratio) of the chopper.

The effective resistance of the ballast load changes by varying the duty ratio of the switch. Figure 4.16 shows the voltage across the ballast load and the ac-side current when the consumer load is only 50% of the rated load of the machine. Due to the machine inductance, the current drawn from the generator is nearly sinusoidal with a superimposed high-frequency ripple component.

In order to run the machine with a single-phase loading for minimum unbalance, the power output of the generator and the value of the excitation capacitor (C_1) connected across the load should have the relationship depicted by Equation 4.21:

$$P = V_g i_L = \sqrt{3} V_g i_{C1} = \sqrt{3} V_g^2 \omega C_1 \qquad (4.21)$$

where

V_g is the generated voltage
i_L is the load voltage
i_{C1} is the current through the capacitor

Power constraints of Equation 4.21 assume that the induction generator is operating near to a balanced condition. Therefore, the family of curves of the terminal voltage, which is a characteristic of the induction generator, and called power–voltage curves, is found for different capacitor values by using the conventional induction generator steady-state equivalent circuit. This model includes the saturated magnetizing inductance of the generator in addition to considering the loading and operating speeds of the machine. The intersections of the terminal voltage characteristics and the voltage versus power are obtained from Equation 4.21. It is noticeable on these characteristics that both operating points satisfy the machine equivalent circuit in steady state. This design method can be easily used in practice for sizing a small hydro scheme using an induction machine. Figure 4.17 shows a typical plot of characteristics for three different values of the excitation capacitor versus the supply voltage i_s. One can observe that if the value of the excitation

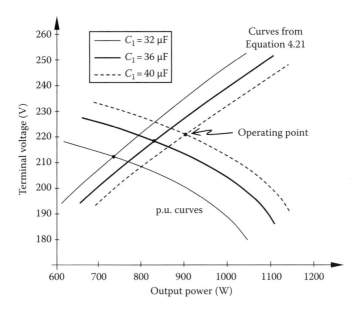

FIGURE 4.17   Voltage versus power characteristics for variable C_1.

capacitor C_1 is 40 µF, the generator produces 900 W at 220 V. Typically, this is a derating of 80%–90% of the nominal machine power.

Under lightly load conditions, or under no load, if the ballast load fails, then only a small part of the mechanical power input is converted into electrical power. Since the turbine is of the run-of-the-river type, the mechanical power into the turbine cannot be controlled, and therefore, the turbine and the generator will accelerate to a runaway speed within a few seconds. The runaway speed depends on the chosen turbine. When runaway occurs, the turbine mainly governs the torque-speed characteristics, and the speed increases to the corresponding torque, which is barely enough to overcome the friction and windage system losses.

For the commonly used cross-flow turbine, runaway speed is around 175% of the optimum speed. Under runaway conditions, the voltage of the generator will increase, which can cause extensive damage to the generator (insulation failure), in addition to the connected loads and excitation capacitors. In order to prevent such overall damage, an overvoltage protection circuit is used. When the generated voltage rises above a certain threshold-preset limit, the overvoltage protection isolates the excitation capacitors and the loads from the generator, thus allowing the induction machine to run freely without any current circulation.

From everything, we have seen in this chapter, the biggest advantage of the induction generator is its simplicity in not needing a separate field circuit as the synchronous generator and not needing to work at a constant speed. It is sufficient that it works at a speed above the synchronous speed, which is the stator magnetic field speed. The higher the torque is, the higher the output power and, therefore, the speed. The distribution network, as cited previously, controls the voltage, and the PF should be compensated by capacitors. These factors make the induction generator advisable for wind energy, heat recovery, and other small sources of electrical energy. The great majority of the wind and hydro turbines up to 500 kVA use this type of generator.

4.7 SOLVED PROBLEMS

1. How can harmonic voltage and current affect the *PF* of an induction generator?

   Solution

   As we know, assuming undistorted voltage the PF is related to the harmonics in current as given by the equation

$$PF \approx \frac{DPF}{\sqrt{1+THD_{(i)}^2}}$$

   Hence, with increase of harmonics, the *PF* becomes smaller, and we have less active power for the loads. This can be overcome by passive filters across the output of the induction generator.

2. Based on Section 4.2, suggest some specifications for an induction generator being considered for supplying a load of 100 kW for a small group of residential houses.

Solution

Parameters
Operating voltage: three-phase, 110 V/phase
Rated power: 100 kW
Rated frequency: 60 Hz
Power factor: 0.9
Rotor speed: 1200 rpm
Insulation class: *F*
Capacity of acceleration: 25% of synchronous speed

3. Specify the self-exciting capacitance for a three-phase, four-pole, 1700 rpm, Y-stator connection induction generator supplying 110 V, 1.0 kW at 60 Hz, whose equivalent circuit is given as $R_1 = 0.20\ \Omega$, $R_2 = 0.42\ \Omega$, $X_1 = 0.15\ \Omega$, and $X_2 = 0.43\ \Omega$; R_2 and X_2 parameters refer to the stator. The excitation current is 8.1 A and is assumed to be constant at the rated values.

Solution

The equation for Q (4.11) is observed in Figure 4.6:

$$X_p = \frac{\left(R_1/F + R_2/F_s\right)^2 + \left(X_1 + X_2\right)^2}{X_1 + X_2}, \text{ taking } F = 1 \text{ and } f_s = \frac{pn_s}{120}$$

$$\Rightarrow n_s = \frac{120 \times 60}{4} = 1800 \text{ rpm}$$

$$X_p = \frac{\left(0.2 + 0.42/0.05\right)^2 + \left(0.15 + 0.43\right)^2}{\left(0.15 + 0.43\right)} = 128.1\ \Omega$$

$$X_{lp} = \frac{Z_l^2}{F^2 X_l} \rightarrow \infty$$

$$X_m = \frac{110}{8.1} = 13.58\ \Omega$$

So, $\dfrac{V_f^2}{X_p} + \dfrac{V_f^2}{X_m} - \dfrac{V_f^2}{X_l} + \dfrac{V_f^2}{X_{lp}} = \sum Q = 0 \Rightarrow \dfrac{1}{X_c} = \dfrac{1}{128.1} + \dfrac{1}{13.58} = 0.08\ \text{S}$

Now, $X_c = 12.28\ \Omega \Rightarrow \dfrac{1}{\omega C} = 12.28 \Rightarrow C = \dfrac{1}{2\pi 60\left(12.28\right)} = 216.03\ \mu F$

4.8 SUGGESTED PROBLEMS

4.1 Give the effective self-exciting capacitance at 1900 rpm and 60 Hz of an induction generator having a rated self-exciting capacitance of 180 μF and a load connected across its terminals of $R_L = 50$ Ω and $L = 0.05$ H. What would the effective capacitance be if the frequency went up to 65 Hz? What would the new value of the real capacitance be to keep the self-exciting point at its original value?

4.2 Using the data given in Tables 5.2 and 5.3, calculate the self-exciting capacitance of the induction generator to give the nominal electrical values for the specified machine. If this self-exciting capacitance was electronically controlled, what should its maximum and minimum values be to keep the voltage regulation within a ±10% range at the rated speed?

4.3 Calculate the induction generator's total internal current to feed the load specified in Exercise 4.2, separating the self-exciting current to the capacitor and the current to the load itself. What is the active power actually supplied by the generator in comparison to that supplied to the load?

REFERENCES

1. Doxey, B.C., Theory and application of the capacitor-excited induction generator, *The Engineer Magazine*, pp. 893–897, November 1963.
2. Murthy, S.S., Nagaraj, H.S., and Kuriyan, A., Design-based computer procedures for performance prediction and analysis of self-excited induction generators using motor design packages, *IEEE Proc.*, 135(1), 8–16, January 1988.
3. Stagg, G.W. and El-Abiad, A.H., *Computer Methods in Power Systems Analysis*, McGraw-Hill Book Company, New York, 1968.
4. Tabosa, R.P., Soares, G.A., and Shindo, R., *The High Efficiency Motor (Motor de alto rendimento), Technical Handbook of the PROCEL Program*, edited by Eletrobrás/Procel/CEPEL, Rio de Janeiro, Brazil, August 1998.
5. Grantham, C., Steady-state and transient analysis of self-excited induction generators, *IEEE Proc.*, 136(2), 61–68, March 1989.
6. Seyoum, D., Grantham, C., and Rahman, F., The dynamics of an isolated self-excited induction generator driven by a wind turbine, *The 27th Annual Conference of the IEEE Industrial Electronics Society, IECON'01*, Denver, CO, December 2001, pp. 1364–1369.
7. Singh, S.P., Bhim Singh, M., and Jain, P., Performance characteristics and optimum utilization of a cage machine as capacitor excited induction generator, *IEEE Trans. Energy Conver.*, 5(4), 679–684, December 1990.
8. Hallenius, K.E., Vas, P., and Brown, J.E., The analysis of a saturated self-excited asynchronous generator, *IEEE Trans. Energy Conver.*, 6(2), 336–345, June 1991.
9. Barbi, I., *Fundamental Theory of the Induction Motor (Teoria fundamental do motor de indução)*, UFSC University Press-Eletrobrás, Florianópolis, Brazil, 1985.
10. Langsdorf, A.S., *Theory of the Alternating Current Machines (Teoría de máquinas de corriente alterna)*, McGraw-Hill Book Co., New York, 1977, p. 701.
11. Chapman, S.J., *Electric Machinery Fundamentals*, 3rd edn., McGraw-Hill, New York, 1999.
12. Hancock, N.N., *Matrix Analysis of Electric Machinery*, Pergamon, New York, 1964, p. 55.
13. Grantham, C., Steady-state and transient analysis of self-excited induction generators, *IEEE Proc.*, 136(2), March 1989.

14. Seyoum, D., Grantham, C., and Rahman, F., The dynamics of an isolated self-excited induction generator driven by a wind turbine, *The 27th Annual Conference of the IEEE Industrial Electronics Society, IECON'01*, Denver, CO, pp. 1364–1369, December 2001.

15. Muljadi, E., Sallan, J., Sanz, M., and Butterfield, C.P., Investigation of self-excited induction generators for wind turbine applications, *IEEE Proc.*, 1, 509–515, 1999.

16. Leses, C.H. and Wang, L., A novel analysis of parallel operated self-excited induction generators, *IEEE Trans. Energy Conver.*, 13(2), June 1998.

17. Murthy, S.S., Bhim Singh, M., and Tandon, A.K., Dynamic models for the transient analysis of induction machines with asymmetrical winding connections, *Elect. Mach. Electromech.*, 6(6), 479–492, November/December 1981.

18. Smith, O.J.M., Three-phase induction generator for single-phase line, *IEEE Trans. Energy Conver.*, EC-2(3), 382–387, September 1987.

19. Bhattacharya, J.L. and Woodward, J.L., Excitation balancing of a self-excited induction generator for maximum power output, *Gener. Transm. Distrib. IEE Proc. C*, 135(2), 88–97, March 1988.

20. Portolann, C.A., Farret, F.A., and Machado, R.Q., Load effects on DC-DC converters for simultaneous speed and voltage control by the load in asynchronous generation, *First IEEE International Caracas Conference on Devices, Circuits and Systems*, Caracas, Venezuela, 1995, pp. 276–280. DOI: 10.1109/ICCDCS.1995.499159.

21. Ekanayake, J.B., Induction generators for small hydro schemes, *Power Eng. J.*, 61–67, April 2002.

22. Farret, F.A., Portolann, C.A., and Machado, R.Q., Electronic control by the load for asynchronous turbogenerators, driven by multiple sources of energy, *Proceedings of Second IEEE International Caracas Conference on Devices, Circuits and Systems*, Isla de Margarita, Venezuela, 1998, pp. 332–337. DOI: 10.1109/ICCDCS.1998.705859.

5 General Characteristics of Induction Generators

5.1 SCOPE OF THIS CHAPTER

Chapters 2 through 4 encompassed an overall discussion of the behavior of the induction generator (IG). Based on standard machine models and on circuit equations relating voltage and current, many characteristics can be derived. In this chapter, mechanical and electrical characteristics are combined in such a way as to give a clear picture of the IG as one of the most important tools for mechanical conversion of renewable and alternative energies into electrical energy. The relationships between load voltage and frequency that result from this electromechanical interaction are made clear as the basic parameters of the turbine as a prime mover and the induction machine as a generator.

5.2 TORQUE–SPEED CHARACTERISTICS OF INDUCTION GENERATORS

Figure 2.3 can be used to determine the torque–speed characteristics of the IG. It is repeated here as Figure 5.1 for convenience. From this figure, and from Equations 2.19, 2.23, and 2.26, neglecting R_m, we can obtain the expressions for air-gap power, converted power, and converted torque, which are

$$P_{airgap} = \frac{3I_2^2 R_2}{s} = \frac{3V_{ph}^2 R_2/s}{\left(R_1 + R_2/s\right)^2 + \left(X_1 + X_2\right)^2} \tag{5.1}$$

$$P_{converted} = \left(1 - s\right) P_{airgap} \tag{5.2}$$

$$T_{converted} = \frac{P_{airgap}}{\omega_s} \tag{5.3}$$

Notice that the precision of R_1 and X_1 can be improved by replacing them with the values taken from the Thévenin equivalent in Figure 5.1. This subject is left for the student as an exercise.

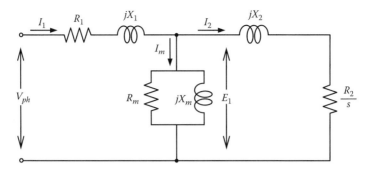

FIGURE 5.1   Stator equivalent circuit of the IG.

Output power may be obtained from Equation 2.33 if there is a way of determining the stray losses, friction, and air resistance as being

$$P_{out} = P_{mec} - P_{losses} = -3I_2^2 R_2 \frac{1-s}{s} - 3I_2^2 R_1 - \frac{3E_1^2}{R_m} - 3I_2^2 R_2 - P_{frict+air} - P_{stray} \quad (5.4)$$

or

$$P_{out} = P_{airgap} - P_{losses} = -3I_2^2 R_2 \frac{1}{s} - 3I_2^2 R_1 - \frac{3E_1^2}{R_m} - P_{frict+air} - P_{stray} \quad (5.5)$$

Substituting Equation 5.1 into 5.3, we get

$$T_{converted} = \frac{3V_{ph}^2 R_2/s}{\omega_s\left[\left(R_1 + R_2/s\right)^2 + \left(X_1 + X_2\right)^2\right]} \quad (5.6)$$

where $s = 1 - (n_r/n_s)$.

Equations 5.5 and 5.6 are plotted in Figure 5.2. It can be observed that (1) there is no torque at the synchronous speed; (2) both the torque–speed and the power–speed curves are almost linear since from no load to full load the machine's rotor resistance is much larger than its reactance; (3) as resistance is predominant in this range, current and the rotor field as well as the induced torque increase almost linearly with the increase of the slip factor s; (4) the rotor torque varies as the square of the voltage across the terminals of the generator; (5) if the speed slows down close to the synchronous speed, the generator motorizes; that is, it works as a motor; (6) as we will show, the generated power has a maximum value for a given current drained from the generator; (7) in the same way, there is a maximum possible induced generator torque called pullout or breakdown torque, and from this torque value on, there will be overspeed.

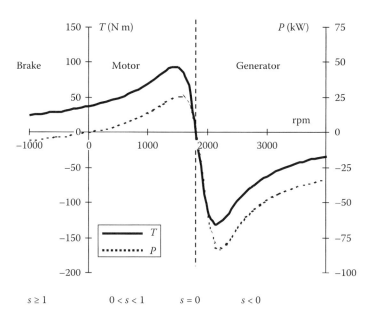

FIGURE 5.2 Torque–/power–speed characteristic of the induction machine working as a brake, motor, or generator.

As can be observed in Figure 5.2, the peak power supplied by the IG happens at a speed slightly different from the maximum torque, and, naturally, no electric power is converted into mechanical power when the rotor is at rest (zero speed). In the same way, in spite of the same rotation, the frequency of the IG varies with the load variation. However, as the torque–speed characteristic is much accentuated in the normal range of operation, the total frequency variation should be limited to not much more than 5%. Such variation is acceptable for most of the loads in stand-alone power plants.[1-3]

In Figure 5.2, it can be observed that, for $0 < s < 1$ (or $0 < n_r < 1800$ rpm), the machine absorbs electric power in the rotor to convert it to a positive mechanical torque. If $s = 0$ (or $n_r = n_s$), there is no electromotive force induced in the rotor, and, therefore, there is no power transfer from the stator to the rotor, and there is no mechanical torque. If $s < 0$ (or $n_r > 1800$ rpm), the mechanical power is converted into electric power by the external application of torque in the shaft to turn the rotor, and the machine works as a generator. If $s \geq 1$ (or $n_r \leq 0$), the machine works as a brake, absorbing mechanical power, which acts negatively on its shaft ($P_{mec} < 0$).

In the characteristic $T \times n_r$, the current I_2 assumes values that depend on the rotation assumed for the rotor. In the characteristic $P_{mec} \times I_2$, in contrast, the slip factor s varies according to I_2.

5.3 POWER VERSUS CURRENT CHARACTERISTICS

Chapter 10 presents optimized methods of control for IGs. Thus, it is important to know the characteristics of mechanical power versus load current ($P \times I$). This relationship may be obtained from Figure 2.7 and from the definition of mechanical power in the rotor. The power dissipation in the rotor is expressed by Equation 2.23 and repeated here for convenience:

$$P_{mec} = 3I_2^2 \left(\frac{R_2}{s} - R_2 \right)$$

(5.7)

Figure 5.1 can be used to make Equation 5.7 independent of slip s and only a function of the rotor current. To do this, I_2 may be given as

$$I_2 = \frac{V_{ph}}{\sqrt{\left(R_1 + R_2/s \right)^2 + \left(X_1 + X_2 \right)^2}}$$

(5.8)

from which the term R_2/s can be isolated to give

$$\frac{R_2}{s} = \sqrt{\left(\frac{V_{ph}}{I_2} \right)^2 - \left(X_1 + X_2 \right)^2} - R_1$$

(5.9)

Substituting Equation 5.9 into 5.7, we finally get

$$P_{mec} = 3I_2^2 \left[\sqrt{\left(\frac{V_{ph}}{I_2} \right)^2 - \left(X_1 + X_2 \right)^2} - \left(R_1 + R_2 \right) \right]$$

(5.10)

Notice in Equation 5.10 that necessarily

$$\left(\frac{V_{ph}}{I_2} \right)^2 \geq \left(X_1 + X_2 \right)^2$$

or

$$I_2 \leq \frac{V_{ph}}{X_1 + X_2}$$

(5.11)

Maximum power is obtained by substituting the inequality (5.11) in (5.10) to give

$$P_{mec(max)} = -I_2^2 \left(R_1 + R_2 \right)$$

(5.12)

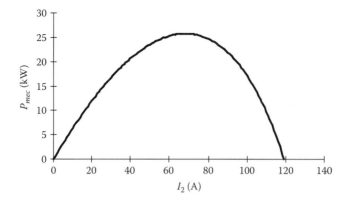

FIGURE 5.3 $P \times I$ characteristic of the IG ($V_{ph} = 220$ V).

Notice that $I_2 \approx I_1$ when we neglect the excitation admittance. The negative sign in Equation 5.12 is due to reversion of the power flow through the terminals of the induction machine. Equation 5.10 is plotted in Figure 5.3, assuming a constant output voltage control.

Knowledge of the characteristic shown in Figure 5.3 is particularly useful in studies of the types of hill climbing and fuzzy control (peak power tracking) discussed in Chapter 10. In this figure, we see that when the load current of the IG increases, two limiting values of current stand out for which the power is null: when there is no load across its terminals ($I_{2min} = 0.0$) and when there is an excess of current over the maximum defined by $P_{mec} = 0.0$ in Equation 5.10. Therefore,

$$I_{2\max} = \frac{V_{ph}}{\sqrt{\left(R_1 + R_2\right)^2 + \left(X_1 + X_2\right)^2}} \tag{5.13}$$

On the other hand, there is also a value for generator current at which generated power is maximized. This value is obtained by the Theorem of the Average Values of Maxima and Minima of Equation 5.10, which is

$$I_{p\max} = \frac{V_{ph}}{\sqrt{2}\left(X_1 + R_2\right)}\left[1 - \frac{R_1 + R_2}{\sqrt{\left(R_1 + R_2\right)^2 + \left(X_1 + R_2\right)^2}}\right] \tag{5.14}$$

5.4 ROTOR POWER FACTOR AS A FUNCTION OF ROTATION

The power factor of an IG is the cosine of the angle represented by the impedance argument seen across the output terminals of the equivalent circuit whose vector diagram is represented in Figure 2.6.[4,5] The angle δ between the stator and rotor

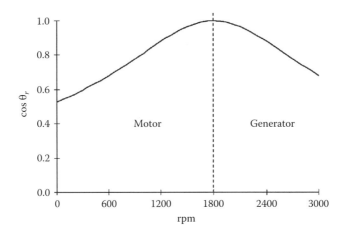

FIGURE 5.4 Rotor power factor as a function of rotation.

magnetic fields can express the rotor power factor by addition of an angle of 90° to θ_r, that is, $\delta = 90° + \theta_r$. As a consequence, $\cos\theta_r = \sin\delta$. In terms of the rotor parameters, the power factor can be expressed from Equation 2.7 as

$$PF = \cos\left(\arctan\frac{sX_{r0}}{R_r}\right) \qquad (5.15)$$

The plot of Equation 5.15 is shown in Figure 5.4 for an induction machine operating as motor and generator. So, for low power factor and high currents, the voltage regulation is deeply affected.

5.5 NONLINEAR RELATIONSHIP BETWEEN AIR-GAP VOLTAGE V_g AND MAGNETIZING CURRENT I_m

Variation of the magnetizing inductance is the main factor in the process of voltage buildup and stabilization in operating self-excited induction generators (SEIGs). Since the magnetization inductance L_m at a rated voltage is derived from the relationship between V_g and I_m, the result is a highly nonlinear function. For better results in terms of representation, a table of experimental values usually gives the magnetization inductance used in laboratory setups as a polynomial series or an analytically adjusted expression.[6–9]

The polynomial representation of the magnetizing inductance taken from experimental results can be reasonably given for practical purposes by a fourth-order curve fit:

$$L_m = a_0 + a_1 V_{ph} + a_2 V_{ph}^2 + a_3 V_{ph}^3 + a_4 V_{ph}^4$$

Another way of representing the magnetization curve is a linear piecewise association within certain ranges of the magnetizing current. The magnetizing reactance $X_m = \omega L_m$ may appear as

$$
X_m = \begin{cases}
a_0 & \text{for } 0 \leq I_m < I_1 \\[2mm]
\dfrac{a_1}{\left(I_m + b_1\right)} & \text{for } I_1 \leq I_m < I_2 \\[2mm]
\dfrac{a_2}{\left(I_m + b_2\right)} & \text{for } I_2 \leq I_m < I_3 \\[2mm]
\dfrac{a_3}{\left(I_m + b_3\right)} & \text{for } I_3 \leq I_m < I_4 \\[2mm]
\dfrac{a_4}{\left(I_m + b_4\right)} & \text{for } I_4 \leq I_m
\end{cases}
\tag{5.16}
$$

The technical literature has reported on several possible ways to relate the air-gap voltage to the magnetizing current,[2-4] showing also that the relationship between V_g and I_m can be established through the following nonlinear equation:

$$
V_g = F I_m \left(K_1 e^{k_2 I_m^2} + K_3 \right)
\tag{5.17}
$$

where
K_1, K_2, and K_3 are constants to be determined
V_g is the air-gap voltage across the magnetizing reactance (without external access)
F is the frequency in p.u., defined as

$$
F = \frac{f}{f_{base}}
\tag{5.18}
$$

where
f is the rotor frequency
f_{base} is the reference frequency used in the tests to obtain the excitation curve (usually, the 60 Hz from the distribution network)

In this case, the magnetizing reactance can be obtained directly from Equation 5.19 as

$$
X_m = \omega L_m = \frac{V_g}{I_m} = F \left(K_1 e^{k_2 I_m^2} + K_3 \right)
\tag{5.19}
$$

The reason for using the magnetizing curve[6] is that the representation of the table points from the lab magnetizing curve test should be adjusted in such a way that it passes through all points. Because of its computational advantages of easy and compact representation, it has been adopted in the discussions in this book.

Another observation that can be made regarding Equation 5.19 is that it can be used to establish the variation of the generator magnetizing reactance, $x_m = dV_g/dI_m$, as well as its boundary limits of validation. For this, taking the derivative of Equation 5.17 with respect to I_m gives

$$\frac{dV_g}{dI_m} = x_m = K_3 + \left(1 + 2K_2 K I_m^2\right) K_1 e^{k_2 I_m^2} \tag{5.20}$$

After some algebraic manipulation from Equations 5.19 and 5.20, we get

$$x_m = \frac{dV_g}{dI_m} = \left(1 + 2K_2 I_m^2\right)\frac{V_g}{I_m} - 2K_2 K_3 I_m^2 \tag{5.21}$$

At the origin ($I_m = 0$), Equation 5.19 yields

$$X_{m0} = \left(K_1 + K_3\right) \tag{5.22}$$

The theoretical maximum of the saturation curve would occur when dV_g/dI_m tends to a minimum value, and with Equation 5.21, we get

$$\frac{V_g}{I_m} = \frac{2K_2 K_3 I_m^2}{1 + 2K_2 I_m^2} = \frac{K_3}{1 + \dfrac{1}{2K_2 I_m^2}} \tag{5.23}$$

As a matter of fact, for the size of induction motor used in asynchronous generation, the magnetizing current in the saturation region is very high, so

$$1 \gg \frac{1}{2K_2 I_m^2}$$

and as a result, for Equation 5.23, we get

$$X_m = \frac{V_g}{I_m} \approx K_3 \tag{5.24}$$

The limit conditions established by Equations 5.22 and 5.24 can be gathered together as follows:

$$K_3 < X_m < K_1 + K_3$$

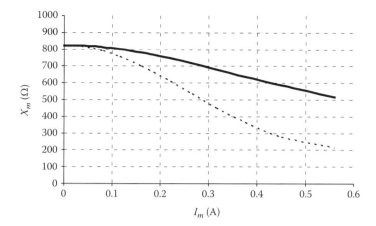

FIGURE 5.5   Variation of the magnetizing reactance with the saturation current.

With the test data obtained from a machine at no load, a graph can be drawn to describe the behavior of the magnetizing reactance with the increase of current I_m. It can be concluded that the value of M decreases steeply as the machine reaches the saturating point as in Figure 5.5. The larger the I_m, the larger is the variation of X_m.

5.5.1   MINIMIZATION OF LABORATORY TESTS

The number of points needed during laboratory tests to obtain the constants K_1, K_2, and K_3, and, therefore, the magnetization curve, can be minimized if it can be taken into account that there are only three of these constants. Therefore, the minimum number of tests necessary to obtain them is three. As quadratic exponents are used to establish these constants, good precision in measuring is fundamental; otherwise, the results can lead to incorrect conclusions.

The theoretical basis for selecting the correct current and voltage values to use in the laboratory tests, for minimization of the number of measurements, comes from Equation 5.17 for $F = 1$ (test frequency). Assume that the following three points were already fitted into Equation 5.17 from the lab measurements and corrected for the residual magnetism according to Equation 5.16:

$$V_{g1} = \left(K_1 e^{K_2 I_{m1}^2} + K_3 \right) I_{m1}$$

$$V_{g2} = \left(K_1 e^{K_2 I_{m2}^2} + K_3 \right) I_{m2} \qquad (5.25)$$

$$V_{g3} = \left(K_1 e^{K_2 I_{m3}^2} + K_3 \right) I_{m3}$$

or, more specifically,

$$a = \frac{V_{g1}}{I_{m1}} = \left(K_1 e^{k_2 I_{m1}^2} + K_3 \right)$$

$$b = \frac{V_{g2}}{I_{m2}} = \left(K_1 e^{k_2 I_{m2}^2} + K_3 \right) \tag{5.26}$$

$$c = \frac{V_{g3}}{I_{m3}} = \left(K_1 e^{k_2 I_{m3}^2} + K_3 \right)$$

Manipulation of Equation set 5.26 can produce

$$\frac{c - K_3}{a - K_3} = e^{k_2 (I_{m3}^2 - I_{m1}^2)}$$

$$\frac{b - K_3}{a - K_3} = e^{k_2 (I_{m2}^2 - I_{m1}^2)}$$

and a constant k is proposed to make the following identity:

$$\frac{c - K_3}{a - K_3} = \left(\frac{b - K_3}{a - K_3} \right)^k \tag{5.27}$$

To make Equation 5.27 more useful, the value of k should be any real value not equal to 1. However, to establish a sequence of multiple integer values easy to memorize and to minimize the current to be used in laboratory tests, it is done as follows:

$$k = \frac{I_{m3}^2 - I_{m1}^2}{I_{m2}^2 - I_{m1}^2} = 2$$

Therefore,

$$I_{m3}^2 - I_{m1}^2 = 2 \left(I_{m2}^2 - I_{m1}^2 \right)$$

or

$$I_{m2}^2 = \frac{\left(I_{m1}^2 + I_{m3}^2 \right)}{2} \tag{5.28}$$

If it is done in Equation 5.28, $I_{m3} = n I_{m1}$, and n is an integer, we get

$$I_{m2} = I_{m1} \sqrt{\frac{\left(n^2 + 1 \right)}{2}} \tag{5.29}$$

To have a sequence of minimum integer numbers, it is important to choose the smallest convenient value of n; in this case, it could be $n = 7$. So

$$I_{m3} = 7I_{m1} \tag{5.30}$$

Also, for $n = 7$ in Equation 5.29, I_{m2} becomes equal to $5I_{m1}$.

Notice that the advisable value for I_{m3} is that of the motor rated current so as not to overload it during tests.

Equation 5.27 has an infinite set of solutions for integers $k > 1$. Therefore, except for trivial solutions, the next smallest integer is $k = 2$, the simplest solution. So

$$(c - K_3)(a - K_3) = (b - K_3)^2 \tag{5.31}$$

or, simplifying,

$$b^2 - 2bK_3 + K_3(a + c) - ac = 0$$

Isolating K_3, we get

$$K_3 = \frac{b^2 - ac}{2b - (a + c)} \tag{5.32}$$

For K_1 and K_2, the following derivations can be used.

If Equation 5.32 is replaced in expressions $a - K_3$ and $c - K_3$, we get

$$a - K_3 = \frac{(a - b)^2}{a + c - 2b} \tag{5.33}$$

$$c - K_3 = \frac{(c - b)^2}{a + c - 2b} \tag{5.34}$$

With further manipulation of Equations 5.33 and 5.34, and taking into account that $I_{m1} = I_{m3}/7$, we get

$$K_2 = \frac{49}{24} \frac{\ell n \left(\dfrac{b - c}{a - b} \right)}{I_{m3}^2} \tag{5.35}$$

Observe that a, b, and c represent the slope of the curve at points 1, 2, and 3, respectively, of the straight line parallel to the saturation curve tangent. The highest slope in decreasing order will always be a, b, and c and the negative sign of the squared argument in the logarithmic expression of K_2 must be rectified.

TABLE 5.1

Table of Measurements for Determination of K_is

Order	Measured Current	Measured Voltages	X_m	Formulas for K_is
1	$I_{m1} = I_{m1}$	V_{g1}	$a = V_{g1}/I_{m1}$	$K_1 = (c - K_3)\left(\dfrac{a-b}{b-c}\right)^{49/24}$
2	$I_{m2} = 5I_{m1}$	V_{g2}	$b = V_{g2}/I_{m2}$	$K_2 = \dfrac{49}{24}\dfrac{\ell n\left(\dfrac{b-c}{a-b}\right)}{I_{m3}^2}$
3	$I_{m3} = 7I_{m1}$	V_{g3}	$c = V_{g3}/I_{m3}$	$K_3 = \dfrac{b^2 - ac}{2b - (a+c)}$

After some algebraic transformations,

$$K_1 = (c - K_3)\left(\frac{a-b}{b-c}\right)^{49/24} = \frac{(b-c)^2}{a+c-2b}\left(\frac{a-b}{b-c}\right)^{49/24} \tag{5.36}$$

However, this is a very complex form in which to express K_1, which introduces high sensitivity to the values obtained in the lab measurements. Table 5.1 gathers the approaches for best numerical results.

Notice that the excitation curve represented by Equation 5.17 can be used to predict the terminal voltage of IGs as a function of the excitation capacitance. For that, the point of intersection between the straight line $V_t = I_t/F\omega C$ and the excitation curve can be determined by substituting I_m in Equation 5.17 by $I_t = F\omega C V_t$ to give

$$\frac{1}{\omega C} = F^2\left[K_1 e^{K_2(\omega C F V_t)^2} + K_3\right] \tag{5.37}$$

Expressing Equation 5.37 in terms of V_t, we get

$$V_t = \frac{1}{\omega C F \sqrt{K_2}}\sqrt{\ell n\left[\frac{1}{\omega C F^2 K_1} - \frac{K_3}{K_1}\right]} \tag{5.38}$$

Equation 5.38 imposes limits to the existence of V_t, that is,

$$\frac{1}{\omega C F^2 K_1} > \frac{K_3}{K_1} \quad \text{or} \quad C < \frac{1}{\omega F^2 K_3} \tag{5.39}$$

5.6 EXAMPLE FOR DETERMINING MAGNETIZING CURVE AND MAGNETIZING REACTANCE

In Table 5.2 are the plate data of an induction motor used for obtaining some lab results with formulas from Section 5.5. The asynchronous motor was made to rotate at no load by a dc motor with the stator current progressively increased by an increase of the voltage across its terminals. The corresponding voltage and current values were measured in the three phases of the motor, and the average of these values is shown in Table 5.3. Eventual differences among the excitation voltage measured in the laboratory and the theoretical values obtained by Equation 5.17 are attributed to imprecision in the reading since no theory predicts such oscillations on the magnetizing curve.

TABLE 5.2
Plate Data of the Tested Motor

Power	10 hp	Δ connection	220 V; 26 A
Isolation	B. IP54	Y connection	380 V; 15 A
Service	S1	Frequency	60 Hz
I_p/I_n	8.6	Rotation	1765 rpm
Category	N	FS	1.0

TABLE 5.3
Excitation Curves

Measured Current (A)	Exciting Voltage (V)	Theoretical Voltage (V)
0.000	0	0.0000
0.020	15	16.4538
0.060	45	49.0341
0.080	65	65.0013
0.115	90	92.1422
0.160	115	125.0500
0.185	140	142.1579
0.220	165	164.5150
0.265	190	190.3397
0.315	215	215.0967
0.375	240	239.6143
0.450	265	263.5572
0.560	290	289.9996
Lab Tests	V_g (V)	I_m (A)
1	65.00	0.08
2	248.33	0.40
3	290.00	0.56

To appreciate the precision of the values given by Equation 5.17 and of the method suggested to obtain the magnetizing curve in the laboratory, the data presented in Table 5.3 can be used. The maximum current of the table was taken as the reference, being the highest possible value of current without putting the integrity of the motor in danger. The other values were obtained by interpolation to satisfy the relationship 1:5:7 foreseen in the method described in Section 5.5 for obtaining K_1, K_2, and K_3.

The values for the formulas in Table 5.1 are filled out in Table 5.4. The no-load curves foreseen by Equation 5.17 for $F = 1$, and the values obtained in laboratory tests (marked with an "*x*") are plotted in Figure 5.6.

TABLE 5.4

Test Data of the Motor Described in Table 5.2

Order	Measured Current	Measured Voltages	X_m	Formulas for K_is
1	$I_{m1} = 0.08$	65.00	$a = V_{g1}/I_{m1}$ $= 812.5$	$K_1 = (c - K_3)\left(\dfrac{a-b}{b-c}\right)^{49/24}$ $= 425.05$
2	$I_{m2} = 0.40$	248.33	$b = V_{g2}/I_{m2}$ $= 620.8$	$K_2 = \dfrac{49}{24}\dfrac{\ell n\left(\dfrac{b-c}{a-b}\right)}{I_{m3}^2}$ $= -4.0455$
3	$I_{m3} = 0.56$	290.00	$c = V_{g3}/I_{m3}$ $= 517.9$	$K_3 = \dfrac{b^2 - ac}{2b - (a+c)}$ $= 398.33$

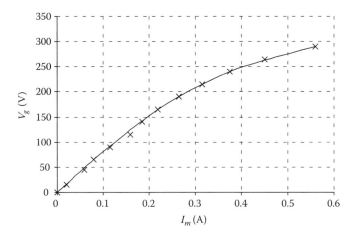

FIGURE 5.6 Practical and theoretical magnetizing curves.

5.7 VOLTAGE REGULATION

The type of load connected to the self-excited IG is the most serious problem for voltage regulation because highly resistive loads (high current), and inductive loads in general, can vary the terminal voltage over a very wide range. Figure 5.7 displays the regulation curves where the dashed line suggests the desired values of rated voltage for a purely resistive load. On top of the secondary-load resistive effects on the generator voltage drops, an inductive load in parallel with the excitation capacitor reduces the effective voltage as given by Equation 4.9.

  This change in the effective self-excitation capacitance increases the slope of the straight line of the capacitive reactance, reducing the terminal voltage. This phenomenon is even more drastic with higher inductive loads. A capacitance for individual compensation can be connected across the terminals of each inductive load in such a way that the IG sees the load as almost resistive. This capacitance should be naturally added or removed with the connection or disconnection of the load. This solution is particularly useful when loads like refrigerators, whose thermostats turn their compressors on and off countless times to maintain their cold temperature at a certain level, are connected to micropower plants.

  Figure 5.8 illustrates the influence of the exciting frequency on the air-gap voltage, which almost doubles when the frequency changes from 45 to 75 Hz. An interesting characteristic of the IG is the variation of the stator frequency with the load, which is remarkably affected by the power factor, shown in Figure 5.9.

5.8 CHARACTERISTICS OF ROTATION

Rotation also influences the terminal voltage of the IG, as would be expected.[9] As faster is the rotation as higher is the induced voltage. However, it should not move too far away from the machine rated limits, not only by mechanical limitations

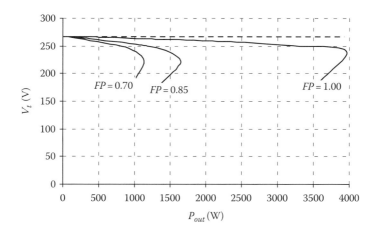

FIGURE 5.7 Influence of the load power factor on the voltage regulation of the IG.

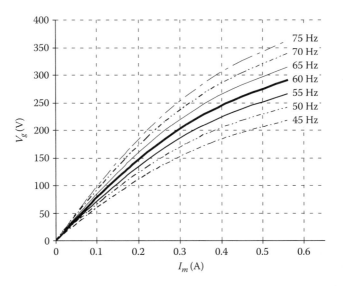

FIGURE 5.8 Influence of the exciting frequency on the air-gap voltage.

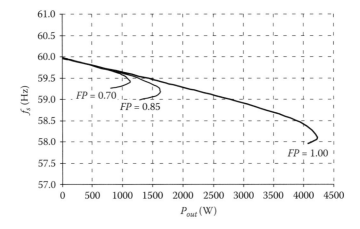

FIGURE 5.9 Frequency variation with the load (kW).

but also by the low terminal voltage induced, as can be observed in Figure 5.10 for a 4-pole IG. From Equations 5.1 and 5.2, the converted power is calculated as follows:

$$P_{conv} = (1-s)P_{airgap} = \frac{3V_{ph}^2 R_2 \dfrac{1-s}{s}}{(R_1 + R_2/s)^2 + (X_1 + X_2)^2} \tag{5.40}$$

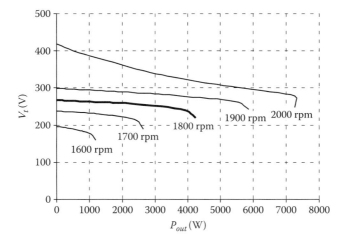

FIGURE 5.10 Voltage–rotation characteristics.

Isolating V_{ph} from the rest, we get

$$V_{ph} = \sqrt{\frac{P_{conv}\left[\left(R_1 + \dfrac{R_2}{s}\right)^2 + \left(X_1 + X_2\right)^2\right]}{3R_2 \dfrac{1-s}{s}}} \qquad (5.41)$$

Finally, substituting Equation 2.2 into Equation 5.41 after some simplification, we get

$$V_{ph} = \sqrt{\frac{P_{conv}\left(\dfrac{n_s}{n_r} - 1\right)\left[\left(R_1 + \dfrac{R_2}{1 - \dfrac{n_r}{n_s}}\right) + \left(X_1 + X_2\right)^2\right]}{3R_2}} \qquad (5.42)$$

Equation 5.42 should be conveniently interpreted; it does not have a finite solution at $n_r = n_s$. It is important to emphasize that in Equation 5.42 when the rotor is turning at a speed faster than synchronous, there must be a sign change on the converted power. Equation 5.42 is plotted in Figure 5.10 illustrating the influence of rotor speed on the terminal voltage. As expected, the induced voltage is approximately proportional to the speed. Rotor speed also affects frequency and, as a consequence, the terminal voltage, as depicted in Figure 5.10. The slip factor can also be displayed with the power factor, as in Figure 5.11.

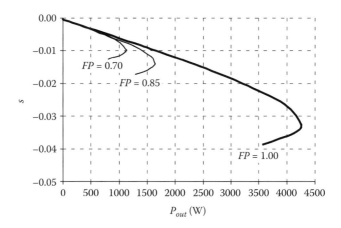

FIGURE 5.11   Slip factor–power factor.

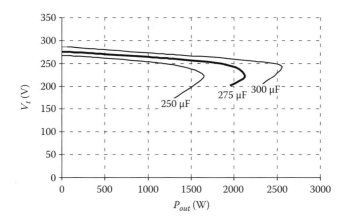

FIGURE 5.12   Self-exciting capacitance–terminal voltage.

Another decisive factor affecting frequency is the generator's load, even at constant speed, as in Figures 5.11 and 5.12, which show the influence of the self-exciting capacitance on the terminal voltage. The voltage increases with the amount of reactive power being supplied to the terminals.

On the other hand, the efficiency of the IG varies widely for small loads, reaching close to 90% for larger loads not very far from the rated value, as one can see in Figure 5.13. Rotation has little influence on efficiency.

As one can easily observe, it is very difficult to start up an induction motor in a network supplied only by IGs; special techniques must be used to temporarily increase the effective capacitance at start-up time, later reducing it gradually to normal. With small loads—for example, the motor impedance of a refrigerator—one of these techniques is the simple connection of a small reactor to limit surge currents during the start-up of the machine (soft start). For heavier loads, modern static reactive compensators are a good practical solution.

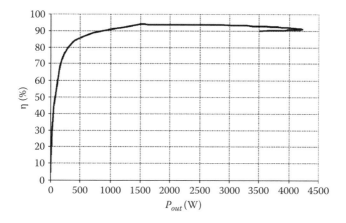

FIGURE 5.13 Relationships between efficiency and load for $n_r = 1700$ rpm.

5.9 COMPARISON OF INDUCTION GENERATORS WITH OTHER GENERATORS

The IG may be used for several applications, including high-speed turbines. When permanent magnet (PM) and switched reluctance (SR) generator controllers are compared with the IG controller, it is important to take into account doubly fed induction generator (DFIG) benefits: (1) it is not necessary to have special sensors; (2) it is possible to achieve very high rotor speeds; and (3) it has fractional converter rating, in addition to several other considerations (such as not being dependent on importation of PMs from countries controlling the market of rare earth materials). For example, the slip control of IG requires a precise measurement of speed, which is easy to implement in practice. In contrast, the control of the SR generator requires very precise measurements of the rotor position involving high technological, precise, and expensive components. For the SR generator, the operating frequency is extremely high, in the range of 6 kHz at 60,000 rpm requiring high-speed power transistors at very high switching rates. On the other hand, the usual operating frequency for the IG is in the range of 1–2 kHz at high speeds depending upon the generator number of poles allowing the switching rate to be lowered within a reasonable range.[10–13]

In the case of PM generators, the power rating of the converters has to cope with several complexities in the controller, requiring voltage boost mechanisms due to the wide variation in the output voltage. The power electronic components must function at high stress levels. In the SR generator controller, high change rates of currents and voltages result in high stress levels for the power electronic devices. In compensation, the IG has naturally a well-regulated sinusoidal output that can be conditioned without using highly stressed electronic components.[14,15]

In Table 5.5 are listed some criteria of comparison between the IG and other generators. These criteria are classified in economical, technical, and commercial aspects, and they can contribute for the final selection of the type of generator to be used in some particular application.

TABLE 5.5

Induction Generator versus Other Generators

	Type of Generator				
Criterion	SCIG	DFIG	EESG	PMSG	SR
Active and reactive power control	Needs compensation	Full	Moderate	Full	Full
Voltage regulation	Poor	Medium	Good	Voltage boost mechanisms	Good
Speed regulation	Full $(10\% > n_s)$	±30%	Full	Full	Full
Frequency regulation	Low	Low	High	High	High
Need of power converter	Lower switching frequency	Small PWM	Small for field plus full scale	Full scale	Full scale
Losses in power converter	Small	High	Minimum	Very high stress levels	Extremely high stress levels
Technology	Proven	In evolution	Traditional	Fast evolution	Fast evolution
Construction	Simple and robust	Complex	Complex	Simple and robust	Robust
Losses	Medium	Medium	Low, field losses	None in the field	None in the field
Thermal dissipation	Medium	Medium	Good	Minimum losses	Minimum losses
Brushes and slip rings	None	Necessary	May have	None	None
Ride-through capab.	Complex	Complex		Less complex	Less complex
Manufacture	Easy	Involving	Complicated	Moderate	Moderate
Acquisition costs	Low	Medium	High	Still high	Still high
Adaptability	High	High	Low	High	High
Capital return	Fast	Medium	Medium	—	—
Commercial availability	Highest	High	Medium	Low	Low
Commissioning	Medium	Medium	Medium	Medium	Medium
Costs	Overall low	Power converter	Highest, field	In the PMs	In the PMs
Connectivity	Medium	Medium	Medium	Medium	Medium
Controllability	Difficult for constant speed	Wide range	For rotor excitation	Needs voltage boost controller	Wide range
Efficiency	89%	88%	high	91.6%	91.6%
Power ranges	3 MW	2 MW	4.5 MW	3 MW	3 MW
Environment impacts	Low	Low	Low	Low	Low
Expertise to run it	Low	High	High	Medium	Medium

(Continued)

TABLE 5.5 (*Continued*)
Induction Generator versus Other Generators

Criterion	Type of Generator				
	SCIG	DFIG	EESG	PMSG	SR
Flexibility	High	High	Medium	High	High
Mains synchronism	Easy	Easy	Complex	Complex	Complex
Maintenance costs (O&M)	Low	High	High	Low	Low
Torque oscillations	Flat	Wavy	Wide	Flat	Flat
O&M	Lowest	Low	Moderate	Average	Average
Pollution levels	Low	Low	Bulky size	Low	Low
Portability	High	Medium	Low	Medium	Medium
Practical application range	Up to 50 kW	Up to 500 kW	800 MW	3 MW	3 MW
Performance	Variable frequency	Allows PF control	Good PF control	Only external PF control	Only external PF control
Reliability	High	Top medium	Proven	High	High
Response speed to surges	Slow	Fast	Fast	Fast	Fast
Robustness	High	Medium	Medium	High	High
Self-curing	No	No	No	No	No
Service interruption	Low	Low	Low	Low	Low
Simplicity (easy to cope)	Easy	Involving	Involving	New	New
Today standardization	High	Medium	High	Low	Low
Synchronization for connection	Loose	Loose	Tight	Tight	Tight
Turbine–generator coupling	Gear box	Direct coupling	Gear box	Possibly direct	Possibly direct
Useful life	30 years	15 years	30 years	20 years	20 years
Speed regulation	Difficult	Loose	Tight	Tight	Tight
Power per weight and volume	Low	Medium	May be heavy and large	May be light	May be light
Specific problems	Demands high reactive power	Requires precise slip measurement	Needs field excitation	High costs and PM demagnetization at high temp.	Requires precise measurement of the rotor position

SCIG, Squirrel cage IG; DFIG, Doubly fed IG; PMSG, Permanent magnet synchronous generator; EESG, Electrically excited synchronous generator; SR, Switched reluctance.

5.10 SOLVED PROBLEM

1. Using the experimental values from the magnetization curve obtained in laboratory tests for an IG,
   a. Estimate the values K_1, K_2, and K_3 (interpolation) of the nonlinear relationship for I_m given in Equation 5.17.
   b. Plot the magnetization curve with the parameters obtained in (a) and with the values from the lab tests. What is the maximum percentage difference?
   c. Calculate the excitation capacitance to have as output terminal voltage the following values: 110, 127, and 220 V.
   d. With the capacitors calculated in item (c), calculate the voltage regulation for the rated power.
   e. What are the respective efficiencies of the IG in item (d)?

Solution

The intention of this laboratory is to understand the operation of a self-excited IG. We have done this lab to get a practical hands-on experience on the induction machine operated as a self-excited generator. Besides, we wanted to get an idea about the magnetization curve for the machine, the size of the capacitor needed for the machine having 1.5 kW rating. We also wanted to know the range of capacitors that could self-excite the generator. Finally, we wanted to look into the voltage regulation of the self-excited IG for pure resistive loads and for resistive-inductive loads.

Experiment and interpretation of results

The first part of this laboratory experiment was to get the magnetization curve of the induction machine. Therefore, we set up the machine as a motor and measured the voltage and current in each phase. Then we averaged them up to get the experimental data shown in Table 5.6.

TABLE 5.6
Practical Excitation Curve

Test #	V_g (V)	I_m (A)
1	131.4	7.57
2	125.8	7.00
3	120.0	6.44
4	114.0	6.07
5	109.0	5.70
6	105.0	5.44
7	98.5	5.00
8	87.6	4.39
9	79.0	3.92
10	57.0	2.91
11	37.0	2.06
12	4.8	0.48

The next job was to calculate the functional relationship between V_g and I_m. As we know, these two variables are related to each other by equation 5.17 as follows:

$$V_g = FI_m(K_1 e^{K_2 I_m^2} + K_3)$$

and for a no-load case, $F = 1$, our job became to find K_1, K_2, and K_3.

We first tried the 1:5:7 method, and we obtained the given in Table 5.7.

Now when we tried to get K_1, K_2 and K_3, we faced a problem. The value of "a" should be bigger than the value of "b"; otherwise, we cannot calculate K_2, which involves $\ln[(b - c)/(a - b)]$. Therefore, the first few readings taken were wrong. To get a functional relationship, we now tried some other values instead of the 1:5:7 rule. We used different value for "a" and used reading #10 as in Table 5.8.

Finally, we calculated the values K_1, K_2, and K_3 from the values of Table 5.8.

$$K_3 = \frac{b^2 - ac}{2b - (a + c)} = 19.630$$

$$K_2 = \frac{49}{24} \frac{\ln\left(\frac{b-c}{a-b}\right)}{I_{m3}^2} = 0.070$$

$$K_1 = (c - K_3)\left(\frac{a-b}{b-c}\right)^{\frac{49}{24}} = -0.036$$

TABLE 5.7

Test Data (Attempt #1)

Lab Tests	V_g (V)	I_m (A)	X_m (Ω)
1	17.03	1.08	$a = 15.77$
2	104.40	5.40	$b = 19.33$
3	131.40	7.57	$c = 17.36$

TABLE 5.8

Test Data (Attempt #2)

Lab Tests	V_g (V)	I_m (A)	X_m (Ω)
1	57.0	2.91	$a = 19.59$
2	104.4	5.40	$b = 19.33$
3	131.4	7.57	$c = 17.36$

TABLE 5.9

Excitation Curves

Measured Current (A)	Exciting Voltage (V)	Theoretical Voltage (V)
7.57	131.4	133.5500
7.00	125.8	129.6290
6.44	120.0	122.1900
6.07	114.0	116.2730
5.70	109.0	109.8960
5.44	105.0	105.2330
5.00	98.5	97.1142
4.39	87.6	85.5667
3.92	79.0	76.5359
2.91	57.0	56.9338
2.06	37.0	40.3380
0.48	4.8	9.4048

Then, we got the following equation for the magnetization curve:

$$V_g = I_m(-0.036e^{0.07\,I_m^2} + 19.630)$$

from which was fitted Table 5.9.

Figure 5.14 shows a comparison between the practical and theoretical magnetization curves.

From the theoretical curve, we got that for self-excitation we needed minimum value of the capacitance as 139.0 μF. After that point, the capacitance line will not intersect the magnetization curve at some stable point. The practical result gave as the minimum capacitance as 141.2 μF. If the rated terminal voltage is 120 V, the rated capacitance for self-excitation will be 141 μF. The rated capacitance we got from the laboratory experiment was 142 μF. The machine can go at most to 130 V;

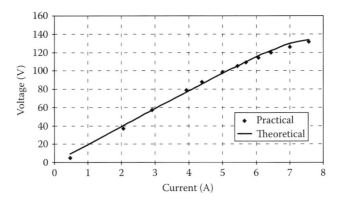

FIGURE 5.14 Practical and theoretical magnetization curves.

hence, the maximum capacitance can be calculated from this value. The maximum capacitance from the theoretical curve is 150 μF, whereas the maximum capacitance from the experimental result was 153 μF.

Therefore, the results from the theoretical excitation curve matches the experimental curve results within considerable accuracy. Now, if we increase the speed of the prime mover and keep the capacitance fixed, the terminal voltage will increase up to certain point. With further increase, the capacitance line will intersect the excitation curve in the plateau; hence, the output voltage will become constant, but the output line current will increase. Therefore, the output line current is the deciding factor for getting maximum speed of the prime mover. We saw the same kind of characteristics during the laboratory test whose data are given in Table 5.10 and plotted in Figure 5.15.

When load resistances are added to the self-excited IG, the output voltage starts to drop down as shown in Table 5.6. This voltage droop follows the excitation curve. After we add a sufficient amount of load resistances, the voltage drops below the knee of the excitation curve and the generator loses excitation and finally the output voltage completely collapses to zero. In the lab, we found that if we increase the load resistance value beyond 145 Ω, the output voltage collapsed. To maintain the rated terminal voltage with the loads, we have to increase the capacitance value.

TABLE 5.10

Changing Capacitance

Frequency (Hz)	V_{out} (V)	I_{out} (A)	C (μF)
61.6	84.5	4.4	0.13
60.0	115.0	6.8	0.15
59.3	137.6	9.7	0.18

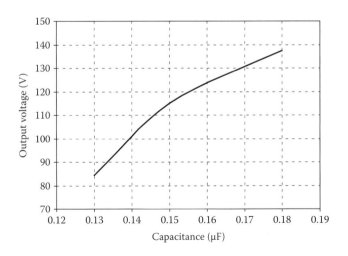

FIGURE 5.15 Output voltage versus capacitance.

TABLE 5.11

Adding a Load Resistance

Frequency (Hz)	V_{out} (V)	I_{out} (A)	I_{load} (A)	R_{load} (Ω)	C (μF)
60.0	120.7	7.3	0.91	132.64	0.15
58.5	75.0	4.2	0.53	141.51	0.15

TABLE 5.12

Adding a Resistive-Inductive Load

Frequency (Hz)	V_{out} (V)	I_{out} (A)	I_{load} (A)	R_{load} (Ω)	C (μF)
59.8	98.2	5.55	0.68	132.64	0.15
59.3	133.2	9.50	0.94	132.64	0.18

Again, as we add the inductance along with the load resistance, two things start to happen at the same time. The terminal voltage starts to drop, and due to inductance, the effective capacitance value starts to drop as observed in Table 5.11. Therefore, the machine starts to lose its excitation faster with smaller impedance value than that of the pure resistive load. This was also visible during lab tests. If we look at Tables 5.11 and 5.12, we see that for a capacitance of 0.15 μF and for a load resistance of 132.64 Ω, the output voltage is considerably lower with the resistive-inductive load. To increase the output voltage, we have to increase the capacitance value.

Comments for this experiment

This laboratory is a good experience to understand the principles of operation of the self-excited IG. One important issue is that we should be careful taking the readings because in our case we have some wrong readings for the low currents; that is why we cannot apply the 1:5:7 rule. Again, proper care should be taken so that the machine output voltages do not exceed too much the rated value that may deteriorate the insulation of the windings. The capacitance value and the speed of the prime mover should be chosen in a way such that the armature current does not exceed the full load current of the machine. When connecting the load, proper care must be taken so that the generator does not lose its excitation. If all these considerations are taken care of, this lab method is good for getting the magnetization curve and for understanding the self-excited operation of the IGs. Here, we had not conducted the efficiency calculation, as it was difficult to get the input power to the IG. If there are precise torque and angular speed measurements by which the prime-mover torque and mechanical power can be found, it is possible to easily find the efficiencies.

5.11 SUGGESTED PROBLEMS

5.1 Using Excel or a similar program, plot the following relationships of the IG you have used in the lab tests:
a. Output power × efficiency
b. Output power × power factor
c. Rotor speed × load frequency
d. Voltage and frequency regulation for 10%, 50%, and the rated load (100%)

5.2 Calculate the power injected by your IG into the grid for speeds 5%, 10%, and 15% above synchronous speed. What are the respective efficiencies in each case?

5.3 Explain why in cases of loads sensitive to frequency variation it is not recommended to connect an IG in parallel with a synchronous generator of comparable rated output power.

5.4 Calculate the necessary torque of a primary machine and the maximum rated power of an IG fed at 6.0 kV, 60 Hz, 4 poles, $R_1 = 0.054\ \Omega$, $R_2 = 0.004\ \Omega$, $X_1 = 0.852\ \Omega$, and $X_2 = 0.430\ \Omega$. What is the rated mechanical power? Calculate the distribution transformer ratings if the subtransmission voltage is 69 kV.

5.5 Using the data given in Problem 5.5, estimate the difference in your calculations if the Thévenin equivalent circuit was used rather than the more approximate model given by the straight values when $R_m = 29.82\ \Omega$ and $X_m = 52.0\ \Omega$. Express the final differences in percentage terms.

5.6 Using Excel or any other automatic calculator, plot the turbine torque and the corresponding output power of the IG according to the mechanical speed variation from 0.0 to 2000 rpm and the data given in Exercise 5.4.

5.7 A small three-phase IG has the following nameplate data: 110 V, 1.140 rpm, 60 Hz, Y connection, and 1250 W. The blocked rotor test at 60 Hz yielded the following: $I = 69.4$ A, $P = 4200$ W. The turn ratio between stator and rotor is 1.13, and the rotor resistance referred to the stator is 4.2 Ω. What are the full load torque and air-gap power?

5.8 In a small generating system, there are two generators in parallel. One of them is a synchronous generator, and the other, an IG. The common load is 3.0 kW with a power factor of 0.92. If the IG supplies only 1.0 kW of the total load at a power factor of 0.85, at what power factor would the synchronous generator be running?

5.9 There is a three-phase IG of 110 V, 60 Hz, 4 poles, 1900 rpm, Y stator connection, and an equivalent circuit given as $R_1 = 0.20\ \Omega$, $R_2 = 0.42\ \Omega$, $X_1 = 0.15\ \Omega$, and $X_2 = 0.43\ \Omega$; R_2 and X_2 refer to the stator. The excitation current is 8.1 A and is assumed constant at the rated values. Due to a reduced flux density, the excitation current is neglected at starting conditions. Evaluate
a. The maximum torque in N m
b. The slip factor at the maximum torque
c. The torque to generate the rated voltage
d. Full-load torque

REFERENCES

1. Lawrence, R.R., *Principles of Alternating Current Machinery*, McGraw-Hill Book Co., New York, 1953, p. 640.
2. Kostenko, M. and Piotrovsky, L., *Electrical Machines*, Vol. II, Mir Publishers, Moscow, Russia, 1969, p. 775.
3. Smith, I.R. and Sriharan, S., Transients in induction motors with terminal capacitors, *Proc. IEEE*, 115, 519–527, 1968.
4. Chapman, S.J., *Electric Machinery Fundamentals*, 3rd edn., McGraw-Hill, New York, 1999.
5. Liwschitz-Gärik, M. and Whipple, C.C., *Alternating Current Machines (Maquinas de corriente alterna)*, Companhia Editorial Continental, D. Van Nostrand Company, Inc., Princeton, NJ, 1970, p. 768.
6. Murthy, S.S., Malik, O.P., and Tandon, A.K., Analysis of self-excited induction generators, *Proc. IEEE*, 129(6), 260–265, 1982.
7. Watson, D.B. and Milner, I.P., Autonomous and parallel operation of self-excited induction generators, *Int. J. Electr. Eng. Educ.*, 22, 365–374, 1985.
8. Murthy, S.S., Nagaraj, H.S., and Kuriyan, A., Design-based computational procedure for performance prediction and analysis of self-excited induction generators using motor design packages, *Proc. IEEE*, 135(1), 8–16, January 1988.
9. Grantham, C., Sutanto, D., and Mismail, B., Steady-state and transient analysis of self-excited induction generators, *Proc. IEEE*, 136(2), 61–68, 1989.
10. Benelghali, S., Benbouzid, M.E.H., and Charpentier, J.F., Comparison of PMSG and DFIG for marine current turbine applications, *XIX International Conference on Electrical Machines—ICEM*, Rome, Italy, 2010.
11. Li, H. and Chen, Z., Overview of different wind generator systems and their comparisons, *IET Renew. Power Gen.*, 2(2), 123–138, 2008. doi: 10.1049/iet-rpg: 20070044.
12. Kushare, B.E. and Kadam, D.P., Overview of different wind generator systems and their comparisons, *Int. J. Eng. Sci. Adv. Technol.*, 2(4), 1076–1081, July–August 2012.
13. Wang-Hansen, M., Wind power dynamic behavior—Real case study on Linderödsåsen wind farm, Master's thesis, Department of Energy and Environment, Chalmers University of Technology, Göteborg, Sweden, 2008.
14. Polinder, H., Bang, D.-J., Li, H., Mueller, M., McDonald, A., and Chen, Z., Integrated Wind Turbine Design, Part 1 and 2, concept report on Generator Topologies, Mechanical & Electromagnetic Optimization. Project up wind funded by the European Commission under the 6th (EC) RTD, Framework Programme 2002–2006, Delft, the Netherlands.
15. Kumar, A. and Sharma, V.K., Implementation of self-excited induction generator (SEIG) with IGBT based electronic load controller (ELC) in wind energy systems, *Int. J. Res. Eng. Technol.*, 2(8), 188–193, August 2013.

6 Construction Features of Induction Generators

6.1 SCOPE OF THIS CHAPTER

Induction machines are of robust construction and relatively low manufacturing cost. They are more economical than synchronous machines. For high-power applications, this difference is less perceptible because those units are custom made. However, for medium and small sizes, the difference in price is dramatic, as much as 80%. For the same kVA rating, though, induction machines are larger than synchronous machines because their magnetizing current circulates through the stator winding. Induction generators connected to the grid need no voltage regulation, less constraints on the turbine speed control, and practically no maintenance. These advantages plus the fact that induction machines are readily available from several manufacturers make them very competitive for just-in-time installation. The U.S. National Electrical Manufacturers Association (NEMA) has standardized the variations in torque–speed characteristics and frame sizes, assuring physical interchangeability between motors of competing manufacturers, thereby making them a commercial success and available for integral horsepower ratings with typical voltages ranging from 110 to 4160 V. As discussed in Chapter 2, the only perceived weaknesses of induction machines are lower efficiency—the rotor dissipates power—and the need for reactive power in the stator. This chapter will discuss the physical construction of various types of induction machines, emphasizing their features in generator mode. An electrical generator used for a wind turbine system has efficiency imposed by three main issues: (1) stator losses, (2) converter losses, and (3) gearbox losses, as depicted in Figure 6.1. Stator losses are considered in this chapter by the proper design of the machine for the right operating range; the converter losses are given by proper design of the power electronic circuits. On-state conduction losses of transistors and diodes, plus their frequency proportional switching losses, are not considered in this book because they must be approached in a specific power electronic topology that best fits the final design. The third main factor responsible for a noticeable power loss is the use of a gearbox, as indicated in Figure 6.1 for a typical wind turbine system.

It can be considered that mechanical viscous losses due to a gearbox are proportional to the operating speed, as indicated by

$$P_{gear} = P_{gear,rated} \frac{\omega_{mech}}{\omega_{mech,rated}} \qquad (6.1)$$

where
$P_{gear,rated}$ is the loss in the gearbox at rated speed (on the order of 3% of rated power)
ω_{mech} is the rotor speed (rotation/min)
$\omega_{mech,rated}$ is the rated rotor speed (rotation/min)

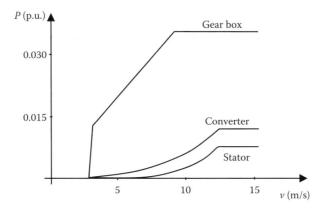

FIGURE 6.1 Losses for a high-power doubly fed induction generator.

Losses in the gearbox dominate the efficiency in most of the wind turbine systems, and simple calculations show that a significant annual energy dissipation in the generator system is due to the gearbox. Therefore, a good motivation for a more efficient generator would be the elimination of such mechanical device. While for large wind turbines a good gearbox system is employed, it is a common sense that for small wind turbines, for inexpensive applications, the gearbox is not optimized and it is expected higher losses than indicated in the literature because those small turbines must have a very competitive low price.

6.2 ELECTROMECHANICAL CONSIDERATIONS

An induction generator is made up of two major components: (1) the stator, which consists of steel laminations mounted on a frame so that slots are formed on the inside diameter of the assembly as in a synchronous machine, and (2) the rotor, which consists of a structure of steel laminations mounted on a shaft with two possible configurations: wound rotor or cage rotor. Figure 6.2 shows an exploded view of all the major parts that make up a rotor-cage induction generator. Figure 6.3 shows a schematic cut along the longitudinal axis of a typical wound-rotor induction machine. Figure 6.3a shows the external case with the stator yoke internally providing the magnetic path for the three-phase stator circuits. Bearings provide mechanical support for the shaft clearance (the air gap) between the rotor and stator cores. For a wound rotor, a group of brush holders and carbon brushes, indicated on the left side of Figure 6.3a, allow for connection to the rotor windings. A schematic diagram of a wound rotor is shown in Figure 6.3b. The winding of the wound rotor is of the three-phase type with the same number of poles as the stator, generally connected in Y. Three terminal leads are connected to the slip rings by means of carbon brushes.

Wound rotors are usually available for very large power machines (>500 kW). External converters in the rotor circuit, rated with slip power, control the secondary currents providing the rated frequency at the stator. For most medium power applications, squirrel cage rotors, as in Figure 6.3c, are used. Squirrel cage rotor windings

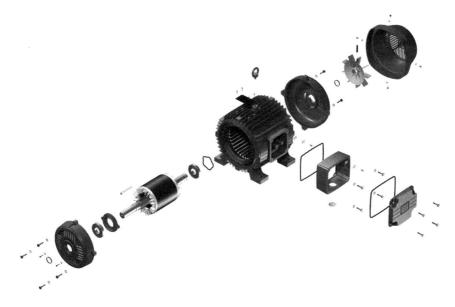

FIGURE 6.2 **(See color insert.)** Exploded view of induction generator major parts. (Courtesy of WEG Motores, Brazil.)

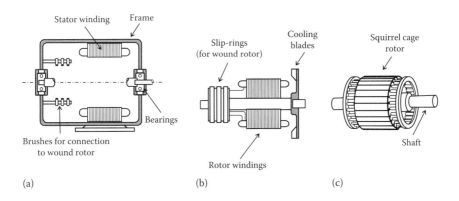

FIGURE 6.3 Induction machine longitudinal cut: (a) stator, (b) wound rotor, and (c) cage rotor.

consist of solid bars of conducting material embedded in the rotor slots and shorted at the two ends by conducting rings. In large machines, the rotor bars may be of copper alloy brazed to the end rings. Rotors sized up to about 20 in. in diameter are usually stacked in a mold made by aluminum casting, enabling a very economical structure combining the rotor bars, end rings, and cooling fan as indicated in Figure 6.4. Such induction machines are very simple and rugged with very low manufacturing cost. Some variations on the rotor design are used to alter the torque–speed features.

Figure 6.5 shows a cross-sectional cut indicating the distributed windings for three-phase stator excitation. Each winding (*a*, *b*, or *c*) occupies the contiguous slots within a 120° spatial distribution. The stator or stationary core is built up from

FIGURE 6.4 (See color insert.) Finished squirrel cage for large power induction motor.
(Courtesy of Equacional Motores, Brazil.)

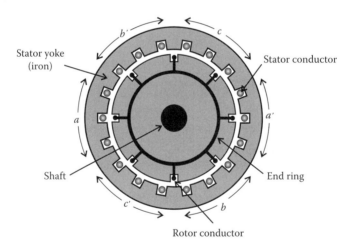

FIGURE 6.5 Cross-sectional cut for an induction machine.

silicon steel laminations punched and assembled so that it has a number of uniformly
spaced identical slots, in integral multiples of six (such as 48 or 72 slots), roughly
parallel to the machine shaft. Sometimes, the slots are slightly twisted or skewed in
relation to the longitudinal axis, to reduce cogging torque, noise, and vibration, and
to smooth up the generated voltage. Machines up to a few hundreds of kW rating and
low voltage have semiclosed slots, while larger machines with medium voltage have
open slots. The coils are usually preformed in a six-sided diamond shape as shown

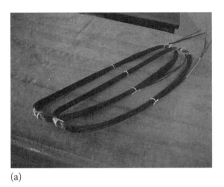
(a)

(b)

FIGURE 6.6 **(See color insert.)** Stator coils: (a) random-wound coil and (b) form-wound coil.

in Figure 6.6, where Figure 6.6a indicates a random-wound coil with several turns of wire and Figure 6.6b is a preformed-wound coil.

The conventional winding arrangement has as many coils as there are slots, with each slot containing the left-hand side of one coil and the right-hand side of another coil. Such top and bottom coils may belong to different phases with relatively high voltage between them requiring reinforced insulation. Those coils and insulation elements are inserted into the slots, and the finished cored and winding assembly are dipped in insulating varnish and baked, making a solid structure.

The number of poles defines how the magnetic flux is distributed along the stator core. An elementary two-pole stator winding is depicted in Figure 6.7, where the pole created by one-phase winding links half of the other phases. Two-pole machines are usually constructed only for motoring applications despite the fact they have large stator conductors with considerable return paths at the end of the windings, resulting in high leakage reactance and more core material to withstand higher flux per pole.

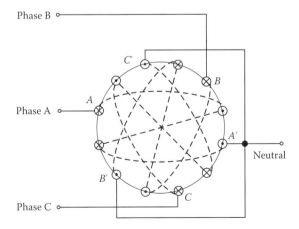

FIGURE 6.7   Concept of a two-pole stator winding.

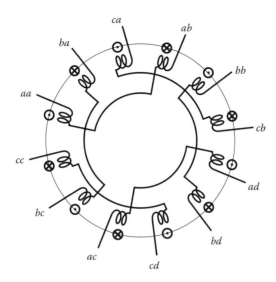

FIGURE 6.8 A four-pole stator winding.

For generator operation, a large number of poles are most common because the prime movers to which they are coupled are usually low-speed turbines.

A four-pole machine-winding diagram is outlined in Figure 6.8. In order to obtain a rotating field with p number of poles, each phase of the stator winding must create those magnetic poles by splitting the coils into $3p$ groups, known as pole-phase groups or phase belts and connected in series with their mmf's adding to each other. Sometimes in large machines, the pole-phase groups are also connected in parallel circuits. The objective is to divide the phase current in several paths as a means of limiting the coil wire diameter, as well as to adjust the required flux per pole for a given voltage and number of turns. Another feature of the parallel connection of phase groups is the control of asymmetric magnetic pull between stator and rotor cores, limiting vibration and noise of electromagnetic origin, particularly in large machines with a great number of poles. Figure 6.9 shows that several different connections of the phase groups are possible. Star and delta connections with series and parallel combinations offer flexibility in altering the rated voltages for the generator. No matter how the phase belts are distributed, the adjacent pole-phase groups will have opposite magnetic polarities. Eventually, the adjacent phase belts may have equal polarities in small consecutive-pole machines or machines with a very large number of poles and a limited number of stator slots.

6.3 OPTIMIZATION OF THE MANUFACTURING PROCESS

There are several factors to be considered in the design of an induction generator. There must be a minimum of materials in the machine with optimized manufacturing costs and the design should be compatible with machining and assembly equipment.

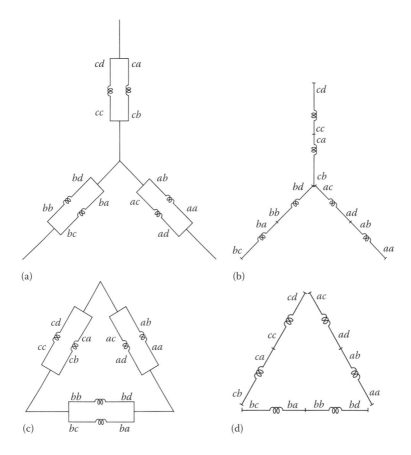

FIGURE 6.9 Possible stator winding connections: (a) star for high current, (b) star for high voltage, (c) Y for high current, and (d) Y for high voltage.

Materials should be available without time-consuming and costly delays. The cost of losses and maintenance over the life of the machine must also be minimal, and there must be a reasonable trade-off between manufacturing and efficient operating costs. Total costs include material costs, fabrication and selling costs, capitalization of losses over generator life, plus maintenance costs.

The design of electrical machines, particularly induction generators, is closely constrained by the performance, dimensions, limits, and technical and economic properties of materials. Progress in magnetic-steel laminations, in copper and aluminum for windings, and in mechanical devices has been continuously affecting the design of machines. Insulating materials in particular play a very important role in electrical machines and have been continually developed, looking forever higher maximum operating temperature limits. Such developments have permitted better utilization of the materials, reducing the size and cost of induction generators. Metallurgical technologies in casting, molding, plastic forming, cutting, and surfacing will strongly affect the design. In recent years, several advances in efficient higher-flux amorphous nanocrystalline materials promise, in the near future, new improvements in the performance and cost of electrical machinery. Maximum ambient temperature, altitude, and cooling systems, especially for high-rated machines, must be considered in the design and optimization of induction generators. These factors strongly affect the size of the machine and consequently the cost of the machine itself and its associated systems.

6.4 TYPES OF DESIGN

Rotor lamination design is particularly important in defining the torque–speed characteristic of an induction machine, in both motor and generator mode. Figure 6.10 shows a cross section of a typical rotor with distinct features. The equivalent rotor-leakage reactance defines the rotor flux lines that do not couple with the stator windings. If the bars of a squirrel cage rotor are placed near the surface, as in Figure 6.10a, there is a small reactance leakage and the pullout torque will be high and closer to the synchronous speed. This occurs at the expense of a small amount of torque for low-speed and high starting current in motoring operation and a hard self-excitation process in generator mode. On the other hand, if the bars of a squirrel cage are deeper in the rotor, as in Figure 6.10e, the leakage reactance is higher, the torque curve is not so steep close to the synchronous speed, and the maximum torque is not so high. There is a lower starting torque with a better self-excitation process and easier closed loop current control. However, the leakage flux at the air gap is very significant, and the output voltage of the generator fluctuates more when operating as an isolated machine or off-grid generator. Double-cage rotors, shown in Figure 6.10c, or deep-bar with skin effect rotor, Figure 6.10b, combining both features, lead to better starting characteristics that are more appropriate for motoring operation. They are the most expensive.

The design, performance, and materials used in squirrel cage rotors are subject to standards issued by technical or standardization organizations such as IEEE, IEC, and the National Electrical Manufacturers Association (NEMA). These organizations have set a plethora of standards—for example, the NEMA Standard MG-1 defines torque, slip, and current requirements to meet design classifications. Figure 6.11 shows typical torque–speed characteristics for a squirrel cage induction machine operating in the motor range. For the generator range, that is, rotating above synchronous speed, the characteristics are essentially the same, with torque in the negative quadrant and breakdown values slightly higher.

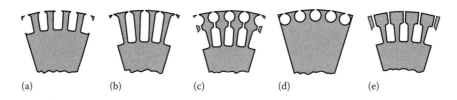

(a)   (b)   (c)   (d)   (e)

FIGURE 6.10 Laminations for squirrel cage induction generators: (a) class A, (b) class B, (c) class C, (d) class D, and (e) deep bars.

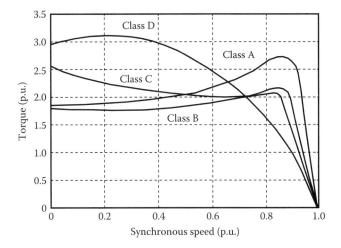

FIGURE 6.11   Torque–speed characteristics for NEMA standard machines.

The following general characteristics, presented here for reference only, are those of the various classes of machines operating in motor mode defined by NEMA:

Design Class A: The full-load slip of a class A machine must be less than 5%, lower than that of a class B machine of equivalent rating. The push-out torque is greater than that of class B. Specified starting torques are the same as those for class B. The starting current is greater than that of class B and may be more than eight times the rated full-load current. Class A machines have single-cage, low-resistance rotors. In class A, there are large rotor bars placed near the rotor surface, leading to little slip at full load, high efficiency, and high starting current.

Design Class B: This class provides a full-load slip of less than 5% (usually 2%–3%). The breakdown torque is at least 200% of full-load torque (175% for motors of 250 hp and above). The starting current usually does not exceed five times rated current. Starting torque is at least as great as full-load torque for motors of 200 hp or less and is more than twice the rated torque for four-pole machines of less than 3 hp. These characteristics may be achieved by either deep-bar with skin effect or double-cage rotors. Class B machines have large narrow rotor bars placed deep in the rotor and lower starting current than class A. They became the standard for industrial applications and probably the least expensive choice for rural and home applications.

Design Class C: These machines have a full-load slip of less than 5%, usually slightly larger than class B. The starting torque is larger than that of class B, running 200%–250% of rated torque. The pull-out torque is a few percent less than for class B, and the starting current is about the same. These excellent overall characteristics are achieved by means of a double-cage rotor with large low-resistance bars buried deep in the rotor and small high-resistance bars near the rotor surface. Class C is the most expensive.

Design Class D: These machines are designed so that maximum torque occurs near zero speed; so they have a very high starting torque (about 275% of full-load value) at relatively low starting current (the starting current is limited to that of class B). The term pullout or breakdown torque does not apply to this design since there may be no maximum in the torque–speed curve between zero and synchronous speed. For this reason, these are often called "high-slip" motors. Class D motors have a single, high-resistance rotor cage. Their full-load efficiency is less than that of the other classes. These motors are used in applications where it is important not to stall under heavy overload, like loads employing flywheel energy storage or lifting applications.

Insulation is very important in defining a motor for overall electrical insulation and resistance to environmental contamination. Motors are built with a full class F insulation system using nonhygroscopic materials throughout to resist moisture and extend thermal endurance. Air gap surfaces are protected against corrosion and windings are impregnated with polyester resin to provide mechanical strength.

Operating as a generator, an induction machine may exhibit some different and important characteristics. As a rule, induction generators do not have to start directly on line, like motors, but are accelerated by means of the prime movers they are coupled to. For this reason, starting torque and starting current are of minor importance. Efficiency is a very important issue in induction generators; keeping losses as low as possible implies a low slip machine. Moreover, to improve dynamic stability, a large maximum transient power is desired, especially when on-grid induction generators are used. To achieve this characteristic, a high breakdown torque is important, which implies a small leakage-reactance rotor design. In the same way, to minimize the amount of reactive power needed for the excitation of induction generators, small leakage and small magnetizing reactance are required. In a general way, to achieve the characteristics mentioned earlier, a cage rotor resembling that of design class A is usually the most appropriate for an induction machine designed for generator operation.

As already stated, a rotor design with low leakage reactance presents a steeper torque–slip characteristic near synchronous speed. Therefore, a relatively precise and fast responding speed regulator is required for the prime mover. When this is not possible or desirable, as in induction generators used with wind turbines or small low-cost hydropower plants, a squirrel cage rotor induction machine is not the best choice. In that case, a possible solution is the use of a wound-rotor induction generator, associated to a secondary converter. This electronic device, rated for slip power only, injects into the rotor circuit current with the slip frequency, enabling the generator to be connected directly to the grid, although the rotor speed may vary through a relatively wide range.

In some applications, special considerations may override some preconceived issues. For example, aircraft generators require minimum weight and maximum reliability, while military generators and motors require reliability and ease of servicing. Large hydropower generators require better starting torque per ampere;

induction generators for renewable energy applications require good steady and dynamic active and reactive capabilities and better efficiency. However, if they work self-excited, a trade-off with a consistent saturation curve might be important.

6.5   SIZING THE MACHINE

When designing an induction generator, the developed torque depends on the following equation:

$$T = \frac{\pi}{4} D^2 l_a J_{sm} \sin(\phi)$$ (6.2)

where
   D is the stator bore diameter
   l_a is the stator core length
   T is the output torque
   ϕ is the rotor power factor

For a given air gap volume and the stator winding J_{sm} current density, as well as the assumed air gap B_{rm} flux density, the following is true[1,2]:

$$\frac{D^2 l_a \phi}{T} \cong \text{constant}$$ (6.3)

Since power is the product of torque and speed, the following coefficient (C_0) can be used for sizing ac generators:

$$C_0 = \frac{kVA}{D^2 l_a \omega}$$ (6.4)

In order to optimize size, the values of current and flux densities are maintained at the limits allowed by cooling capability and saturation of the ferromagnetic material utilized in the laminations. The output sizing coefficient C_0 is calculated from those considerations and put into Equation 6.4 to support the decision on diameter and stack length with respect to output power and rated speed. A fairly complex design procedure is undertaken to decide the ratio of core length to pole pitch.

The major issues in designing induction generators are in the following five areas:

1. *Electrical*: Electrical design involves the selection of the supply voltage, frequency, number of phases, minimum power factor, and efficiency; and in a second stage, the phase connection (delta or Y), winding type, number of poles, slot number, winding factors, and current densities. Based on the output coefficient, power, speed, number of poles, and type of cooling, the

rotor diameter is calculated. Then, based on a specific current loading and air gap flux density, the stack length is determined.

2. *Magnetic*: Magnetic design implies establishing the flux densities in various parts of the magnetic path and calculating the magnetomotive force requirements. Slots are sized to optimize stack length and find a suitable arrangement of the winding conductors. While aiming to design machines with large diameters to benefit from increased output coefficients, the designers of induction generators need to ensure that the machine operates with a satisfactory power factor. The magnetizing current decreases (and the power factor improves) as the diameter grows, but it becomes larger as the number of poles is increased. A generator having a large pole pitch and a small number of poles will have a higher power factor than one with a small pole pitch. The diameters and lengths are selected for minimum cost, while these values are proportioned to provide a good power factor at a reasonable cost.

3. *Insulation*: Insulation design is concerned with determining the composition of slot and coil-insulating materials and their thickness for a given voltage level. End connection insulation and terminal leads depend upon voltage, insulation type, and the environment in which the generator operates.

4. *Thermal*: Thermal design depends on the calculation of losses and temperature distribution in order to define the best cooling system to keep the windings, core, and frame temperatures within safe limits.

5. *Mechanical*: Mechanical design refers to selection of bearings, inertia calculations, definition of critical rotating speed, noise, and vibration modes that may cause stresses and deformations. Forces on the windings during current transients must be calculated and centrifugal stresses on the rotor must not surpass safe limits. Some isolated generator systems require pole changing winding stator procedures and the right parallel star or series-star connection will influence the design.

The design of induction generators is generally divided into two distinct levels: less than and greater than 100 kW. In general, below-100 kW generators have a single winding stator built for low voltages and a rotor with an aluminum cast squirrel cage. Induction generators above 100 kW are built for 460 V/60 Hz or even higher voltages, 2.4–6 kV, and sometimes 12 kV. For such high voltages, the design is more constrained with regard to insulation issues and more appropriate for air or water-cooling techniques. At such high-power levels, wound rotor generators are frequently used. The configuration and construction of wound-rotor machines are different from squirrel cage ones, basically in the design of the rotor. They incorporate brushes and slip rings, and skin effect must be considered for higher power generators. Unlike cage rotors, wound rotors need an insulating system for coils similar to their stators and also require banding tapes or retaining rings to support the end windings against centrifugal forces. Figure 6.12 shows typical flowcharts used in designing induction generators comprising stator, rotor, and mechanical design. More details on the design equations can be found in Refs. [4,5].

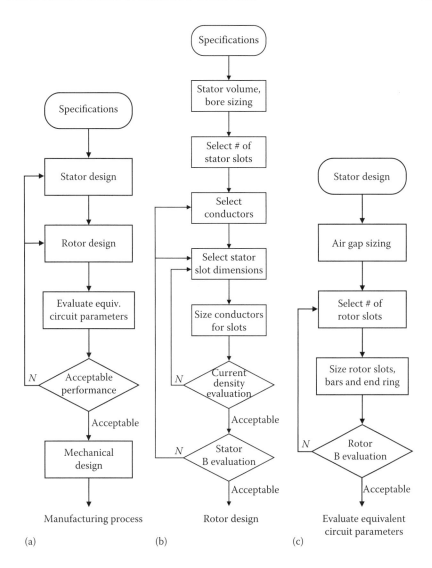

FIGURE 6.12 Flowchart for design: (a) overall process, (b) stator design, and (c) rotor design.

6.6 EFFICIENCY ISSUES

An induction generator is a long-term investment. An appropriate design should consider maximizing the machine efficiency. There are four different kinds of losses occurring in a generator: electrical losses, magnetic losses, mechanical losses, and stray load losses. Those losses can be reduced by using better quality materials, as well as by optimizing the design. Electrical losses increase with generator load. Increasing the cross section of the stator and rotor conductors can reduce them, but more copper is required, at a higher cost. Magnetic losses occur in the steel laminations of the stator and at a much lower level in the rotor. They are due to hysteresis and eddy currents and vary with the flux density and the frequency. These losses can be reduced by increasing the cross-sectional area of the iron paths mainly in the stator. They can also be limited by using thinner laminations and improved magnetic materials. If rotor and stator core length is increased, the magnetic flux density in the air gap of the machine is reduced, limiting the magnetic saturation and core losses, again at a higher cost.

Mechanical losses are due to friction in the bearings, power consumption in the ventilating fans, and windage losses. They can be decreased by using more bearings that are expensive and a more elaborate cooling system. The magnetic field inside a generator is not completely uniform. Thus as the rotor turns, a voltage is developed on the shaft longitudinally (directly along the shaft). This voltage causes microcurrents to flow through the lubricant film on the bearings. These currents, in turn, cause minor arcing, heating, and eventually bearing failure. The larger the machine is, the worse the problem. To overcome this problem, the rotor side of the bearing body is often insulated from the stator side. In most instances, at least one bearing will be insulated, usually the one farthest from the prime mover for generators and farthest from the load for motors. Sometimes, both bearings are insulated.

Efficient generators also need improved fan design and more steel used in the stator of the machine. This enables a greater amount of heat to be transferred out of the machine, reducing its operating temperature. The fan and the rotor surface are then redesigned to reduce windage losses. Stray load losses are due to leakage flux, nonuniform current distribution, mechanical imperfections in stator and rotor surfaces, and irregularities in the air gap flux density. Contributing to these losses is the skin effect in stator and rotor conductors, flux pulsation due to slotting, and harmonic currents. Stray load losses can be reduced by optimization of the electromagnetic design and careful manufacturing, eventually leading to a larger volume machine.

Figures 6.13 through 6.18 portray some pictures of actual medium- and large-size machines. Figure 6.13 shows a medium-sized eight-pole induction machine,

FIGURE 6.13 (**See color insert.**) Medium-sized eight-pole induction machine, with form wound stator coils. (Courtesy of Equacional Motores, Brazil.)

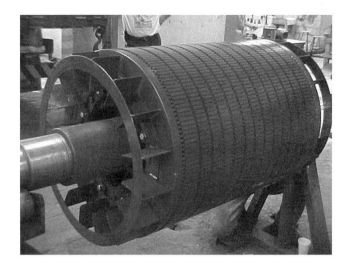

FIGURE 6.14 **(See color insert.)** Squirrel cage rotor for a large induction machine showing the brazed bars to the end rings. (Courtesy of Equacional Motores, Brazil.)

FIGURE 6.15 **(See color insert.)** Stator and rotor punchings for a large induction machine. (Courtesy of Equacional Motores, Brazil.)

with wound rotor coils, and Figure 6.14 is a squirrel cage rotor showing the brazed bars to the end rings. Figure 6.15 depicts the stator and rotor punchings for a large induction machine. Figure 6.16 shows the stator core of a medium-sized induction machine prepared for winding. Figure 6.17 shows a medium-sized squirrel cage rotor. Figure 6.18 portrays a random stator winding of a small induction machine.

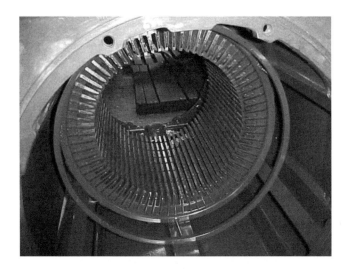

FIGURE 6.16   **(See color insert.)** Stator core of a medium-sized induction machine pre-pared for winding. (Courtesy of Equacional Motores, Brazil.)

FIGURE 6.17   **(See color insert.)** Medium-sized squirrel cage rotor winding. (Courtesy of Equacional Motores, Brazil.)

6.7   COMPARISON OF INDUCTION GENERATORS, PM, AND FERRITE MACHINES

A large power range for wind turbine conversion leads to an overdesigned genera-tor with correspondent higher losses. Consequently, maximizing the power range area is not usually done in wind turbine generator design. The methodology pro-posed[7] is a new way to handle the design of very low-speed generators as a com-promise between acceptable losses level and the power range. Table 6.1 shows the

FIGURE 6.18 **(See color insert.)** Random stator winding of a small induction machine. (Courtesy of Equacional Motores, Brazil.)

TABLE 6.1
Comparison of Mass and Energy Cost for the Designed Machines

	Ferrite	Bonded NdFeB	Sintered NdFeB
Iron mass	67 kg	37 kg	32 kg
Copper mass	38 kg	22 kg	20 kg
Magnet mass	27 kg	23 kg	21 kg
Total mass	132 kg	82 kg	73 kg
Energy	1.7 kWh	2.1 kWh	2.3 kWh
Mass/energy	80 kg/kWh	38 kg/kWh	32 kg/kWh
Cost/energy	0.6 k€/kWh	2.2 k€/kWh	1.9 k€/kWh

optimization results for three machines, where one is made of ferrites, while two others are from different NdFeB magnets.

The level of flux density is lower for a ferrite magnet machine. Therefore, the total mass is increased to obtain the same performance. In this case, the energy stored is less than in the other two cases. The best set of characteristics in terms of mass, energy stored and consequently (the mass–energy ratio) are reached for the sintered NdFeB (1.2 T) magnet configuration due to the high efficiency of the permanent magnet. However, when the cost–energy ratio is computed using economics and cost marked data for all components (iron, copper, and permanent magnet), the cost–energy ratio criteria show a different scenario where the ferrite magnet configuration is the smallest ratio. Although the mass is the highest one,

TABLE 6.2

Comparison of a Chosen Optimal Design of a Ferrite versus a PM Machine

	PM Machine	Ferrite
Turbine power	6.8 kW	5 kW
Maximum torque	1049 N m	1000 N m
Efficiency	79%	77%
Total mass	[a]82 kg	132 kg
External radius	795 mm	240 mm
Air gap radius	689 mm	208 mm
Active length	55 mm	476 mm
Wind speed	N/A	6 m/s
Rotor speed	N/A	30 rpm
Turbine radius	N/A	1 m

[a] The mass given by manufacturer datasheet.

the price of the ferrite magnets is about 20 times lower than the price of NdFeB magnets, showing that total cost is smaller than the NdFeB magnet-based design. Neodymium is a strategic material, and its price will most probably increase in the future. Thus, it is expected that the cost–energy ratio will further increase in near future. Although the mass–energy ratio is the lowest for ferrite magnets, it is indeed still a good alternative when compared with rare earth permanent magnet options. Table 6.2 shows a comparison of a generator with Bonded NdFeB magnet and a high torque generator taken in the literature; the machine selected in the literature has the same electromagnetic characteristics (torque and speed) and the same type of permanent magnet that the machine designed in [7] and can be used for comparison.

The flux-switching electrical generator designed in [7] is an option for very low speed (eliminating the need of gearbox). In addition, it can be constructed of low-cost standard ferrite magnets.

While it is certain that permanent magnet generators can be constructed for very low-power and cheap applications, or maybe for PM synchronous machines, the future economics for commercialization of rare-earth magnets might be a problem, because very few countries have those minerals. In order to design induction genera-tors, our schools and future electrical engineers must use the principles established by Nikola Tesla. In order to manufacture induction machines, one requires only cop-per and iron. The fabrication techniques have been completely available with current knowledge. Therefore, the induction generator is a choice for now and forever, it is the best solution for small-scale to medium-power wind turbines, typically used in rural systems, small farms and villages, as well as for small and medium hydropower applications.

6.8 SOLVED PROBLEMS

1. Torque production by an induction generator depends on the air-gap volume. Write a summary of the practical limitations on magnetic flux and conductor current densities that constrain this physical fact.

Solution

From Equation 6.2, it is clear that the torque production is dependent on air gap volume, the stator winding current density, and flux density. Therefore, to produce the same torque with the smaller size machine, we need higher current density and/or flux density. However, current density is directly related to more losses, which in turn increases temperature rise of the machine and this adversely affects the insulation of the machine. On the other hand, higher flux density drives the machine more into the saturation, hence causing more magnetizing current requirements.

2. Since most renewable energy prime movers are very low speed, discuss the design characteristics of an induction generator for those applications.

Solution

For low-speed application, some important requirements of the induction generator are

Need for gearbox
Higher number of poles (20/40 poles)
Generally, diameter of the machine is huge compared with its length to contain physically higher number of poles
High inertia of the machine
Salient pole construction of the stator

6.9 SUGGESTED PROBLEMS

6.1 Gas turbine and compressed air systems will serve as prime movers at a very high speed. Discuss the design characteristics of an induction generator for those applications.

6.2 Explain in detail why induction generators of the same rating as synchronous generators are larger.

6.3 What are the differences between a high standard induction generator and the corresponding induction motor?

6.4 Research about the mining, manufacturing, and production of permanent magnets, which countries produce, and the ones that import them. Compare the economics and sustainability of permanent magnet-based machines versus induction generators.

REFERENCES

1. Hamdi, H.S. and Hamdi, E.S., *Design of Small Electrical Machines*, John Wiley & Sons, New York, 1994.
2. Cathey, J.J., *Electric Machines: Analysis and Design Applying MATLAB*, McGraw-Hill, New York, 2000.

BIBLIOGRAPHY

Boldea, I. and Nasar, S.A., *The Induction Machine Handbook*, CRC Press, Boca Raton, FL, 2001.

Nasar, S.A., *Handbook of Electric Machines*, McGraw-Hill, New York, 1987.

Ojeda, J., Godoy Simões, M., Li, G., and Gabsi, M., Design of a flux switching electrical generator for wind turbine systems, *IEEE Trans. Ind. Appl.*, 48(6), 1808–1816, November–December 2012, doi: 10.1109/TIA.2012.2221674.

Say, M.G., *Alternating Current Machines*, John Wiley & Sons, New York, 1976.

7 Power Electronics for Interfacing Induction Generators

7.1 SCOPE OF THIS CHAPTER

Power electronics is the branch of electronics that studies application systems ranging from less than a few watts to more than 2 GW. These systems encompass the entire field of power engineering, from generation to transmission and distribution of electricity and its industrial use, as well as transportation, storage systems, and domestic services. The progress of power electronics has generally followed the microelectronic device evolution and influenced the current technological status of machine drives. The amount of power produced by renewable energy devices like photovoltaic (PV) cells and wind turbines varies significantly on an hourly, daily, and seasonal basis due to variations in the availability of the sun, the wind, and other renewable resources. This variation means that sometimes power is not available when it is required and sometimes there is excess power.

Figure 7.1 shows how power electronics interfaces energy sources with loads. The variable output from renewable energy devices also means that power conditioning and control equipment are required to transform this output into a form (voltage, current, and frequency) that can be used by electrical appliances. This chapter will present power electronic semiconductor devices and their requirements for interfacing with renewable energy systems: ac–dc, dc–dc, dc–ac, and ac–ac, conversion topologies as they apply to the control of induction machines used for motoring and generation purposes.

7.2 POWER SEMICONDUCTOR DEVICES

After the transistor was invented, the emergence of thyristor also called silicon-controlled rectifier (SCR), initiated the first generation of power electronics. In the 1960s, inverter-grade thyristors enabled the introduction of force-commutated thyristor inverters like the McMurray inverter, the McMurray-Bedford inverter, the Verhoef inverter, the ac-switched inverter, and the dc-side commutated inverter. Many other commutation techniques were discussed in the literature. This class of inverters gradually faced obsolescence because of the advent of self-commutated gate turn-off thyristors (GTOs) and improved bipolar transistors, ending the first generation of power electronics by the middle of the 1970s.

Thyristors and GTOs continue to grow in power rating (most recently 6000 V, 6000 A) for multimegawatt voltage-fed and current-fed converter applications (with

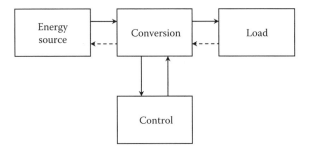

FIGURE 7.1   Power electronics interface energy sources with load.

reverse blocking devices). The device is slow switching, which causes large switching losses, and keeps its switching frequency low (a few hundred hertz) in high-power applications. Large dissipative snubbers are essential for SCR and GTO converters. It is important to note that as the evolution of the new and advanced devices continued, the voltage and current ratings and electrical characteristics of those devices dramatically improved the performance of power electronic circuits, making possible a widespread deployment of new applications. Even today, thyristors are indispensable for handling very high power at low frequency for applications such as high-voltage DC (HVDC) converters, phase control type static VAR compensators, cycloconverters, and load-commutated inverters. It appears that the dominance of thyristors in high power handling will not be challenged, at least in the near future.

The introduction of large Darlington and Ziklay, bipolar-junction-transistors (BJTs) in the 1970s brought great expectations in power electronics. However, at the same time the metal–oxide–semiconductor field-effect transistors, or power MOSFETs were introduced. Although the conduction drop is large for higher voltage devices, the switching loss is small. MOSFET is the most popular device for low-voltage high-frequency applications, like switching mode power supplies (SMPS) and other battery-operated power electronic apparatuses. Bipolar junction transistors have been totally ousted in new applications in favor of power MOSFETs (higher frequency) and the insulated-gate bipolar transistors (IGBT) for higher voltage.

The introduction of the IGBT in the 1980s was an important milestone in the history of power semiconductor devices, marking the second generation of power electronics. Its switching frequency is much higher than that of the BJT, and its square safe operating area (SOA) permits easy snubberless operation. The power rating (currently 3500 V, 1200 A) and electrical characteristics of the device are continuously improving. IGBT intelligent power modules are available with built-in gate drivers, control, and protection for up to several hundred kilowatt power ratings. The present IGBT technology has conduction drops slightly larger than that of a diode and a much higher switching speed. A MOS-controlled thyristor (MCT) is another MOS-gated device, which was commercially introduced in 1992. The present MCT (P-type with 1200 V, 500 A), however, has limited reverse-biased SOA (RBSOA), and its switching speed is much lower than that of IGBTs. MCTs are being promoted for soft-switched converter applications where these limitations are not barriers.

The integrated gate-commutated thyristor (IGCT) is the newest member of the power conductor family at this time. It was introduced by ABB in 1997. It is a high-voltage, high-power, hard-driven, asymmetric-blocking GTO with unity turn-off current gain (currently 4500 V, 3000 A), turning on with a positive gate current pulse and turning off with a large negative gate current pulse of about 1 µs. It is fully integrated with a fiber-optic-based gate drive. By means of a *hard-drive* gate control with a unity gain turn-off, the element changes directly from thyristor mode to transistor mode during turn-off. This allows operation without any snubber. The gate driver required is part of the IGCT, and no special gating requirements are needed. The claimed advantages over GTO are smaller conduction drop, faster switching, and a monolithic bypass diode that combines the excellent forward characteristics of the thyristor and the switching performance of the bipolar transistor. It is claimed that it acts as a short circuit in failure mode (advantageous in converter series operation), while IGBT modules usually act as an open circuit.

The third generation of power electronics is currently under way through the integration of power devices with intelligent circuits. Isolation, protection, and interfacing are already incorporated in several commercial modules. Digital computation–intensive embedded systems are making it easier to integrate pulse width modulation (PWM) techniques, signal estimation, and complex algorithms for control of machines. Although silicon has been the basic raw material for power semiconductor devices for a long time, several other raw materials, like silicon carbide and diamond, are showing promise. These materials have a large band gap, high carrier mobility, high electrical and thermal conductivity, and strong radiation hardness. Therefore, devices can be built for higher voltage, higher temperature, higher frequency, and lower conduction drop and are likely to be commercially available soon. Devices capable of withstanding voltages and currents in both directions (ac-switches) are also under development and will make possible the implementation of direct frequency changers (matrix converters) in the fourth generation of power electronics. Table 7.1 depicts the most important power electronic devices.

7.3 POWER ELECTRONICS AND CONVERTER CIRCUITS

There are different circuit topologies that can be used to handle variations in supply or change the electrical current into a form that can be used by industrial, rural, or household loads. The following classes of power electronic equipment are frequently used in renewable energy systems: regulators for battery charge controllers, inverters for interfacing dc with ac, and protection/monitoring units.

7.3.1 REGULATORS

A battery charge controller or regulator should be used to protect the battery bank from overcharging and overdischarging. The simplest method of charge control is to turn off the energy source as the battery voltage reaches a maximum and turn off the load when the battery voltage reaches a preset minimum. There are three main types of regulators: shunt, series, and chopper.

1. *Shunt regulators*: Once the batteries are fully charged, the power from the renewable source is dissipated across a dump load. These are commonly used with wind turbines.

TABLE 7.1

Power Electronic Device Capabilities

Device	Symbol	Applications	Ratings
SCR	Anode A i_A v_{AK} Gate G K Cathode	Very high power, multimegawatt power systems	100 MW to 1 GW, frequency < 100 Hz
GTO	Anode Gate Cathode	Very high power, multimegawatt traction, and controlled systems	1–100 MW, frequency <500 Hz
IGCT	Anode Gate Cathode	High-voltage/high-current megawatt systems	1–100 MW, frequency < 500 Hz
TRIAC	MT_2 Gate MT_1	Low power household ac–ac control	Up to 250 W, 60 Hz
Power MOSFET	Drain D i_D Gate G v_{DS} S Source	Switching mode power supplies and small power actuators/drives	Up to 10 kW, frequency < 1 MHz
IGBT	Collector C Gate G E Emitter	Medium power industrial drives, machine control, inverters, converters, and active filter	Up to 500 kW, frequency < 100 kHz
MCT	Anode A Gate G K Cathode	Resonant-based topologies	Up to 10 kW, frequency < 100 kHz

2. *Series regulators*: Once the batteries are fully charged, the power from the renewable source is switched off in the simplest series regulator.
3. *Chopper regulators*: These regulators use a high-frequency switching technique. The regulator switches the control device on and off quickly. When the batteries are discharged, the unit will be fully switched on. As the battery reaches a fully charged state, the unit will start switching the control device on and off in proportion to the level of charging required. When the battery is fully charged, no current will be allowed to flow to the battery.

7.3.2 INVERTERS

Renewable energy systems often provide low voltage, direct current (dc) from batteries, solar panels, or wind generators with rectification. An inverter is an electrical device that changes direct current (dc) into alternating current (ac). The inverter enables the use of standard household appliances. Inverters often incorporate extra electronic circuits that control battery charging and load management. If inverters are used only to supply power from a generator to a load, they are unidirectional. On the other hand, if they need to transfer power to an electrical machine to operate in motoring mode (such as for start-up, braking, or pumping), they need to be bidirectional. Inverters must produce power of a similar quality to that in the main electricity grids. Therefore, power quality and compliance with harmonic distortion standards are very important. Inverters can operate in stand-alone or connected mode to the main grid. In the latter case, interconnection guidelines must be negotiated with the local public services company.

7.3.3 PROTECTION AND MONITORING UNITS

In systems with a number of power sources, sophisticated system controllers are required. These controllers are usually computer controlled, with inputs indicating the state of the system being fed into the controller. The microprocessor makes changes to the system operation if necessary. System controllers can measure energy demanded in a house and communicate with the utility company to achieve contracted load management—for example, water heaters and furnaces—or to dispatch to a local generator. The functions performed by system controllers include the following:

- Disconnecting or reconnecting renewable energy sources
- Disconnecting or reconnecting loads
- Implementing a load management strategy
- Starting diesel generators when battery voltage is too low or if the load becomes too heavy
- Synchronizing ac power sources (e.g., inverters and diesel generators)
- Shutting systems down if overload conditions occur
- Monitoring and recording key system parameters

Renewable energy systems can be classified as stationary and rotatory. Stationary systems usually provide direct current; PV arrays and fuel cells (FCs) are the main renewable energy sources in this group. Rotatory systems usually provide alternating current; induction, synchronous, and permanent magnet generators are the main drivers for hydropower, wind, and gas turbine energy sources. Of course, dc machines are of rotatory type, but not usually employed due to their high cost, size, and maintenance needs.

Renewable energy sources in the stationary group do not absorb any power; energy only flows outwards to the load. Sources in the rotatory group, on the other hand, require bidirectional power flows, either to rotate in motoring mode (for pumping, braking, or starting up) or to absorb reactive power (for induction generators (IGs)). The storage systems can also be either stationary (batteries, supercapacitors, magnetic) or rotatory (connection of generators to the grid, flywheels, hydropumping). A generalized concept of power electronic needs for renewable energy systems is portrayed in Figure 7.2, embracing three categories:

1. Static systems with a dc input converter to a dc or ac output and unidirectional power flow (no moving parts, lower time constant, silent, no vibrations, minimally pollutant, lower power up to now)
2. Rotating systems with an ac input converter to a dc or ac output and bidirectional power flow (angular speed, noisy, mechanical inertia, weight and volume, inrush characteristics)
3. Storage systems—necessarily bidirectional with requirements of ac and dc conversion (dependent on the application)

A converter uses a matrix of power semiconductor switches to convert electrical power at high efficiency. The standard definition of a converter assumes a source to be connected to a load as indicated in Figure 7.1. The inherent regenerative capabilities of renewable energy systems require less restrictive terms as input/output or

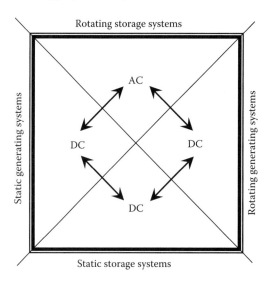

FIGURE 7.2   Power electronic needs for renewable energy conversion systems.

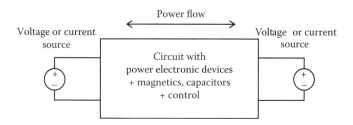

FIGURE 7.3   Power electronic converter for the connection of two sources.

supply/load when referencing to the connection of two sources. The links between a power converter and the outside world can be perceived as having the possibility of power reversal, where the sink can be considered as a negative source, in accordance with Figure 7.3. Therefore, it is a better fit to define the terms *voltage source* (VS) and *current source* (CS) when describing the immediate connection of converters to its entry and exit ports. The VS maintains a prescribed voltage across its terminals, irrespective of the magnitude or polarity of the current flowing through the source. The prescribed voltage may be constant dc, sinusoidal ac, or a pulse train. The CS maintains a prescribed current flowing between its terminals, irrespective of the magnitude or polarity of the voltage applied across these terminals. Again, the prescribed current may be constant dc, sinusoidal ac, or a pulse train.

Current source inverters (CSIs) have several distinct advantages over variable voltage inverters. They provide protection against short circuits in the output stage; they can handle oversized motors. In addition, CSIs have relatively simple control circuits and good efficiency. Their disadvantages are that CSIs produce torque pulsations at low speed. They cannot handle undersized motors, and they are large and heavy. The phase-controlled bridge rectifier CSI is less noisy than its chopper-controlled counterpart. It does not need high-speed switching devices, but it cannot operate from direct dc voltage. The chopper-controlled CSI can operate from batteries and produces more noise because of its need for high-speed switching devices. With the advent of fast high-power-controlled devices, VS inverters became very popular. Table 7.2 summarizes a comparison between VS and CS inverters. The following discussion in this chapter covers only VS inverters.

7.4   DC TO DC CONVERSION

Electrical power in dc form is readily available from batteries and other inputs like PV arrays and FCs. Batteries are a common energy storage device for portable applications ranging from handheld drills and portable computers to electric vehicles. Another typical dc source is the output of rectifiers. When using power electronic devices to synthesize ac or variable dc waveforms, it is usually easier to begin with a known, fixed source rather than a varying source.

A fundamental form of dc–dc converter is the buck converter, sometimes called a chopper, shown in Figure 7.4. A buck converter can create an average dc output less than or equal to its dc source. The switch used in a buck converter must be fully controllable, that is, a MOSFET or an IGBT, so that the switch can forcibly turn off

TABLE 7.2

Summarized Comparison between VS Inverters and CS Inverters

VSI	CSI
Output is constrained voltage	Output is constrained current
DC bus is dominated by shunt capacitor	DC bus dominated by series inductor
DC bus current proportional to motor/generator power and hence dependent on motor/generator PF	DC bus voltage proportional to motor/generator power and hence dependent on motor/generator PF
Output contains voltage harmonics varying inversely as harmonic order	Output contains current harmonics varying inversely as harmonic order
Prefers motor/generator with larger leakage reactance	Prefers motor/generator with lower leakage reactance
Can handle motor/generator smaller than inverter rating	Can handle motor/generator larger than inverter rating
DC bus current reverses regeneration	DC bus voltage reverses in regeneration
Immune to open circuit	Immune to short circuit

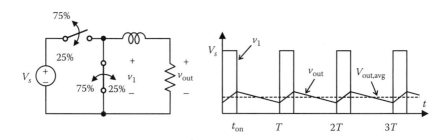

FIGURE 7.4   DC–DC converter.

its current. On the other hand, it is rarely desirable to extinguish rapidly the current in an inductor. In fact, one would like the inductor to store the excess power from the source when the switch is off and then surrender the stored energy to the load while the switch is on. For the inductor to power the load while the main switch is open, the designer must add a second switch to provide another current path.

PWM is a technique of digitally encoding analog signal levels. Power circuits need to be turned on and off in order to modulate a voltage or current level. PWM can be implemented through old-fashioned operational amplifier-based analog circuits or, recently, through timers available in most microcontrollers, including on-chip PWM firmware. Figure 7.4 shows the principle of PWM, where by turning on and off the switches, a variable average voltage can be synthesized at the output.

This secondary current path is referred to as the freewheeling path. The second switch must be off when the primary switch is on, and vice versa, as shown in Figure 7.5. Intuitively, a bigger filter inductor stores more energy at a given current level than a smaller one. For a given switching frequency, therefore, more inductance

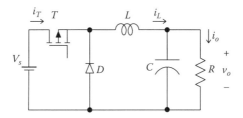

FIGURE 7.5   Buck converter.

provides more filtering. Although the main switch in a buck converter must be fully controllable, the secondary switch does not have to be. As the secondary switch functions as the logical complement of the controllable switch, the most economical implementation of the auxiliary switch is a diode, as shown in Figure 7.5. Such a diode is called a freewheeling diode. When the main switch is conducting, current flows from the source through the inductor to the load. The diode points upward, so there is no danger of shorting the source. When the controlled switch turns off, the inductor Ldi/dt voltage forces the diode to conduct the current.

Some switching power supplies are based on the same idea. However, transformers are more frequently used, both for the help from their turn ratios and for electrical isolation. These dc–dc converters actually convert dc to ac, which is placed across the transformer. The secondary of the transformer is then rectified, converting ac back to the desired dc. The details are not included here. By rearranging the elements in the buck converter, a designer can create a boost converter as shown in Figure 7.6. One must add the additional capacitor to support the load while the active switch is on. When the active switch is on, the source is directly across the inductor, causing ramp up to the inductor current. When the active switch is off, the diode feeds the inductor current into the load and the capacitor. If the converter is in steady state, the inductor current must go back down during this mode; thus, the voltage must be higher than the source voltage. As usual, the ripple on the output voltage is determined by the switching frequency and the size of the filter elements, especially the capacitor in this case.

Buck converters and boost converters can be considered the building blocks for the structure of inverters. The buck-boost converter shown in Figure 7.7 can generate either a higher voltage at the output for a duty cycle greater than 50% or a voltage lower than the input for a duty cycle less than 50%. Figures 7.8 through 7.10,

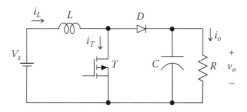

FIGURE 7.6   Boost converter.

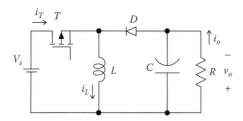

FIGURE 7.7   Buck-boost converter.

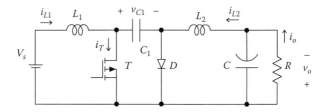

FIGURE 7.8   Ćuk converter.

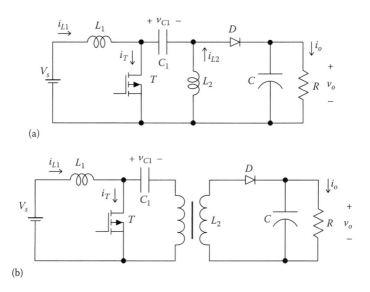

(a)

(b)

FIGURE 7.9   Sepic converter: (a) nonisolated topology and (b) isolated topology.

respectively, show a Ćuk converter, a Sepic converter, and a Zeta converter; all of them are used for situations where the input current must be smooth. The dc–dc converters discussed in this chapter are relevant for FCs, PV applications, and battery chargers. There are other isolated topologies of converters—forward, flyback, half-bridge, and full-bridge—that are very common for SMPS applications. Detailed information on the operation of dc–dc converters can be found in the references.[1–3]

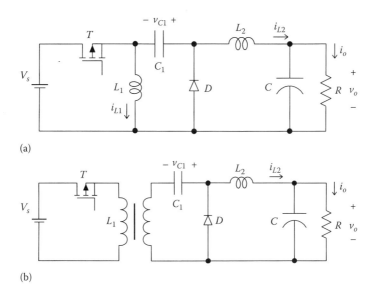

FIGURE 7.10   Zeta converter: (a) nonisolated topology and (b) isolated topology.

7.5   AC TO DC CONVERSION

For ac–dc conversion, the following parameters are required to design the circuit:

- Evaluation of ac input limitations
- Average output voltage (V_{dc})
- Output ripple voltage (V_{ripple})
- Efficiency (η)
- Circuit load, output power
- Regulation to input voltage variation
- Regulation to output load variation
- Power flow direction if required, ac–dc and dc–ac

Half-wave rectifiers are not used for medium and high power due to the severe constraints imposed on the devices and the amount of harmonics generated. In addition, due to the low power factor (PF) in half-wave rectifiers, a bulky and inefficient transformer limits their application to the range of less than 100 W. The following single-phase full-wave rectifiers are used in practice.

7.5.1   Single-Phase Full-Wave Rectifiers, Uncontrolled and Controlled Types

Single-phase rectifiers can be constructed with a transformer with one simple secondary or with a center-tapped secondary.

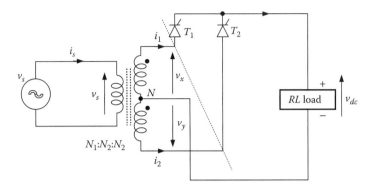

FIGURE 7.11   Center-tapped secondary single-phase rectifier.

7.5.1.1   Center-Tapped Single-Phase Rectifier

Figure 7.11 shows the topology of a center-tapped single-phase rectifier circuit. The center tap splits the voltage into 180° out-of-phase voltages (v_x and v_y) with respect to the neutral terminal N. Therefore, each diode conducts for each half-cycle of the input voltage $v_s = V_m \sin(\omega t)$. The circuit operation can be phase-controlled by thyristors (SCRs), where α is the firing angle for the SCR. For ordinary diodes, it is sufficient to make $\alpha = 0$ in the following description. The average output voltage is given by

$$V_{dc} = 2 \times \frac{1}{2\pi} \int_\alpha^{\pi+\alpha} V_m \sin\theta d\theta = \frac{2V_m}{\pi}\cos\alpha \tag{7.1}$$

This equation is valid for very high reactance ωL, a condition that imposes continuous conduction mode at the output. Under these conditions, each device, T_1 or T_2, will continue to conduct current until the next one takes over; that is, a natural commutation occurs. Figure 7.12 shows the main waveforms. The average voltage can be controlled by commanding the firing angle, that is, by a time delay with respect to the voltage zero crossing by the application of a gate pulse i_g; for $\alpha = 0°$, the average voltage (V_{dc}) is maximum, and for $\alpha = 90°$, the average voltage is minimum. The ripple increases with the decrease of the average voltage. As the output current becomes flatter for highly inductive loads, the thyristor currents become rectangular. The displacement factor (DPF) on the input side is approximately given by $\cos \alpha$. The distortion factor can be roughly bounded by a square wave current.

7.5.1.2   Diode-Bridge Single-Phase Rectifier

Figure 7.13 shows a diode-bridge single-phase rectifier. It is an uncontrollable converter since the diodes turn on as soon as the impressed voltage becomes positive. With a pure resistive load, the current waveform takes the same shape as the rectified semicycles. For a light inductive load where the current is continuous, the following solution for the load current can be used:

$$L\frac{di_L}{dt} + Ri_L = \sqrt{2V_s}\sin\omega t \quad 0° \le \omega t \le 180° \tag{7.2}$$

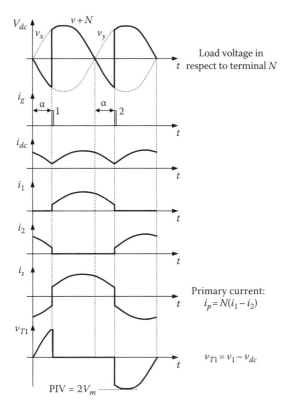

FIGURE 7.12 Main waveforms for center-tapped secondary single-phase rectifier.

and the solution for this differential equation is

$$i_L = \frac{\sqrt{2V_s}}{Z}\sin(\omega t - \theta) + A_1 e^{-Rt/L} \tag{7.3}$$

where

$$Z = \sqrt{R^2 + (\omega L)^2}$$
$$\theta = \tan^{-1}(\omega L/R)$$

Figure 7.14 shows the output voltage and input current waveforms for a light inductive load.

7.5.1.3 Full-Controlled Bridge

Figure 7.15 shows a full-controlled single-phase converter. The thyristors T_1 and T_2 must be triggered at the same time—for example, by using a pulse transformer with double secondary for the positive half-cycle and trigger T_3 and T_4 for the negative half-cycle. Assuming an inductive load, T_1 and T_2 keep conducting even after $\omega t = \pi$

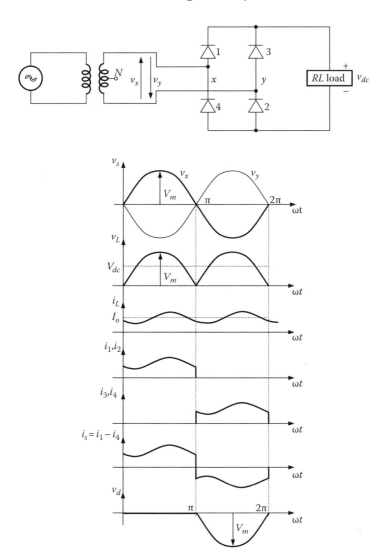

FIGURE 7.13 Circuit topology and waveforms for diode-bridge single-phase rectifier.

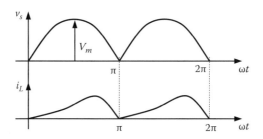

FIGURE 7.14 Output voltage and input current waveforms for light inductive load.

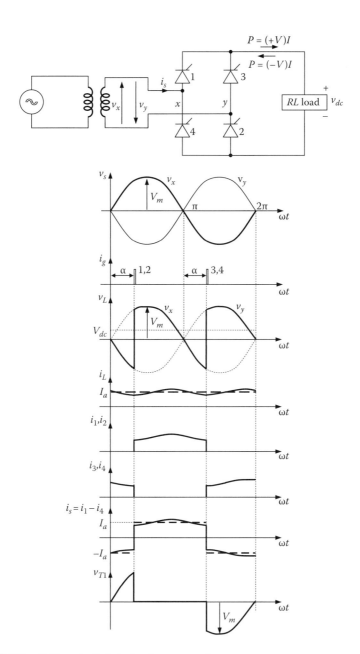

FIGURE 7.15 Full-controlled single-phase converter.

with negative voltage; devices T_3 and T_4 during this half-cycle are forward-biased and ready to commute the current when triggered at angle α. This type of single-phase converter is typical for power up to 15 kW. However, the harmonic pollution on the utility side and consequent neutral loading in three phases and a very low PF inhibits its use for higher power. The average load voltage is the same as the full-wave topology, considering two voltage diode drops of the series devices. The following equation applies for highly inductive loads:

$$V_{dc} = \frac{2}{2\pi} \int_\alpha^{\pi+\alpha} V_m \sin \omega t d\omega t = \frac{2V_m}{2\pi}\left[-\cos \omega t\right]_\alpha^{\pi+\alpha}$$

$$= \frac{2V_m}{\pi}\cos \alpha = V_{do}\cos \alpha \qquad\qquad 0 \le \alpha \le \pi \qquad (7.4)$$

V_{dc} can be varied from $-2V_m/\pi$ to $+2V_m/\pi$. Therefore, the main advantage of this full-controlled converter topology is the possibility of power regeneration to the primary side. If the thyristors are fired to apply a negative voltage for a constant positive current at the load, the power is regenerated to the primary side.

The averaged normalized output voltage is as follows:

$$V_n = \frac{V_{dc}}{V_{do}} = \cos \alpha \qquad\qquad (7.5)$$

The rms output voltage is

$$V_{rms} = \left[\frac{2}{2\pi}\int_\alpha^{\pi+\alpha} V_m^2 \sin^2 \omega t d\omega t\right]^{1/2}$$

$$\left[\frac{V_m^2}{2\pi}\int_\alpha^{\pi+\alpha}\left(1-\cos 2\omega t\right)d\omega t\right]^{1/2}$$

$$= \frac{V_m}{\sqrt{2}} = V_s \qquad\qquad (7.6)$$

The total harmonic distortion is $THD = \sqrt{\left(I_{rms}/I_{rms1}\right)^2 - 1} = 48.34\%$ with a DPF $DF = \cos \phi \cong \cos \alpha$. The PF is $PF = I_{rms1}\cos \alpha/I_{rms} = 2\sqrt{2}\cos \alpha/\pi$. In this converter, the fundamental value of the input current is always 90.03% of I_a, and the harmonic factor is constant at 48.34%.

A simplification of the full-controlled converter is achieved by the half-converter portrayed in Figure 7.16. Only two controlled devices are used; the bottom devices are diodes, and there is a freewheeling diode across the load. The freewheeling diode increases the average voltage across the load and improves the PF on the input side.

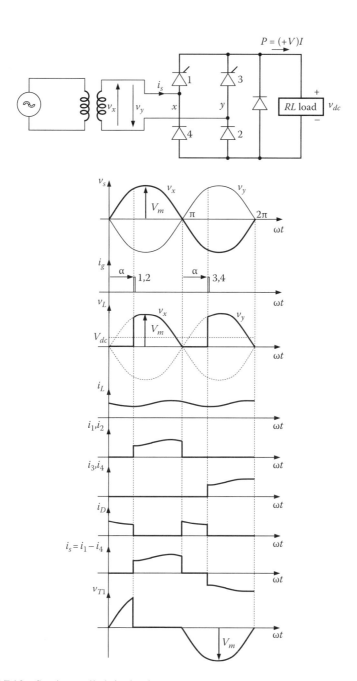

FIGURE 7.16 Semicontrolled single-phase converter.

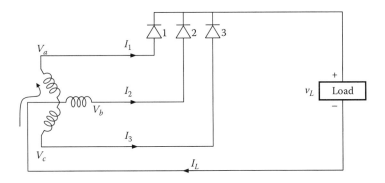

FIGURE 7.17   Half-wave three-phase bridge.

Of course, the characteristics of bidirectionality in the previous converter are lost because the half-converter only impresses positive voltages across the highly inductive load.

Three-phase rectifiers are commonly used in high-power applications; there are more rectifying pulses resulting in more effective filtering than single-phase rectifiers; in addition, the transformer utilization factor for three-phase rectifiers is improved. There are several topologies for three-phase rectifiers.

7.5.1.4   Half-Wave Three-Phase Bridge

The half-wave three-phase bridge is depicted in Figure 7.17. It is a basic circuit used for understanding most polyphase circuits. Balanced three-phase voltages are assumed to be available. Every period that a phase voltage has the largest instantaneous voltage, the correspondent diode is on as indicated by the period a in Figure 7.18. For a half-wave three-phase bridge, there are three pulses at the rectified voltage across the load.

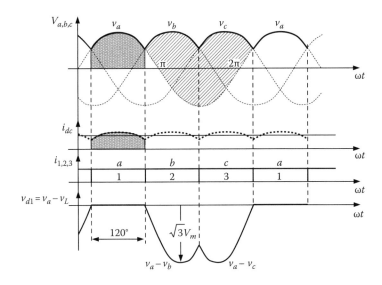

FIGURE 7.18   Waveforms for voltages and currents in a half-wave three-phase bridge.

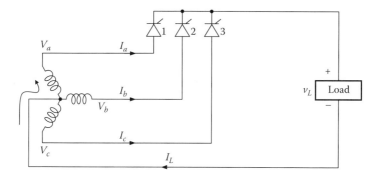

FIGURE 7.19 Controlled three-phase half-wave converter.

The average value (neglecting the diode voltage drops) is

$$V_{dc} = \frac{1}{2\pi/3} \int_{\pi/6}^{5\pi/6} V_m \sin\theta d\theta = \frac{3\sqrt{3}}{2\pi} V_m \qquad (7.7)$$

where V_m is the peak value of the phase voltage $= \sqrt{2}V_{rms}$.

The dashed line indicating the current i_{dc} represents the current through a resistive load. For an RL load, this waveform would be smoothed. The circuit indicated in Figure 7.19 is the version of the controlled three-phase half-wave converter. Figure 7.20 shows the typical waveforms for a controlled three-phase half-wave converter.

Unlike single-phase rectifiers where the phase angle α is defined by the zero crossing for three-phase rectifiers, the phase angle α is defined by the crossing of phase voltages, that is, where the diodes would naturally commute. The average voltage is calculated by

$$V_{dc} \frac{1}{2\pi/3} \int_{(\pi/6)+\alpha}^{(5\pi/6)+\alpha} V_m \sin\theta d\theta = \frac{V_{do}}{2} \cos\alpha \qquad (7.8)$$

where $V_{do} = 3\sqrt{3}/\pi V_m$

The primary side of the transformer is often Δ-connected, imposing an alternating primary current and avoiding saturation. In addition, triple harmonics $(3p)$ are kept inside the windings and do not circulate through the line.

7.5.1.5 Three-Phase Full-Wave Bridge Rectifier (Graetz Bridge)

The three-phase full-wave bridge rectifier shown in Figure 7.21 is in very wide use in industrial systems; it is also called the Graetz converter. It can operate with or without a transformer and gives a six-pulse-ripple waveform at the output. The output voltage is the subtraction of two half-wave three-phase voltages with respect to the neutral N.

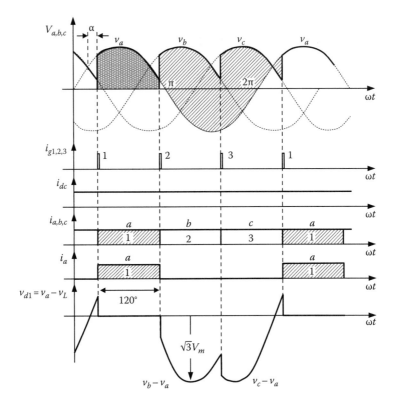

FIGURE 7.20　Waveforms for controlled three-phase half-wave converter.

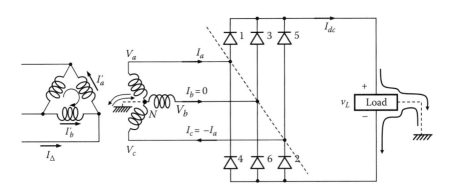

FIGURE 7.21　Graetz rectifier.

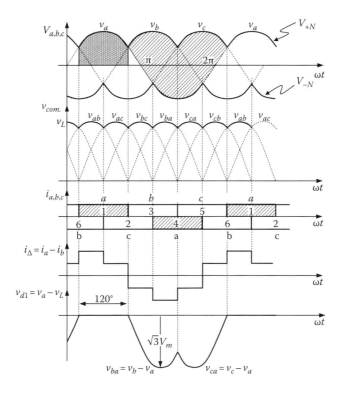

FIGURE 7.22 Main waveforms for Graetz rectifier.

The waveforms are shown in Figure 7.22. The line current i_Δ is a six-step waveform resembling a sine wave with an improved PF on the input side. The average voltage is given by

$$V_{dc} = 2\frac{3\sqrt{3}}{2\pi} V_{m(phase)} = \frac{3}{\pi} V_{m(line)} \qquad (7.9)$$

7.5.1.6 Three-Phase Controlled Full-Wave Bridge Rectifier

The waveforms of the three-phase controlled full-wave bridge rectifier are similar to those of the Graetz bridge shifted by α (Figure 7.23). The sequence of thyristors is indicated in Figure 7.24. The average voltage is computed by

$$V_{dc} = \frac{3}{\pi} V_{m(line)} \cos\alpha = \frac{3\sqrt{2}}{\pi} V_{(line, rms)} \cos\alpha$$

$$= \frac{3\sqrt{6}}{\pi} V_{rms(phase)} \cos\alpha = V_{do} \cos\alpha \qquad (7.10)$$

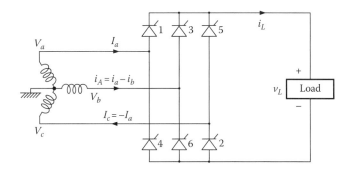

FIGURE 7.23　Three-phase controlled full-wave bridge rectifier.

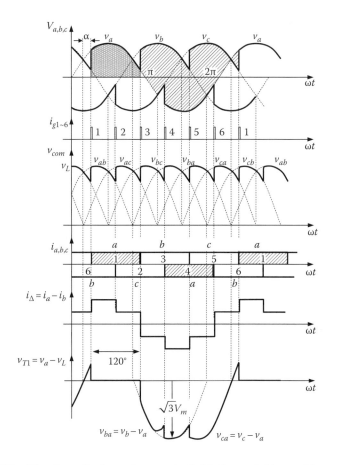

FIGURE 7.24　Waveforms for three-phase controlled full-wave bridge rectifier.

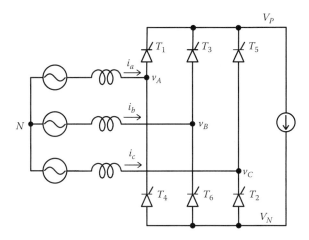

FIGURE 7.25 Graetz converter with ac-side input inductance.

The rms fundamental component of the line current is given by

$$I_{\ell 1} = \frac{\sqrt{6}}{\pi} I_{dc} \tag{7.11}$$

7.5.1.7 Effect of Series Inductance at the Input AC Side

In practical converters, it is necessary to include the effect of the ac-side inductance L_s, as depicted in Figure 7.25. Now for a given delay angle α, the current commutation takes a finite commutation interval μ. Figure 7.26 shows the commutation process for the situation where thyristors 5 and 6 have been conducting previously, and at $\omega t = \alpha$, the current begins to commute from thyristor 5 to 1.

The average voltage is reduced by this series inductance to

$$V_{dc} = \frac{3\sqrt{2}}{\pi} V_{(line,rms)} \cos\alpha - \frac{3\omega L_s}{\pi} I_d$$

Knowing α and I_d, we can calculate the interval μ from

$$\cos\left(\alpha + \mu\right) = \cos\left(\alpha\right) - \frac{2\omega L_s}{\sqrt{2V_{(line,rms)}}} I_d$$

7.5.1.8 Three-Phase Half-Converter

Three-phase half-converters are used in industrial applications up to 120 kW, where unidirectional power flow (one-quadrant) is required. A freewheeling diode is used across the load, and the bottom devices are ordinary diodes, as indicated by Figure 7.27.

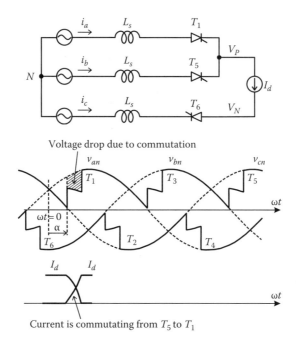

FIGURE 7.26 Commutation effect due to ac-side input inductance.

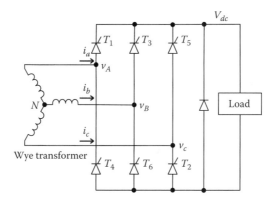

FIGURE 7.27 Three-phase semiconverter.

7.5.1.9 High-Pulse Bridge Converters

There are several other types of converters: 12-pulse and higher-pulse-number bridge converters, 6-pulse bridge converters with star-connected transformers, 6-pulse bridge converters with interphase transformers, and so on. The choice of converter topology depends on the application. Figures 7.28 and 7.29 show a series connection for high voltage and a parallel connection for high-current applications. Figure 7.30 shows a typical scheme of HVDC power transmission line,

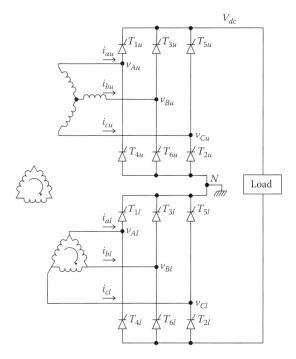

FIGURE 7.28 Series connection for high-voltage applications.

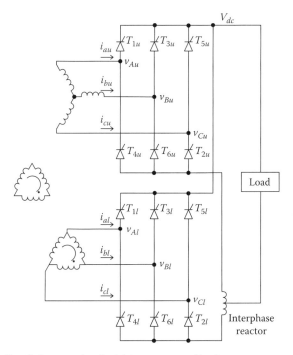

FIGURE 7.29 Parallel connection for high-current applications.

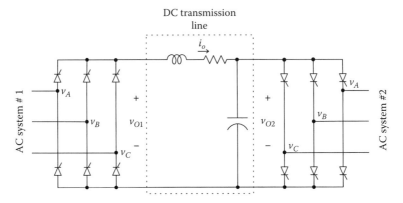

FIGURE 7.30   A six-pulse unipolar dc transmission system.

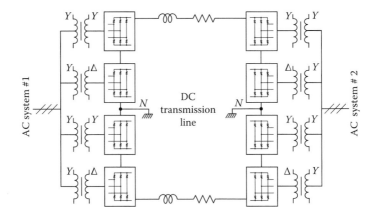

FIGURE 7.31   A 12-pulse bipolar dc transmission system.

and Figure 7.31 shows a fault-tolerant scheme for energy transmission where in emergencies, one pole of the line can operate without the other pole with current returning through the ground path. For a detailed description of these converters, see Kimbark.[4,5]

7.6   DC TO AC CONVERSION

Inverters perform conversion from dc to ac. A general description of the operation of inverter topologies is valid for any power electronic device; the differences will be in hardware implementation, switching frequency, gate drivers, and, of course, power ratings. Inverters require electronic control of the pulses applied to their gates. Such control needs a prescribed pulse train that will command the frequency and voltage of the output voltage. There are several PWM techniques and space vector techniques to synthesize the output voltage.[5–8]

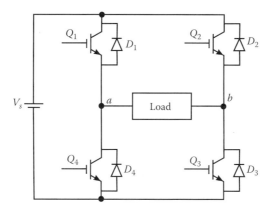

FIGURE 7.32 H-bridge inverter.

7.6.1 SINGLE-PHASE H-BRIDGE INVERTER

The schematic diagram of a single-phase H-bridge inverter is shown in Figure 7.32. It is called an H bridge because of the two vertical legs connecting the horizontal load. The inverter contains four switches: Q_1, Q_2, Q_3, and Q_4.

Q_i is a power device with an antiparallel diode. The load, which in this case is assumed a passive RL load, is connected between the two legs of the inverter. The inverter is supplied by a dc source with a voltage of V_{dc}. The switches of each leg usually have complementary values; that is, when Q_1 is on, Q_4 is off and vice versa. Furthermore, the switches are operated in pairs. When switches Q_1 and Q_2 are on, Q_3 and Q_4 are off. Similarly, when Q_3 and Q_4 are on, Q_1 and Q_2 are off. However, a minor delay time is provided between the turning off of a pair of switches and the turning on of the other pair. This period, called the blanking period, is provided to prevent the dc source from being short-circuited. Consider, for example, the transition when Q_3 and Q_4 are turned off and Q_1 and Q_2 are turned on. During this period if Q_1 gets turned on before Q_4 turns off completely, then it will connect the two leads of the dc source directly. It is therefore mandatory that Q_4 turns off completely before Q_1 is turned on. To ensure this, a blanking period (dead time) is used. The continuity of the current during the blanking period is maintained by the antiparallel diodes. Although the blanking period is very important in practice, for evaluation of operation, it is usually neglected.

7.6.2 THREE-PHASE INVERTER

The schematic diagram of a three-phase VS inverter is shown in Figure 7.33. It contains six switches: Q_1, Q_2, Q_3, Q_4, Q_5, and Q_6; each one is made up of a power semiconductor device and an antiparallel diode. The switches of each leg are complementary: that is, when Q_1 is on, Q_4 is off and vice versa. The inverter is connected to a dc link, and a three-phase load is connected at the output of the inverter. For a floating neutral point, the phase currents will add up to zero, and no triple harmonic current will flow in the load. Each phase leg must be driven with dead-time embedded

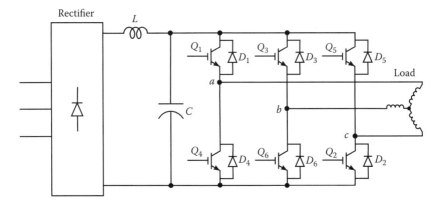

FIGURE 7.33 Three-phase VS inverter.

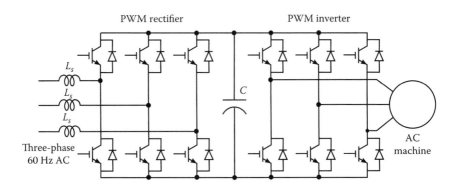

FIGURE 7.34 Double PWM converter.

signals to avoid the shoot-through fault. The inverter can be controlled by hysteresis PWM with sinusoidal tracking or by space vector modulation. Double PWM converters connected back-to-back have been used for ac–ac conversion with an intermediate dc-link voltage that must be boosted to a higher voltage level capable of imposing linear operation to both converters. Double PWM converters (shown in Figure 7.34) have been used for bidirectional power controllers and in applications dominated by electronic transformers, that is, ac–ac transformation without magnetic coupling.

7.6.3 MULTISTEP INVERTER

Multistep inverters have been around for several years for GTO applications. They are constructed using a series of six-step inverters and a magnetic circuit to provide phase shift between the inverters. For example, by providing a phase shift of 30° between two inverters with a transformer connection, it is possible to generate a 12-step waveform with a spectrum that contains 11th and higher harmonics only. By combining two of such inverters, a 24-step inverter is achieved. With a series–parallel operation, the devices require matching, and some amount of voltage or current derating

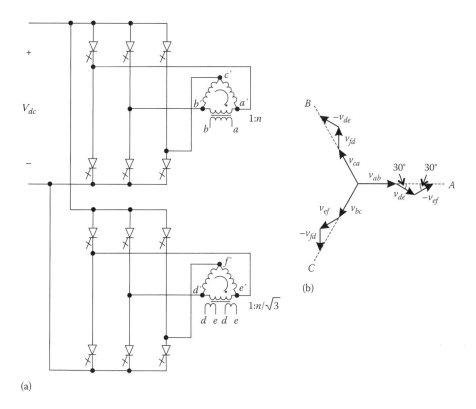

(a)

(b)

FIGURE 7.35 Example of a 12-step inverter: (a) circuit topology and (b) synthesis of voltages.

of the devices is essential. A possible solution for a higher power rating is a parallel connection of three-phase inverters through center-tapped reactors at the output. For large power applications, it is desirable that the inverter output wave be multistepped (approaching a sine wave) because the filter size can be reduced on both the dc and ac sides. The lower order harmonics of a 6-stepped wave (the fifth and seventh) can be neutralized by synthesizing a 12-stepped waveform as depicted in Figure 7.35. Multistep inverters are used for bulk power transmission systems.

7.6.4 MULTILEVEL INVERTER

The inverter configuration shown in Figure 7.36 is that of a three-level inverter as the output voltage can take only two values, $+V_{dc}/2$ and $-V_{dc}/2$. Multilevel inverters take more levels at the output through a special configuration of diodes and split voltages in the dc link. A very straightforward and typical configuration is depicted in the three-level phase leg shown in Figure 7.37, capable of generating $+V_{dc}/2$, 0 and $-V_{dc}/2$. It contains four switches (S_{ua}, S_{ub}, $S_{\ell a}$, and $S_{\ell b}$), each consisting of a power device plus a freewheeling diode and two diodes to impose the zero level (D_1 and D_2). A three-phase inverter is made of three similar structures as in Figure 7.37. Table 7.3 shows the output voltage of a three-level inverter with respect to the switch command.

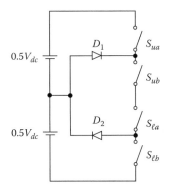

FIGURE 7.36   Phase-leg circuit for a three-level inverter.

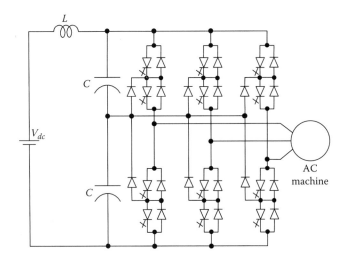

FIGURE 7.37   Three-level three-phase inverter.

TABLE 7.3
Synthesis of Three-Level Voltage

Switch Command				Phase Voltage
S_{ua}	S_{ub}	$S_{\ell a}$	$S_{\ell b}$	V_{AN}
Off	Off	On	On	$-V_{dc}/2$
Off	On	On	Off	0
On	On	Off	Off	$+V_{dc}/2$

Multilevel inverters are constructed with capacitors splitting the dc-link voltage and with corresponding command signals for the devices that synthesize the required output voltages. Multilevel inverters demonstrated a potential for harmonic cancellation, but they have not been widely accepted due to their high-voltage wiring complexity and cost.

7.7 DIRECT AC TO AC CONVERSION

Direct ac–ac conversion was a strong solution using a thyristor for phase-controlled regulation and a phase-controlled cycloconverter during the 1960s. With the fall of power semiconductor prices, dc-link cascaded converters became a reality during the 1990s. Recently, the high-frequency matrix converter for direct conversion has been receiving worldwide attention. The matrix converter offers an all-silicon solution for ac–ac power conversion, removing all need for the reactive energy storage components used in conventional inverter-based converters like capacitors and inductors. An n-phase-to-m-phase matrix converter consists of an array of $n \cdot m$ bidirectional switches as shown in Figure 7.38. The power circuit is arranged in such a fashion that any of the output lines of the converter can be connected to any of the input lines. Thus, the voltage at any input terminal can be made to appear at any output terminal or terminals while the current in any phase of load can be drawn from any input phase or phases. A line filter is included to circulate high-frequency switching harmonics. The switches are modulated to generate the desired output-voltage waveform.

The matrix converter concept was first introduced in 1976; it was then considered to be a cycloconverter where the devices were fully controllable. Hence, the matrix converter is sometimes called a forced commutated cycloconverter. A lot of research interest followed the publication of Venturini's first matrix converter paper

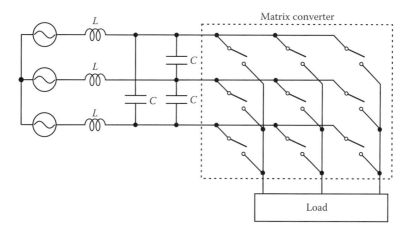

FIGURE 7.38 Three-phase to three-phase matrix converter.

in 1980, which put the matrix converter control algorithm on a strong mathematical foundation. The advantages of this converter are given as follows:

- No dc-link capacitor or inductor. Therefore, noise, electromagnetic interference (EMI), size, and weight are greatly reduced.
- Less maintenance, more durable.
- Inherently bidirectional, so it can regenerate energy back to the utility.
- High efficiency as the number of devices connected in series is less.
- Four quadrants of operation.
- Depending on modulation technique, sinusoidal input/output waveforms.
- Controllable input DPF independent of output load current.
- No electrolytic capacitors, hence can be used in high-temperature surroundings. An ideal topology to utilize future technologies such as high-temperature silicon carbide (SiC) devices.
- Small weight and volume.

A matrix converter requires a bidirectional switch capable of blocking voltage and conducting current in both directions. Unfortunately, there are no such devices currently available to fulfill the need. So a combination of available discrete devices is generally used to get the bidirectional switches. There are some common arrangements for getting these types of switches; each has advantages and disadvantages.

7.7.1 Diode-Bridge Arrangement

The diode-bridge arrangement consists of an IGBT at the center of a single-phase diode bridge shown in Figure 7.39. The main advantage of this configuration is only one gate driver per commutation cell. The direction of current through the switch cannot be monitored, which is a disadvantage, as many highly reliable commutation methods described in the literature cannot be used.

7.7.2 Common-Emitter Antiparallel IGBT Diode Pair

The common-emitter antiparallel IGBT diode pair arrangement is shown in Figure 7.40. It consists of two diodes and two IGBTs connected in antiparallel. Here it is possible to control the direction of the current in each of the devices independently. Each of the switch cells needs an isolated power supply for the gate drives.

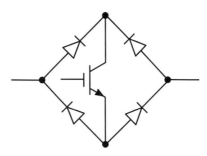

FIGURE 7.39 Diode-bridge bidirectional switch.

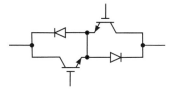

FIGURE 7.40 Common-emitter back-to-back switch.

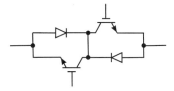

FIGURE 7.41 Common-collector back-to-back switch.

7.7.3 COMMON-COLLECTOR ANTIPARALLEL IGBT DIODE PAIR

The common-collector antiparallel IGBT diode pair, shown in Figure 7.41, is similar to the previous structure, but here, the IGBTs are arranged in the common-collector configuration. Here, fewer isolated gate drives are required than with the common-emitter arrangement. However, practical implementation of this configuration makes the common-emitter configuration the preferred one. The matrix converter requires very intensive digital signal computation to enable the correct connection of the power devices and very specialized modulation techniques that are beyond the scope of this book. It is expected that future systems will pervasively incorporate matrix converters for controlling ac machines and ac systems.

7.8 POWER ELECTRONICS TO REDUCE SELF-EXCITATION CAPACITANCE

Converters and power electronic circuits can be useful in decreasing the size of capacitors for self-excitation. Features such as low cost, robustness, ruggedness, high power density, commercial availability, and easy replacement make the IG a viable alternative for small stand-alone energy systems. However, an induction machine working as a self-excited induction generator (SEIG) requires reactive capacitance for providing the necessary reactive power during its operation.[9] In stand-alone mode operation, any SEIG has poor voltage and frequency regulation during load and primary source variations more evident when feeding inductive loads. For rated output power, say, above 50 kW, the cost of IG plants may become vastly affected by the self-excitation capacitor bank limiting the use of SEIG in high-power applications.[10] With regards of that, many studies have been conducted to simultaneously improve the voltage and frequency operating ranges. Among these studies of cost reduction is the establishment of simple techniques using reactive switched banks for voltage regulation.[11–13]

Early developments in IGs were possible with fixed capacitors used for excitation, since suitable active power devices were not available. This resulted in unstable power output since the excitation could not be adjusted as the load or speed deviated from the nominal values. This approach became possible only for large power systems with very firm voltage and frequency. With the availability of high-power switching devices, the IG was provided with appropriate controls for an adjustable excitation in stand-alone operation.

The IG has two main electromagnetic components: (1) the rotating magnetic field constructed using high conductivity, high strength bars located in a slotted iron core to form a squirrel cage, (2) and the stationary windings. Figure 7.42 shows the construction of a typical squirrel cage IG in a cross-sectional view.

The output multiple-phase ac voltage from the IG can be regulated using power converters. The control of the voltage is accomplished in a closed loop operation where the excitation current is adjusted to generate constant output voltage regardless of the variations of speed and load current. The excitation current is supplied to the stationary armature winding from which it is induced into the short-circuited squirrel cage secondary winding in the rotor. The excitation current, its magnitude, and its frequency are determined by the control system.

Switched power converters seem to be one way to reduce the self-capacitance size. The current technological alternatives for reactive load compensation to improve voltage and frequency regulation using static compensators (STATCOM) are dealt in the technical literature.[14,19] However, such compensators are based on a three-phase fully controlled converter, where the capacitive reactive power is supplied to the IG by a converter control using PWM modulation techniques. The generated voltage and the adjustment of the IG self-excitation are regulated by modulating controls according to the generator instantaneous speed and load. Even so, this method uses large electrolytic capacitors with auxiliary loads, increasing costs again for a reduced reliability. Another possibility is to provide a stable inverter output voltage at a fixed frequency by connecting the IG to a cascaded uncontrolled bridge rectifier and inverter.[17,18] Such systems are seen as a nonlinear load for the IG, draining currents with large harmonic contents, which also causes voltage distortion across the

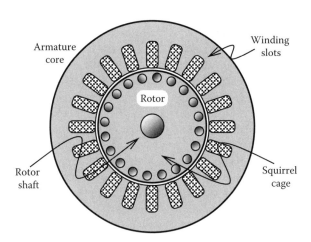

FIGURE 7.42 Squirrel cage IG cross-sectional view.

IG terminals. Thus, larger self-excitation capacitor banks are used, and they eventually increase core losses and they cause mechanical vibrations and increased costs.

Reactive compensation may improve the IG energy quality with minimization of the installation costs. One simplified method to maintain the regulated load voltage and frequency with small self-excitation capacitor is the use of an indirect matrix converter (IMC) between the SEIG and the load.[19–22] Such topology allows for load voltage regulation, frequency control, and minimization of the self-excitation capacitors, especially when the SEIG is feeding inductive loads, which is the most ordinary case. This association increases the IG capacity to supply inductive loads without increasing the capacitor bank and keeps the load frequency quite close to its rated value. Such a method works well only with stand-alone loads of resistive and inductive characteristics, and it is not recommended to supply nonlinear and unbalanced loads because they degrade the generated voltage and, consequently, the voltage supplied to the load.

The dynamic model of the induction machine as developed by Szczesny and Ronkowsky[20] has been used in minimization studies of self-excitation capacitance. Also, inclusion of the saturation effects in the magnetic core can be added to this mathematical model as proposed by Marra and is discussed in the following.[21]

7.8.1 SEIG–IMC Connection

The IMC connected across the SEIG terminals may be used to reduce the size of capacitive banks. It consists of a fully controlled rectifier stage feeding an inverter, without any dc-link energy storage element as depicted in Figure 7.43. The IG feeds the controlled rectifier stage with a reduced self-excitation capacitor bank. The rectifier in its turn will provide power to the dc link, decoupling generator output and load with respect to the frequency. A rectifier dc link voltage controlled by a sinusoidal pulse width modulation (SPWM) strategy to regulate output load voltage and frequency feeds the inverter. The rectifier is controlled by a PWM technique in order to deliver a sinusoidal current to the input with low harmonic distortion and high PF.

When connecting nonlinear loads to IGs, a high harmonic content of current and voltage is injected in the machine, causing losses and high level of magnetic

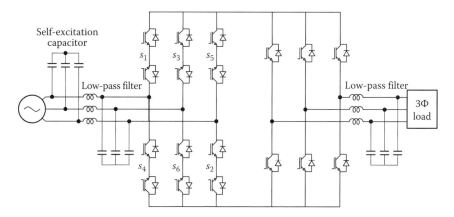

FIGURE 7.43 IMC power circuit connected between SEIG and a load.

irradiation.[19] Therefore, it is important that the IMC compensates the harmonic content producing a nearly sinusoidal current giving resistive characteristic to the rectifier as seen by the IG terminals. This association also allows the inverter control to provide regulated voltage to the load at a constant frequency.

7.8.2 POWER ELECTRONIC CONTROLS OF THE IMC

The rectifier used as a capacitor bank to self-excite the IG must have capacity enough to drain load current with high PF and low harmonic content. Therefore, the IMC control must be designed to provide an IG current in phase with the phase voltage. The presence of an LC low-pass filter across the IMC input causes a current leading the phase voltage, contributing with reactive power to the generated voltage taken from the capacitor filter C_{fi}. The IMC will supply the balanced load with balanced three-phase voltages and low harmonic contents if the rectifier PWM is synchronized with the IG space vector by a phase locked loop (PLL).

Transition states of the IMC bidirectional rectifier switches s_1, s_2, and s_6 at the turning-on instants are illustrated in Figure 7.43. For the circuit illustrated in Figure 7.43, one can observe that there are always three branches conducting current. With this particular timing, current i_a will flow from the bottom rail to the top rail, circulating through the outside circuit (inverter and load subsystems). At this point, current i_a will be split into currents i_b and i_c from the bottom rail. Because of the many possible combinations for all individual switches in this circuit, a particular attention must be taken to avoid short circuit between positive and negative rails (VS) to allow switches to turn on to establish freewheeling paths (CS).

The IMC control can be split into the rectifier-side control and inverter-side control, although the inverter control will be dependent on the dc link voltage. Furthermore, the dc link voltage can be obtained through the IG voltages by PLL synchronization because the IG phase voltages are reflected on the dc link.

Matrix converters are very complex systems and may lead to generator oscillation problems caused, particularly by the filter across the generator's terminals and the lack of dc link energy storage elements. The control scheme must precisely balance the instantaneous load power from the generator. For nonlinear loads, that is, with high harmonic content, a deep degradation of the generated voltage occurs, which is very difficult to compensate in real time. Therefore, this system cannot handle nonlinear loads.[21,22] With an appropriate PWM control, the rectifier and inverter may provide sinusoidal voltage and current outputs at a low total harmonic distortion (THD). However, a correct control design will have to keep the voltage gain and the time constant of the input filter within appropriate limits, thus guaranteeing a stable operation of the matrix converter.[28] The IMC maximum voltage gain (inverter output voltage related to the generated voltage) can be typically around 0.5.

The IMC control has a unit PF and harmonic sinusoidal waveform slightly leading the IG phase voltage. Therefore, the IMC impresses a kind of resistive electrical load to the IG. As a result, the IG voltage regulation is significantly improved for inductive loads. The design also allowed constant voltage across resistive loads, since the inverter stage control compensates generated voltage variations. In addition, the output voltage frequency can synthesize a constant frequency of 60 Hz.

The inductive load used in some practical results made possible a reduction of about 37% in the self-excitation capacitance, including the IMC when compared with the same SEIG voltage drop without IMC.[21–24] The same occurs with the simulated results. With purely inductive loads, this reduction can be even larger.

7.9 STAND-ALONE INDUCTION GENERATOR SCHEMES

The stand-alone operation allows IG-based power supplies in locations away from the public network where there is a primary source sufficient for energy self-genera-tion. An integrated form of these small power plants is suggested in Figure 7.44. This scheme ensures a safe range of voltage and frequency that meet loads commonly found in urban and rural areas. This scheme consists of a battery self-excited IG with the necessary reactive power to be attended by a STATCOM with the following modes of operation: (1) IG directly connected to the load, (2) battery bank directly supplying the load using a voltage inverter, and (3) battery bank acting in parallel with the IG to meet temporarily high loads. There is a circuit breaker between the generator and the load to open the direct connection when the inverter becomes the load on its inverter mode. The battery bank in the dc bus can store the surplus energy generation to meet loads in the temporary absence of wind and improve the stabilization of the dc voltage, the generated voltage, and the speed of the turbine/generator set.[22,25,26]

In the scheme of Figure 7.44 are used two three-phase IGBT fully controlled converters and a small bank of excitation capacitors. The drive connected directly to the generator acts as a STATCOM controlling the excitation of the IG, the amplitude of the generated voltage, frequency, and the level of charge/discharge of the bat-tery bank. The load connected to the converter operates as an inverter to meet the load from the dc bus but also operates as an active filter to compensate for possible unbalanced and nonlinear loads. The differential of this scheme with respect to other stand-alone schemes is that the load remains connected to the generator even under variable voltage and frequency. Another difference is the presence of a converter working in two ways, one as an inverter and the other as an active filter.

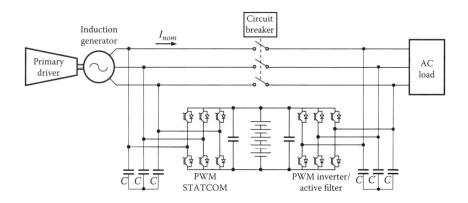

FIGURE 7.44 IG in a wind system connected to the utility grid.

7.10   SOLVED PROBLEMS

1. Research manufacturers' datasheets and prepare a table comparing the devices SCR, GTO, IGCT, TRIAC, Power-MOSFET, IGBT, and MCT with respect to the following parameters: maximum voltage and current ratings, symmetric or asymmetric voltage blocking, voltage or current gating, junction temperature range, conduction drop, drop sensitivity with temperature, SOA, switching frequency, turn-off gain, reapplied *dv/dt*, turn-on time, turn-off time, leakage current, snubber/ protection needs, and main application

Solution

In Table 7.4, a comparison for all devices listed earlier has been made. Of course, such comparison may change over the years, depending on the availability of power electronic devices in the market.

2. Discuss methods of cooling power converters. What do you understand about thermal resistance of a heat sink?

Solution

Heat exchange is a major characteristic of any power generation scheme. As the induction machine is usually a local type of power generation needing conversion of its ac voltage and frequency by power converters, it produces about 10% heat. This heat can be just dissipated away or associated to some useful scheme. Of the three major heat exchanger mechanisms, convection and conduction are the major interest to the power converter designer. Radiation within the supply is generally a nuisance, as heat radiated away by one component is generally absorbed by adjacent components. In high-power conversion, the phenomenon is very much pronounced. Some very common cooling methods for high-power converters are discussed now.

Convection cooling

If a free airflow is available, then convection or forced air-cooling is by far the most cost-effective way of removing unwanted heat. Heat exchangers with good forced air-cooling properties generally have a large effective surface area because of having many cooling fins.

Convection cooling becomes less effective at high altitudes because of the reduction of air density. It should be noted that there is a 20% reduction in cooling efficiency at 10,000 ft.

Further, in nonforced convection cooling, the thermal resistance of the heat exchanger is not linearly proportional to the size. That means larger surfaces will not be as effectively cooled, as the air will be heated as it passes over the surface of the exchanger. For example, the thermal resistance of a vertical finned extrusion decreases with length, and very little improvement is obtained beyond a length of 12 in.

Conduction cooling

Where very little airflow is available, conduction cooling is a viable alternative. For conduction cooling, the finned heat sinks are replaced by thermal shunts (bridges) between the heat-generating components and the chassis. The thermal conduction

TABLE 7.4
Comparison among Devices for Problem # 1

Device	SCR	GTO	IGCT	TRIAC	MOSFET	IGBT	MCT
Part name	5STP50Q1800	5SGA4014501	5SHY3514510	BTA225B800	FGA11N90	FGA15N120AN	MCTG35P60F1
Manufacturer	ABB	ABB	ABB	Philips	Fairchild	Fairchild	Harris
Maximum voltage rating	1800 V	4500 V	4500 V	800 V	900 V	1200 V	600 V
Maximum current rating	6100 A	4000 A	4000 A	25 A	11.4 A	15 A	35 A
Voltage blocking	1880 V	4500 V	Asym. 4500 V_{ac}	800 V	900 V	1200 V	600 V
Gating type	400 mA, pulse	1.2 V, 4 A, pulse	24–40V_{ac}	0.1 A	5 V	5.5 V	
Junction temperature	125°C	125°C	125°C	125°C	150°C	150°C	150°C
Conduction drop	1.04 V	4.4 V	2.7 V	1.2 V	1.4 V	3 V	1.35 V
SOA					Rectangle	Rectangle	
Swtiching frequency	50 Hz	200 Hz	1000 Hz	300 Hz	1 MHz	50 kHz	10 kHz
Reapplied dv/dt	Exp. to 0.67 * V_{max}			4000 V/µs	4 V/ns		50 V/µs
Turn-on time	3 µs (gate)	100 µs	10 µs	3	260 ns	370 ns	1.7 ns
Turn-off time	500 µs (gate)	100 µs	10 µs		500 µs	500 µs	500 µs
Leakage current	300 mA	100 mA	50 mA	0.1 mA	100 µA	3 mA	1.5 mA
Snubber protection	No	Yes	No	No	No	No	
Main application	HVDC	Medium and high power		Lighting	HF SMPS	Motor drive	Resonant converter

properties of the material chosen for the heat shunts are important. The chassis in turn must have a good thermal contact to some external heat exchanger, for example, the case of the equipment.

By using temperature differences and a rate of heat dissipation, a quantitative measure of heat transfer efficiency across two components can be expressed in terms of the thermal resistance R, defined as

$$R = \frac{\Delta T}{\dot{Q}}$$

where ΔT is the temperature difference between the two locations.

This thermal resistance is analogous to the electrical resistance R_e, given by Ohm's law:

$$R_e = \frac{\Delta V}{I}$$

with ΔV being the voltage difference and I the current.

The unit of thermal resistance is usually given in °C/W, indicating the temperature rise per unit rate of heat dissipation.

For conduction cooling, it should be remembered that aluminum is only half as good a thermal conductor as copper, and steel is 25% as good as aluminum.

Radiation cooling

As previously explained, radiation is not usually a very effective method of cooling in power conversion. Radiant heat is an electromagnetic wave phenomenon and as such travels in straight lines, and good *line of sight* free radiant paths are not often available in power converters. Radiant energy from hot spots falling on case or other components either is reflected back or simply raises the temperature of the other components and the environment. However, when a good radiant path can be established, this mode of cooling should be considered.

Again, poor radiators are also good reflectors, and this can be used to advantage in protecting heat-sensitive components from nearby hot components. For example, polished aluminum foil can be placed between a hot resistor and an electrolytic capacitor to reduce the radiant heating of the capacitor. Since the reverse is also true, good radiant heat dissipators are good heat absorbers, and power converters designed for radiant cooling should be kept away of direct sunlight.

Forced-air cooling

Natural convection heat sinks are normally useful only for low power applications where very little heat is involved. Although it is difficult to generalize, most natural convection heat sinks have a thermal resistance (R_e) greater than 0.5°C/W and often exceeding 10°C/W. Probably, the most common cooling method used with converters is the forced convection. When compared with natural convection heat sinks, a substantially better performance can be realized. The thermal resistance of quality forced convection systems typically falls within a range of

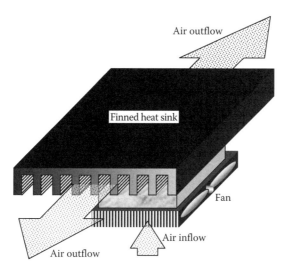

Air outflow

Finned heat sink

Fan

Air inflow

Air outflow

FIGURE 7.45 Forced convection heat sink system showing preferred air flow.

0.02°C/W–0.5°C/W. Many standard heat sink extrusions are available that, when coupled with a suitable fan, may be used to form the basis of a complete cooling assembly. Cooling air may be supplied from a fan or blower and may be passed totally through the length of the heat sink or may be directed at the center of the fins and pass out both open ends. This second airflow pattern, illustrated in Figure 7.45, generally provides the best performance since the air blown into the face of the heat sink creates greater turbulence, resulting in improved heat transfer. For optimum performance, the housing of an axial fan should be mounted a distance of 8–20 mm (0.31–0.75 in.) from the fins. Other configurations may be considered depending on the application.

Apart from the obvious improvement provided by the more rapid exchange of air, the flow direction can be used to advantage by placing hot components in the exhaust port so that heat is carried away from the rest of the unit. Further, airflow can be directed to prevent any static air buildup, an effect that is difficult to avoid in free-flow convection-cooled power converters.

Power converters with outputs of 500 W or more generally have some form of forced-air cooling. The amount of air required for cooling depends on the air density, the power to be dissipated, and the permitted temperature rise. At sea level, the following equation can be applied:

$$\text{Airflow (cfm)} = \frac{1.76 \times W_{loss}}{\Delta T}$$

where
  cfm is the cubic feet per minute
  W_{loss} is the internal loss, W
  ΔT is the manufacturer-permitted internal temperature rise, °C

The fan must overcome the backpressure that results from the restriction to the flow within the converter enclosure. The backpressure depends on the size and packing density, and it will normally be on the order of 0.1–0.3 in. of water per 100 cfm. This parameter is best measured in the finished unit, and a fan is selected to give the required airflow at the measured backpressure.

Liquid-cooled heat sinks

Liquid-cooled heat sinks provide the highest thermal performance per unit volume and, when optimally designed, can exhibit a very low thermal resistance. Although there are many exceptions, the thermal resistance of liquid-cooled heat sinks typically falls between 0.01°C/W and 0.1°C/W.

Today, as more and more compact converters with high current rates are designed in smaller cases, the thermal requirement is the key point to meet these requirements: the heat sink must be integrated as close as possible from heat sources. By integrating heat sinks just under the IGBT chips, cooling can be improved, which leads to an increase in the compactness and lifetime of the module. There are some technical papers showing the effectiveness of liquid-cooled microchannel heat sinks for power microchip modules.

In the final analysis, the mechanical and thermal design of the power converter and enclosure must be capable of removing waste heat from the converter and any load within the enclosure, under all operating conditions. When in doubt, the temperatures of critical components must be measured once thermal equilibrium has been established in the working environment. Briefly, effective cooling methods must be used for long-term reliability of the power converter.

Thermal resistance of a heat sink

With the increase in heat dissipation from microelectronic devices and the reduction in overall form factors, thermal management becomes a more and more important element of electronic product design. Both the performance reliability and life expectancy of electronic equipment are inversely related to the component temperature of the equipment. The relationship between the reliability and the operating temperature of a typical silicon semiconductor device shows that a reduction in the temperature corresponds to an exponential increase in the reliability and life expectancy of the device. Therefore, long life and reliable performance of a component may be achieved by effectively controlling the device operating temperature within the limits set by the device design engineers.

Heat sinks are devices that enhance heat dissipation from a hot surface, usually the case of a heat-generating component, to a cooler ambient, usually air. For the following discussions, air is assumed to be the cooling fluid. In most situations, heat transfer across the interface between the solid surface and the coolant air is the lead efficient within the system, and the solid–air interface represents the greatest barrier for heat dissipation. A heat sink lowers this barrier mainly by increasing the surface area that is in direct contact with the coolant. This allows more heat to be dissipated and/or lowers the device operating temperature. The primary purpose of a heat sink is to maintain the device temperature below the maximum allowable temperature specified by the device manufacturer.

Thermal circuit

Let us define the following parameters:

$\dot{Q}$: Total power or rate of heat dissipation in W, representing the rate of heat dissipated by the electronic component during operation. For selecting a heat sink, the maximum operating power dissipation is used.

T_j: Maximum junction temperature of the device in °C. Allowable T_j values range from 115°C in typical microelectronic applications to as high as 180°C for some electronic control devices; in special military applications, 65°C–80°C are common.

T_C: Case temperature of the device in °C. Since the case temperature of a device depends on the location of measurement, it usually represents the maximum local temperature of the case.

T_S: Sink temperature in °C. Again, this represents the maximum temperature of a heat sink at the location closest to the device.

T_a: Ambient air temperature in °C.

Thermal resistance circuit

Consider a simple case where a heat sink is mounted on a device package as shown in Figure 7.46. Using the concept of thermal resistance, a simplified thermal circuit of this system can be drawn, as also shown in the figure. In this simplified model, heat flows serially from the junction to the case, then across the interface into the heat sink, and finally dissipated from the heat sink to the air stream.

The thermal resistance between the junction and the case of a device is defined as

$$R_{jc} = \frac{\Delta T_{jc}}{\dot{Q}} = \frac{T_j - T_c}{\dot{Q}}$$

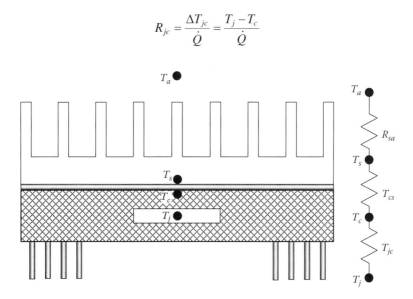

FIGURE 7.46 Heat sink mounted on a device package.

This resistance through R_{jc} is specified by the device manufacturer. Although the value R_{jc} of a given device depends on how and where the cooling mechanism is employed over the package, it is usually given as a constant value. It is also accepted that R_{jc} is beyond the user's ability to alter or control.

Similarly, case-to-sink and sink-to-ambient resistances are defined as

$$R_{cs} = \frac{\Delta T_{cs}}{\dot{Q}} = \frac{T_c - T_s}{\dot{Q}}$$

$$R_{sa} = \frac{\Delta T_{sa}}{\dot{Q}} = \frac{T_s - T_a}{\dot{Q}}$$

Here, R_{cs} represents the thermal resistance across the interface between the case and the heat sink and is often called the interface resistance. This value can be improved substantially depending on the quality of mating surface finish and/or the choice of interface material. R_{sa} is the heat sink thermal resistance.

Obviously, the total junction-to-ambient resistance is the sum of all three resistances:

$$R_{ja} = R_{jc} + R_{cs} + R_{sa} = \frac{T_j - T_a}{\dot{Q}}$$

Required heat sink thermal resistance

In the heat sink selection, the first step is to determine the heat sink thermal resistance required to satisfy the thermal criteria of the component. By rearranging the previous equation, the heat sink resistance can be easily obtained as follows:

$$R_{sa} = \frac{T_j - T_a}{\dot{Q}} - R_{jc} - R_{cs}$$

In the expression, the device manufacturer provides T_j, $\dot{Q}$, R_{jc}, and T_a and R_{cs} are user-defined parameters.

The ambient air temperature T_a for cooling electronic equipment depends on the operating environment in which the component is to be used. Typically, it ranges from 35°C to 45°C, if the external air is used, and from 50°C to 60°C, if the component is enclosed or is placed in a wake of another heat-generating equipment.

The interface resistance R_{cs} depends on the surface finish, flatness, applied mounting pressure, contact area, and, of course, the type of interface material and its thickness. Precise value of this resistance, even for a given type of material and thickness, is difficult to obtain, since it may vary widely with the mounting pressure and other case-dependent parameters. However, data that are more reliable can be obtained directly from material manufacturers or from heat sink manufacturers. Typical values for common interface materials are tabulated in Table 7.5.

TABLE 7.5

Thermal Properties of Interface Materials

Material	Conductivity (W/in.°C)	Thickness (in.)	Resistance (in.²°C/W)
Ther-O-link thermal compound	0.010	0.002	0.19
High-performance thermal compound	0.030	0.002	0.07
Kon-Dux	0.030	0.005	0.17
A-Dux	0.008	0.004	0.48
1070 Ther-A-Grip	0.014	0.006	0.43
1050	0.009	0.005	0.57
1080	0.010	0.002	0.21
1081	0.019	0.005	0.26
A-Pli22o - 20 psi	0.074	0.020	0.27
1897 In-Sil-8	0.010	0.008	0.81
1898 In-Sil-8	0.008	0.006	0.78

A-Pli™ is a "super-soft" low durometer material designed to fill gaps between hot components and their heat sinks or enclosure (plastic or metal). The flexible, elastic nature of this product allows it to blanket uneven surfaces, either individually or as a group. Heat is conducted away from the individual components, or an entire PCB, into metal covers, frames, or spreader plates.

With all the parameters on the right side of the preceding expression identified, R_{sa} becomes the required maximum thermal resistance of a heat sink for the application. In other words, the thermal resistance value of a chosen heat sink for the application has to be equal to or less than the above R_{sa} value for the junction temperature to be maintained at or below the specified T_j.

7.11 SUGGESTED PROBLEMS

To solve the following problems, you will need to consult power electronics books available in your library. You should do such research in order to complement what you have learned in this chapter.

7.1 IGs need dc–ac, ac–dc, and ac–ac converters. Describe one application, relevant for IG-based energy system, for each of those power conversions.

7.2 Where are dc–dc converters used in IG systems?

7.3 Assume $L_s = 0$ and an HIL (high inductive load) for the half-converter indicated in Figure 7.27. Calculate the relationship of the delay angle α with the average voltage across the load V_{dc}. Draw the output voltage waveform across the load and identify the devices that conduct during various intervals. Obtain the DPF and the PF in the input-line current and compare results with a full-bridge converter operating at $V_{dc} = 3\sqrt{2}V_{line,rms}/2\pi$.

7.4 Research the PWM technology and write a summary report.

7.5 Describe the features and characteristics of multilevel converters and multi-step converters.

7.6 Matrix converters require specialized modulation techniques. Conduct research and write a summary report.

7.7 A full-bridge three-phase rectifier supplies energy to a dc load of 300 V and 60 A from a three-phase bridge of 440 V through a ΔY transformer. Select a diode and specify the transformer for a voltage drop across each diode of 0.7 V and continuous current.

7.8 In a conventional thyristor 12-pulse converter, estimate the internal drop of voltage, regulation, and PF for the following data set.

Source	Load
$V = 220$ V	
$f = 60$ Hz	$\alpha = 20°$
$L = 0.1$ mH	$R_{load} = 20\ \Omega$
$R = 0.1\ \Omega$	$L_{load} = 0.2$ Mh

7.9 A single-phase diode bridge is supplied by an ac source of 257 V, 60 Hz. This bridge in turn supplies a dc load of $I_{dc} = 60$ A. Estimate the voltage drops due to (1) source with an inductance of 0.02 mH, (2) each diode with forward voltage drop of $\Delta V_f = 0.6 + 0.0015\ I_{dc}$, and (3) source and wiring resistances of $R_r = 0.0015\ \Omega$. Draw an equivalent circuit to represent such rectifier.

7.10 Research the circuit topologies for implementing HVDC in order to connect multiple offshore wind turbines or multiple tidal generators to the grid with an underwater cable.

REFERENCES

1. Mohan, N., Undeland, T.M., and Robbins, W.P., *Power Electronics: Converters, Applications, and Design*, 3rd edn., John Wiley & Sons, New York, 2002.
2. Erickson, R.W. and Maksimovic, D., *Fundamentals of Power Electronics*, Kluwer Academic Publishers, Dordrecht, the Netherlands, 2001.
3. Rashid, M.H., *Power Electronics: Circuits, Devices, and Applications*, Prentice Hall, New York, 2003.
4. Kimbark, E.W., *Direct Current Transmission*, Vol. 1, Wiley Interscience, New York, 1971.
5. Bose, B.K., *Modern Power Electronics and AC Drives*, Prentice Hall, New York, 2001.
6. Ghosh, A. and Ledwich, G., *Power Quality Enhancement Using Custom Power Devices*, Kluwer Academic Publishers, Dordrecht, the Netherlands, 2002.
7. Chattopadhyay, A., AC-AC converters. In: *Power Electronics Handbook*, edited by M.H. Rashid, Academic Press, San Diego, CA, 2001, pp. 327–331.
8. Youm, J.H. and Kwon, B.H., Switching technique for current-controlled AC-to-AC converters, *IEEE Trans. Ind. Electron.*, 46(2), 309–318, April 1999.
9. Bassett, D.E. and Potter, M.F., Capacitive excitation for induction generators, *Trans. American Inst. Electr. Eng.*, 54, 540–543, June 1935.
10. Ahmed, T., Noro, O., Hiraki, E., and Nakaoka, M., Terminal voltage regulation characteristics by static var compensator for a three-phase self-excited induction generator, *IEEE Trans. Ind. Appl.*, 40(4), August 2004.
11. Muljadi, E., Butterfield, C.P., Sallan, J., and Sanz, M., Investigation of self-excited induction generators for wind turbine applications, *IEEE Industry Applications Society, Annual Meeting*, Phoenix, AZ, October 1999.

12. Kuo, S.C. and Wang, L., Analysis of isolated self-excited induction generator feeding a rectifier load, *IEE Proc. Gener. Transm. Dislrib.*, 149, 90–97, January 2002.
13. Tanady, A.H., Modeling of self-excited induction generator and AC/DC/AC converter for energy conversion systems, School of Electrical and Computer Engineering, Curtin University of Technology, Bentley, Western Australia, Australia, 1987.
14. Marra, E.G. and Pomilio, J.A., Sistemas de geração baseados em gerador de indução operando com tensão regulada e freqüência constante, *SBA Controle & Automação*, 11(1), April 2000.
15. Singh, B., Murthy, S.S., and Gupta, S., STATCOM-based voltage regulator for self-excited induction generator feeding nonlinear loads, *IEEE Trans. Ind. Electron.*, 53(5), October 2006.
16. Iimori, K., Shinohara, K., and Yamamoto, K., A study of dead-time of PWM rectifier of voltage-source inverter without DC link components and its operating characteristics of induction motor, *IEEE, IAS*, 3, 1638–1645, 2004.
17. Kolar, J.W., Schafmeister, F., Round, S.D., and Ertl, H., Novel three-phase AC-AC sparse matrix converters, *IEEE Trans. Power Electron.*, 22(5), September 2007.
18. Chen, Y.-M., Hsieh, C.-H., and Cheng, Y.-M., Modified SPWM control scheme for three-phase inverters, *Proceedings of IEEE International Conference on Power Electronics and Drive Systems*, 2001, pp. 651–656.
19. Simões, M.G. and Farret, F.A., *Renewable Energy Systems: Design and Analysis with Induction Generators*, CRC Press, Boca Raton, FL, 2008, p. 456.
20. Szczesny, R. and Ronkowsky, M., A new equivalent circuit approach to simulation of converter—Induction machine associations, *European Conference on Power Electronics and Applications (EPE'91)*, Florence, Italy, 1991, pp. 4/356–4/361.
21. Marra, E.G., Gerador de indução associado a inversor PWM operando com Freqüência Constante, PhD thesis, Faculty of Electrical Engineering and Computation, State University of Campinas, Campinas, São Paulo, Brazil, 1999.
22. Trapp, J.G., Static excitation of induction generators, active filter and energy storage for standalone wind turbines, PhD thesis, Federal University of Santa Maria, Camobi, Santa Maria, Brazil, March 2013.
23. Vaidya, J. and Gregory, E., Advanced electric generator & control for high speed micro/mini turbine based power systems, patent pending, Electrodynamics Associates, Inc., Orlando, FL, 2009.
24. Trapp, J.G., Parizzi J.B., Farret, F.A., Serdotte, A.B., and Longo A.J., Stand alone self-excited induction generator with reduced excitation capacitors at fixed speed, *IEEE Brazilian Power Electronics Conference (COBEP)*, 2011, pp. 955–962, doi:10.1109/COBEP.2011.6085328.
25. Jayaramaiah, G.V. and Fernandes, B.G., Novel voltage controller for stand-alone induction generator using PWM-VSI, *Industry Applications Conference, 41st IAS Annual Meeting, Conference Record of the IEEE*, Vol. 1, 2006, pp. 204–208.
26. Perumal, B.V. and Chatterjee, J.K., Analysis of a self-excited induction generator with STATCOM/battery energy storage system, *IEEE Power India Conference*, 2006, doi: 10.1109/POWERI.2006.1632596.
27. Billings, K. and Taylor, M., *Switch Mode Power Supply Handbook*, 3rd edn., McGraw-Hill, New York, September 2010.
28. Lee, S., How to select a heat sink, A paper published by Aavid Thermal Technologies Inc., Laconia, NH, http://www.electronics-cooling.com/1995/06/how-to-select-a-heat-sink/, accessed on May 4, 2014.
29. Chakraborty, S., Simões, M.G., and Kramer, W., *Power Electronics for Renewable and Distributed Energy Systems: A Sourcebook of Topologies, Control and Integration*, Springer-Verlag, London, U.K., 2013.

8 Scalar Control for Induction Generators

8.1 SCOPE OF THIS CHAPTER

Scalar control of induction motors and generators means control of the magnitude of their voltage and frequency to achieve suitable torque and speed with an impressed slip. Scalar control can be easily understood based on the fundamental principles of induction-machine steady-state modeling. A power electronic system is used either for a series connection of inverters and converters between the induction generator and the grid or as a parallel path system capable of providing reactive power for isolated operation. This chapter describes such principles, laying out foundations for understanding the more complex vector-controlled systems.

8.2 SCALAR CONTROL BACKGROUND

Scalar control disregards the coupling effect on the generator; that is, the voltage will be set to control the flux and the frequency in order to control the torque.[1] However, flux and torque are functions of frequency and voltage, respectively. Scalar control is different from vector control in which both magnitude and phase alignment of the vector variables are directly controlled. Scalar control drives give an inferior performance, but they are not difficult to implement and are very popular in machine drives for pumping, several industrial applications, and large megawatt systems. The importance of scalar control has diminished recently because of the superior performance of the vector-controlled drives and the introduction of high-performance inverters, but pulse-width modulation (PWM) transistor–based inverters have a maximum power range. Scalar control is very useful with multilevel topologies because of their inherent lack of modulation and its flexibility when compared with PWM, although for three-level inverters, it is possible to implement them also with full PWM capabilities. High-performance inverters offer prices competitive to scalar control–based inverters (V/Hz inverters), although they are cheap and widely available.

The principles of scalar control are presented in this chapter along with a discussion of some simple enhancements that may be retrofit onto existing commercial V/Hz systems.

The main constraint on the use of a scalar control method for induction motors and generators is related to the transient response.[2] If shaft torque and speed are bandwidth-limited and torque varies slowly (within hundreds of milliseconds up to the order of almost a second), scalar control may be a good control approach. Hydropower and wind power applications have slower mechanical dynamics for

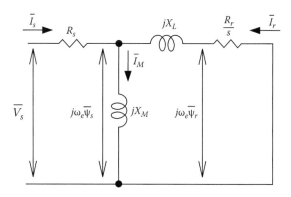

FIGURE 8.1   Γ-stator equivalent model.

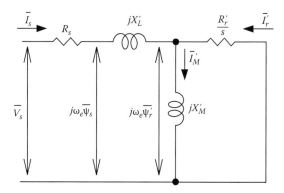

FIGURE 8.2   Γ-rotor equivalent model.

the scalar control method. Therefore, it seems that scalar control is a good control approach for renewable energy applications.

As a background to understanding the scalar control method, a steady-state two-inductance per-phase equivalent circuit for an induction motor can be used.[3] There are two models: the Γ-stator equivalent model, presented in Figure 8.1, and the Γ-rotor equivalent model, presented in Figure 8.2. The Γ-stator equivalent model is related to the ordinary per-phase equivalent model by the transformation coefficient $\gamma = X_s/X_m$, as shown in Table 8.1. The Γ-rotor equivalent model is related to the ordinary per-phase equivalent model by the transformation coefficient $\rho = X_m/X_r$, as in Table 8.2.

In the Γ-stator equivalent model, the stator flux ψ_s is considered to be approximated by the air-gap flux ψ_m, while the stator resistance R_s effects are usually neglected. Considering that $R_{r,\gamma}/s = R_{r,\gamma}\omega_e/\omega_r$, the rotor current in the Γ-stator equivalent model can be expressed as

$$|I_{r,\gamma}| = \frac{\psi_m}{R_{r,\gamma}} \frac{\omega_r}{\sqrt{\left(\tau_r\omega_r\right)^2 + 1}} \tag{8.1}$$

where $\tau_r = L_{l,\gamma}/R_{r,\gamma}$, $L_{l,\gamma}$ is the total leakage inductance in this model; that is, $L_{l,\gamma} = X_{l,\gamma}/\omega$.

TABLE 8.1

Γ-Stator Equivalent Parameters

Rotor resistance referred to stator	$R_{r,\gamma} = \gamma^2 R_r$
Magnetizing reactance	$X_{m,\gamma} = \gamma X_m = X_s$
Total leakage reactance	$X_{l,\gamma} = \gamma X_{ls} + \gamma^2 X_{lr}$
Rotor current referred to stator	$I_{r,\gamma} = \dfrac{I_r}{\gamma}$
Rotor flux	$\psi_{r,\gamma} = \gamma \psi_r$
Angular frequency of rotor currents (rotor frequency)	$\omega_r = s\omega_e$
Rotor angular frequency related to the slip frequency	$\omega_r = \dfrac{p}{2} \omega_{sl}$

TABLE 8.2

Γ-Rotor Equivalent Parameters

Rotor resistance referred to stator	$R_{r,\rho} = \rho^2 R_r$
Magnetizing reactance	$X_{m,\rho} = \rho X_m$
Total leakage reactance	$X_{l,\rho} = X_{ls} + \rho X_{lr}$
Rotor current referred to stator	$I_{r,\rho} = \dfrac{I_r}{\rho}$
Rotor flux	$\psi_{r,\rho} = \rho \psi_r$
Angular frequency of rotor currents (rotor frequency)	$\omega_r = s\omega_e$

When the resistive losses in the rotor $(3R_{r,\gamma}|I_{r,\gamma}|^2)$ is neglected, the electrical power transferred from the mechanical shaft to the rotor circuit is approximated by

$$P_{elec} = 3R_{r,\gamma} \frac{\omega}{\omega} |I_{r,\gamma}|^2 \approx P_{mech} \qquad (8.2)$$

The generator torque shaft can be calculated as the ratio of P_{mech} to the mechanical angular speed. For scalar voltage control, the angular shaft speed is leading the stator angular speed by $\omega_m = 2/P\omega_e(1 + s)$, and the developed electrical torque on the shaft is

$$T_e = 3 \frac{p}{2} \frac{\psi_m^2}{R_{r,\gamma}} \frac{\omega_r}{(\tau_r\omega_r)^2 + 1} \qquad (8.3)$$

For $\omega_r = 1/\tau_r$ the critical slip is $s_{critical} = 1/\tau_r\omega_e$, and the maximum torque developed is given by

$$T_{M,max} = 3 \frac{p}{2} \frac{\psi_m^2}{(L_{l,\gamma})}$$

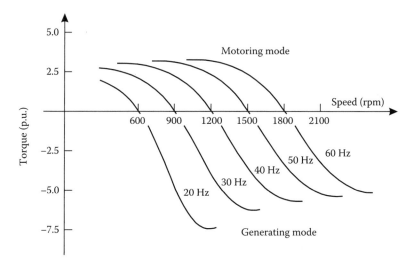

FIGURE 8.3   Torque/speed performance for constant V/Hz.

The machine will operate above (generator) and below (motor) the synchronous frequency ω_s as in Figure 8.3 with its operating slip well below the critical slip. Therefore, the denominator of Equation 8.3 is close to unity, and the torque is virtually proportional to the rotor angular frequency for constant flux. Equation 8.4 is derived from Equation 8.3 and transformed back to the standard machine T model, considering that flux is $\psi_m = V_s/\omega_e$.

$$T_e = 3\frac{p}{2}\frac{1}{R_r}\psi_m^2\omega_{sl} \tag{8.4}$$

Equation 8.4 indicates that for a fixed V/Hz ratio a characteristic curve is locked, and the torque crossover operating point changes in accordance with the impressed stator frequency as shown in Figure 8.3. Therefore, depending on the shaft speed, one point along the torque–speed curve will be developed. For negative slip, the machine will develop negative torque, and the operation will be as a generator. If variable speed is required, the V/Hz ratio must be maintained to keep rated air-gap flux. The firmness of air-gap flux results in good utilization of the motor iron, a high torque per stator ampere, and rated torque is possible for a wide speed range. It is important to note that in the previous discussion we neglected stator resistance and the saturation effect, both playing a very important role for a realistic quantitative evaluation. For motoring operation, the voltage drop across the stator resistance is opposed to the impressed voltage, and a boost scheme is required for low-frequency operation. For generating mode, the stator voltage is reversed, resulting in an increased back electromotive force (emf) with increased air-gap flux. In that case, the torque initially increases with speed reduction, but saturation of the magnetizing inductance offsets such an increase.

8.3 SCALAR CONTROL SCHEMES

The fundamental objective behind the scalar method is to provide a controlled slip operation. The scheme depicted in Figure 8.4 with a grid-connected induction generator is a possible one. The static frequency converter in the figure can be a cycloconverter, a matrix converter (bidirectional in nature), or a rectifier/inverter connection with a dc-link interfacing the 60 Hz grid with the generator stator. The simplified scheme shown in Figure 8.4 requires a programmed slip, which will be dependent on machine parameter variation, temperature variation, and mechanical losses. Therefore, a closed-loop control will improve the drive performance. Since the slip frequency ω_{sl} is proportional to torque, an outer speed control loop will generate a signal proportional to the required slip—for example, through a PI regulator.

The system portrayed in Figure 8.5 has an outer-loop speed control and a PI regulator that generates slip, which is added to the shaft speed to generate the stator frequency. The machine terminal voltage is also programmed through a look-up table. The converter receives both inputs ω_e^* and V_s^*, which command a three-phase sinusoidal generator by PWM in the inverter. The scheme considers that external torque is applied on the generator shaft, and the shaft speed reference ω_r^* is computed in order to seek a shaft operating speed that keeps the slip signal negative ω_{sl}^*, maintaining the machine in generating mode. The three-phase inverter receives a negative phase sequence command, and the power delivered across the battery is indicated in the block diagram. Since most systems are grid connected, a dc-link is

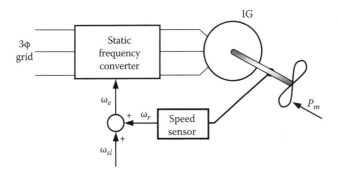

FIGURE 8.4 Principle of slip control for induction generators.

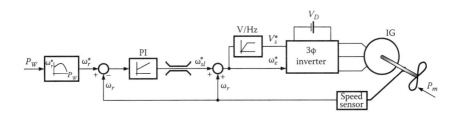

FIGURE 8.5 Closed-loop speed control of IG with V/Hz.

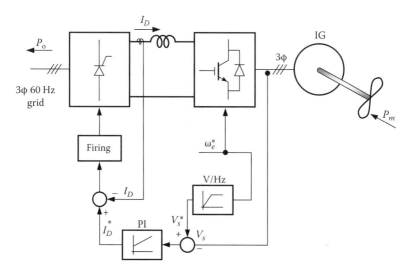

FIGURE 8.6   Current-fed link system for grid connection with V/Hz.

required to interface the machine converter to a grid inverter and pump energy back to the ac side, as indicated in Figure 8.6.

A current-fed link system for grid connection with V/Hz is depicted in Figure 8.6. The dc-link current allows easy bidirectional flow of power. Although the dc-link current is unidirectional, a power reversal is achieved by a change in polarity of the mean dc-link voltage, and symmetrical voltage-blocking switches are required. A thyristor-based controlled rectifier manages the three-phase utility side, and the machine-side inverter can use a transistor with a series diode. The system in Figure 8.6 is commanded by the machine stator frequency reference ω_e^*, and a look-up table for V/Hz sets a voltage reference, which is compared to the developed machine terminal voltage. A PI control produces the set point for the dc-link current, and firing for the thyristor bridge controls the power exchange (P_o) with the grid. The stator frequency reference ω_e^* can be varied in order to optimize the power tracking of the induction generator or may be programmed in accordance with the input power availability at the generator shaft (P_M).

An enhanced voltage-link double-PWM converter is depicted in Figure 8.7. The dc-link capacitor voltage is kept constant by the converter connected to the utility grid. The series inductances that connect the system to the grid keep the converter control within safe limits by inserting a current-source-like feature. When power is transferred from the induction generator, the dc-link voltage will increase slightly, and the feedback control of the grid-side converter will generate a sinusoidal pulse-width modulation (SPWM) in order to pump this power to the grid. The generator-side inverter is current controlled with either PI controllers on the stationary frame (with SPWM) or with hysteresis band controllers. Three-phase reference currents are generated through a programmed sine wave generator that receives the stator electrical angular speed reference ω_e^* and the current peak amplitude $\hat{I}_s^*$ supplied by

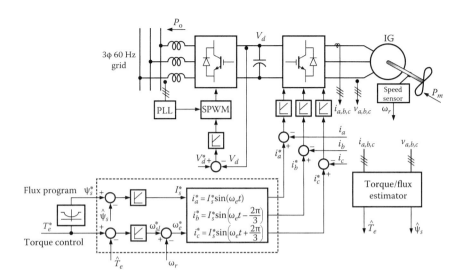

FIGURE 8.7 Voltage-fed link system for grid connection with current control for generator side.

corresponding torque and flux loops. Those loops are closed with the estimation of actual torque flux in the generators by feeding back the generator current and voltage.

In order to optimize the efficiency of the generator, a look-up table reads the torque command and programs the optimum flux reference for the system. A start-up sequence initially boosts the dc-link voltage to a higher voltage than the peak value of the grid, in order for the PWM to work properly. The overall system is very robust due to the current control in the inverters, the dc-link capacitor (with lower losses and faster response than dc-link inductors), the online estimation of generator torque and flux. The system may be implemented with the last generation of microcontrollers since internal computations are not so mathematically intensive. All these scalar-based control schemes can also incorporate speed governor systems on the mechanical shaft in order to control the incoming power for hydropower applications.

If the induction generator is primarily used in stand-alone operation, reactive power must be supplied for proper excitation, and the use of power electronics allows for several approaches such as variable impedance[4,5] and series and shunt regulators. Recently, p–q theory has been used for these applications.[6] The overall scheme is based on the diagram of Figure 8.8. The capacitor bank is a bulk uncontrolled source of reactive current. The static VAR compensator is an inverter providing a controlled source of reactive current. The inverter uses a scalar control approach, as previously discussed; the dashed subsystem of Figure 8.7 constitutes the block called scalar control in Figure 8.8.

Machine torque and flux input will be used in this application to regulate both dc-link voltage at the capacitance C_p and generator voltage supplied for the ac load. These regulators have to reject the disturbances produced by load and speed variations. A dc-link voltage regulator has been implemented to achieve a high-enough

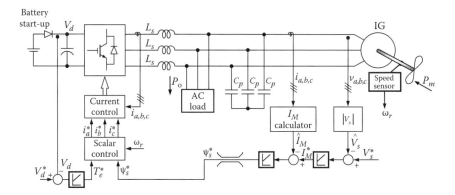

FIGURE 8.8   Static VAR compensator scalar-based control with voltage regulation.

dc-link voltage for proper current-controlled inverter operation. The regulator input is the difference between the dc-link voltage reference and the measured value.

At the generator side, terminal current and voltages are measured to calculate the magnetizing current needed for the generator, and the instantaneous peak voltage is compared to a stator voltage reference V_s^*, which generates a set point for flux through the feedback loops on the inverter side. This system requires a charged battery for start-up. That can be recharged with an auxiliary circuit (not shown) after the system is operational. Since stator voltage is kept constant, the frequency can be stabilized at about 60 Hz for the ac load, but some slight frequency variation is still observed, and the range of turbine shaft variation should be within the critical slip in order to avoid instability. Therefore, ac loads should not be too sensitive for frequency variation in this stand-alone application.

8.4   SOLVED PROBLEMS

1. This chapter showed that there is a critical slip value where machine torque peaks. Discuss why scalar control is in danger of instability due to a particular characteristic and what measures should be taken to avoid unstable operation.

   Solution

   Scalar control means fixed V/Hz to make ψ_m constant. When we fix (V/Hz), we choose to operate in a particular $T_e - \omega$ curve, as indicated in Figure 8.9. For that particular curve, we have a corresponding maximum torque point. But at that point, the operation is unstable, because a slight increase in speed (i.e., $|s| > |s_{max}|$) will drive the machine to a point where the speed will be more and the system will be more unstable. Hence, for the induction generator (IG) where the speed variation is very small, we can operate at a fixed ψ_m. However, when there is a chance of greater speed variation, we have to shift to a different (V/Hz) ratio, which means to different speed–torque characteristics. The figure shows that it is necessary to move from characteristics B to characteristics A in order to incorporate speed increase in the prime mover.

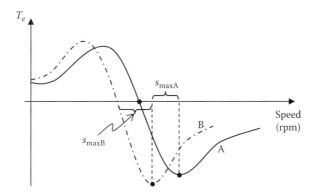

FIGURE 8.9   Torque versus speed for a maximum slip.

2. The current-fed converter scheme explained in Figure 8.6 has a thyristor-based controlled rectifier interfacing with the electrical grid. Describe why this scheme does not operate at unity power factor on the grid side and how the scheme can be improved in this respect.

Solution

The current-fed converter scheme given in Figure 8.6 consists of a thyristor-controlled rectifier interfacing with grid. This thyristor-controlled scheme operates on natural commutation. That means to turn off the thyristor, the current through the device should be zero. This needs a delay between the voltage and the current, which makes circulation of reactive power between grid and the converter. For this reason, one cannot have unity power factor on the grid side. There are some options by which the problem can be overcome. But some of them may not be practically possible if we consider the cost associated:

- Add a capacitor bank at the output of the converter: cheap, easy.
- Change devices to back-back transistors and diodes or MOS-controlled thyristor (MCT) in order to have forced commutation: costlier, depending on the grid voltage level.
- Change topology to voltage-fed converter: costly.

8.5   SUGGESTED PROBLEMS

8.1 Figure 8.3 shows an interesting feature that generating torque tends to increase with decreasing stator frequency. Describe why this trend is desirable for renewable energy conversion systems and what limits the actual performance of induction machines operating at a very low speed.

8.2 The torque equation (8.4) shows that for a fixed air-gap flux machine torque is directly proportional to slip frequency. Variable rotor resistance due to temperature suggests that a derating factor could be applied for correction. Discuss the real implications of parameter variation in the scalar control method.

8.3 Simulate one of the scalar control schemes with block-diagram system simulator software.

8.4 Do a complete research and reading in order to understand how multimegawatt induction generators can be controlled with a scalar control. Assume power electronic devices capable of handling high voltages and high currents, maybe not operating at high-frequency PWM, and probably connected to a subtransmission line voltage (typically greater than 13 kV), write a complete report with your findings.

REFERENCES

1. Boys, J.T. and Walton, S.J., Scalar control: An alternative AC drive philosophy, *IEEE Proc. B, Electr. Power Appl.*, 135(3), 151–158, May 1988.
2. Murphy, J.M.D. and Turnbull, F.G., *Power Electronic Control of AC Motors*, Oxford University Press, New York, 1988.
3. Trzynadlowski, A.M., *Control of Induction Motors*, Academic Press, San Diego, CA, 2000.
4. Brennen, M.B. and Abbondati, A., Static exciters for induction generators, *IEEE Trans. Ind. Appl.*, IA-13, 422–428, September/October 1977.
5. Bonert, R. and Hoops, G., Standalone induction generator with terminal impedance controller and no turbine controls, *IEEE Trans. Energy Convers.*, 5(1), 28–31, 1990.
6. Leidhold, R., Garcia, G., and Valla, M.I., Induction generator controller based on the instantaneous reactive power theory, *IEEE Trans. Energy Convers.*, 17(3), 368–373, September 2002.

9 Vector Control for Induction Generators

9.1 SCOPE OF THIS CHAPTER

Vector control is the foundation of modern high-performance drives. It is also known as decoupling, orthogonal, or transvector control. Vector control techniques can be classified according to their method of finding the instantaneous position of the machine flux, which can be either (1) the indirect or feedforward method or (2) the direct or feedback method. This chapter will discuss field orientation—the calculation of stator current components for decoupling of torque and flux—while laying out the principles for rotor flux– and stator flux–oriented control approaches for induction generator systems. A discussion of parameter sensitivity, detuning, sensorless systems, PWM generation, and signal processing issues can be found in the available technical literature.

9.2 VECTOR CONTROL FOR INDUCTION GENERATORS

The induction machine has several advantages: size, weight, rotor inertia, maximum speed capability, efficiency, and cost. However, induction machines are difficult to control because being polyphase electromagnetic systems they have a nonlinear and highly interactive multivariable control structure. The technique of field-oriented or vector control permits an algebraic transformation that converts the dynamic structure of the AC machine into that of a separately excited decoupled control structure with independent control of flux and torque.

The principles of field orientation originated in West Germany in the work of Hasse[1] and Blaschke[2] at the technical universities of Darmstadt and Braunschweig and in the laboratories of Siemens AG. Varieties of other derived implementation methods are now available, but two major classes are very important and will be discussed in this chapter: direct control and indirect control. Both approaches calculate the online instantaneous magnitude and position of either the stator or the rotor flux vector by imposing alignment of the machine current vector to obtain decoupled control. Therefore, a very broad and acceptable classification of vector control is based on the procedure used to determine the flux vector.

Even though Hasse proposed electromagnetic flux sensors, currently direct vector control (DVC) is based on the estimation through state observers of the flux position. The technique proposed by Blaschke is still very popular and depends on feedforward calculation of the flux position with a crucial dependence on the rotor electrical time constant, which may vary with temperature, frequency, and saturation, as will be discussed in this chapter. The principles of vector control can be employed for control of instantaneous active and reactive power. Therefore, it is probably fair to say that the modern p–q instantaneous power theory has also evolved from the vector control theory.[3] If a totally symmetrical and three-phase balanced induction machine is used, a d–q transformation is used to decouple time-variant parameters, helping to lay out the principles of vector control.

9.3 AXIS TRANSFORMATION

The following discussion draws on voltage transformation, but other variables like flux and current are also similarly transformed. Figure 9.1 shows a physical distribution of the three stationary axes a^s, b^s, and c^s, 120° apart from each other, symbolizing concentrated stator windings of an induction machine. Cartesian axes are also portrayed in Figure 9.1, where q^s is a horizontal axis aligned with phase a^s, and a vertical axis rotated by −90° is indicated in the figure by d^s. Three-phase voltages varying in time along the axes a^s, b^s, and c^s can be algebraically transformed onto two-phase voltages varying in time along the axes d^s and q^s.[3–5]

This transformation is indicated by Equation 9.1. The inverse of such a transformation can be taken by assuming that the zero sequence v_{os}^s is equal to zero for a balanced system and Equation 9.2 is valid for the inverse transformation. Equation 9.1 takes a two-phase quadrature system and transforms it to three phase, while Equation 9.2 takes a three-phase system and converts it to a two-phase quadrature system. Figure 9.2 shows simplified block diagrams representing transformations of this kind commonly used for vector control block diagrams:

$$\begin{bmatrix} v_{as} \\ v_{bs} \\ v_{cs} \end{bmatrix} = \begin{bmatrix} 1 & 0 & 1 \\ -\dfrac{1}{2} & -\dfrac{\sqrt{3}}{2} & 1 \\ -\dfrac{1}{2} & \dfrac{\sqrt{3}}{2} & 1 \end{bmatrix} \begin{bmatrix} v_{qs}^s \\ v_{ds}^s \\ v_{os}^s \end{bmatrix} \qquad (9.1)$$

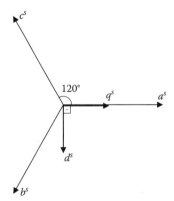

FIGURE 9.1 Physical distribution of three stationary axes and a quadrature stationary axis.

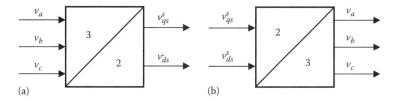

FIGURE 9.2 Simplified block diagrams for stationary frame conversion: (a) three-phase to two-phase conversion and (b) two-phase to three-phase conversion.

$$\begin{bmatrix} v_{qs}^s \\ v_{ds}^s \\ v_{os}^o \end{bmatrix} = \begin{bmatrix} 1 & 0 & 1 \\ 0 & -\dfrac{\sqrt{3}}{3} & \dfrac{\sqrt{3}}{3} \\ \dfrac{1}{2} & \dfrac{1}{2} & \dfrac{1}{2} \end{bmatrix} \begin{bmatrix} v_{as} \\ v_{bs} \\ v_{cs} \end{bmatrix} \tag{9.2}$$

In addition to the stationary d^s–q^s frame, an induction machine has two other frames depicted in Figure 9.3. One is aligned with one internal flux (stator, air gap, or rotor) denoted by d^e–q^e synchronous rotating reference frame and the other aligned with a hypothetical shaft rotating at the electrical speed dependent on the number of poles, denoted by $d^r q^r$ rotor frame. Machine instantaneous variables (voltages, currents, fluxes) can be transformed from the stationary frame $d^s q^s$ to the synchronous rotating reference frame d^e–q^e through Equation 9.3. For convenience of notation, the variables referred to the synchronous reference frame do not take the index e, that is, the voltage along the axis q^e is v_{qs} instead of v_{qs}^e. Equation 9.3 takes a two-phase quadrature system along the stationary frame and transforms it onto a two-phase reference frame, while Equation 9.4 takes a two-phase reference frame and converts it back to the stationary frame. The angle θ_e is considered the instantaneous

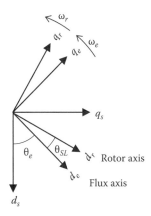

FIGURE 9.3 Stationary, synchronous, and rotor frames.

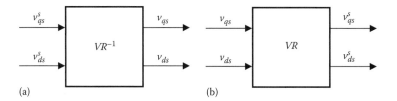

FIGURE 9.4 Block diagrams for vector rotation conversion: (a) from stationary to synchronous frame and (b) from synchronous to stationary frame.

angular position of the flux where the reference frame will be aligned. Therefore, sinusoidal time variations along the stationary frame d^s–q^s will appear as constant values onto the synchronous rotating reference frame d^e–q^e. Those operations are also called vector rotation (VR) and inverse vector rotation (VR^{-1}). Figure 9.4 shows simplified block diagrams representing transformations of this kind usually used for block diagrams of the vector control.

$$\begin{bmatrix} v_{qs} \\ v_{ds} \end{bmatrix} = \begin{bmatrix} \cos(\theta_e) & -\sin(\theta_e) \\ \sin(\theta_e) & \cos(\theta_e) \end{bmatrix} \begin{bmatrix} v_{qs}^s \\ v_{ds}^s \end{bmatrix} \tag{9.3}$$

$$\begin{bmatrix} v_{qs}^s \\ v_{ds}^s \end{bmatrix} = \begin{bmatrix} \cos(\theta_e) & \sin(\theta_e) \\ -\sin(\theta_e) & \cos(\theta_e) \end{bmatrix} \begin{bmatrix} v_{qs} \\ v_{ds} \end{bmatrix} \tag{9.4}$$

In an induction machine, the mutual inductances between stator and rotor windings vary with rotor position. The transformation from three-phase equations for stator and rotor to either reference frame eliminates this time variance. Thus, a set of two equivalent circuits along the d-axis and q-axis can represent the dynamics of the three-phase machine. Figure 9.5 shows the d–q stationary model, that is, onto d^s–q^s stationary frame. In this model, the variables are transformed between a–b–c and d^s–q^s through Equations 9.1 and 9.2, which are valid for both voltages and currents. In this equivalent d–q circuit, the steady-state response of voltages and currents leads to sinusoidal and cosinusoidal waveforms, which represent the machine three-phase 120° phase-shifted variables. Figure 9.6 depicts the d–q synchronous rotating model, that is, onto d^e–q^e reference frame. In this model, the variables are transformed between a–b–c and d^e–q^e through Equations 9.1 through 9.4.

The instantaneous electrical angular speed ω_e is required for the additional terms indicated in Figure 9.6. The vector rotation (VR) and inverse vector rotation (VR^{-1}) transformations are taken after the integration of the given angular speed ω_e signal. In this equivalent d–q circuit, the steady-state response of voltages and currents leads to constant (dc) signals, which represent the machine three-phase 120°-shifted variables. Such equivalent circuit model of the synchronous rotating reference frame runs faster in simulators due to the dc nature of the impressed signals. The synchronous rotating reference frame is easily converted to the stationary frame by imposing $\omega_e = 0$ in the equivalent circuit. In addition, such equivalent circuit concepts are fundamentals for the vector control approach, as will be discussed later in this chapter.

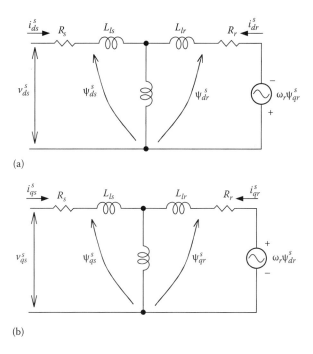

FIGURE 9.5 Induction machine model in stationary frame: (a) direct axis equivalent circuit and (b) quadrature axis equivalent circuit.

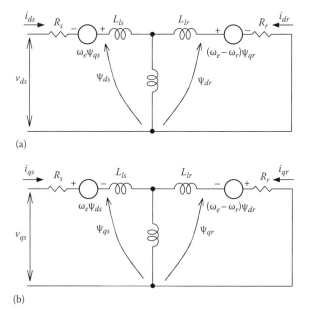

FIGURE 9.6 Induction machine model in synchronous rotating frame: (a) direct axis equivalent circuit and (b) quadrature axis equivalent circuit.

9.4   SPACE VECTOR NOTATION

A three-phase stator voltage excitation impressed on three-phase geometrically distributed windings results in three magnetomotive forces (mmf), which are added to the result in a traveling wave, that is, a magnetic field across the air gap rotating at synchronous speed. Such a spatial conception is implicit in traditional electrical machinery analysis. If we apply the same concept to voltages and currents, the resulting space vectors will be virtual rather than physical, but conveniently express a $d–q$ model through complex variables. For example, the stator current space vector can be computed by Equation 9.5, where

$$\alpha = e^{j\frac{2\pi}{3}}$$

$$\underline{i}_{qds} = \frac{2}{3}\left(i_a + \alpha i_b + \alpha^2 i_c\right) \tag{9.5}$$

The notation used in Equation 9.5 shows an underscore on the variable indicating a complex number. The index s indicates a stator quantity, and indexes d and q in the subscript show the merging of direct and quadrature representation in just one variable, the coefficient 2/3 scales the result in relation to the peak values of the three-phase variable. Transformation from the space vector back to the three-phase original currents is performed through matrix equation (9.6).

The relations of the space vector with the synchronous rotating reference frame depend upon the axis orientation of the $d–q$ frame. In this book, when the q-axis is at horizontal, the d-axis is at $-90°$. Therefore, the current space vector can be expanded as $\underline{i}_{qds} = i_{qs} - ji_{ds}$, where j represents the $\pi/2$ rotational operator. Real and imaginary operators, as indicated by Equations 9.7 and 9.8, can be used to indicate this mapping between the reference frame and the space vector notation. The same space vector approach can be used to handle voltages and fluxes in either stator or rotor circuits. Figure 9.7 indicates the space vector complex algebraic equivalent model of a squirrel cage machine; it is a very compact representation that embeds the transient nature:

$$\begin{bmatrix} i_a & i_b & i_c \end{bmatrix}^T = \mathrm{Re}\begin{bmatrix} 1 & \alpha & \alpha^2 \end{bmatrix}^T \underline{i}_{qds} \tag{9.6}$$

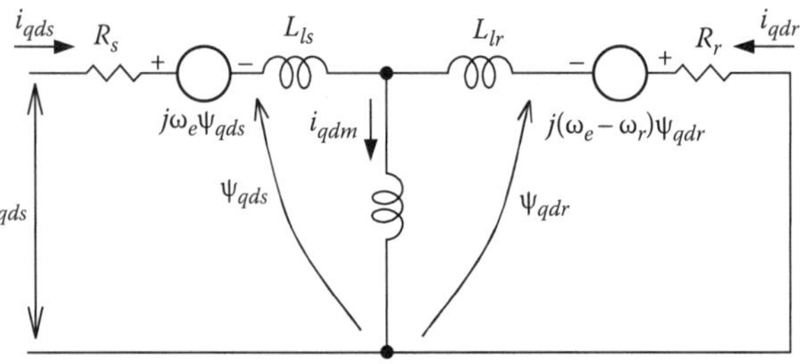

FIGURE 9.7   Space vector equivalent transient model of an induction machine, synchronous rotating reference frame equivalent.

$$i_{qs} = \text{Re}\left[i_{qds}\right] \tag{9.7}$$

$$i_{ds} = \text{Im}\left[i_{qds}\right] \tag{9.8}$$

Space vector Equations 9.9 and 9.10 represent the transient response of an induction machine.[3–6] Each equation is expanded in two-axis equations. Torque is indicated in Equation 9.11 with an expanded real-time equation in Equation 9.12. The machine also needs an electromechanical equation as indicated by Equation 9.13:

$$\underline{v}_{qds} = R_s \underline{i}_{qds} + j\omega_e \underline{\psi}_{qds} + \frac{d}{dt}\underline{\psi}_{qds} \tag{9.9}$$

$$0 = R_r \underline{i}_{qdr} + j\left(\omega_e - \omega_r\right)\underline{\psi}_{qdr} + \frac{d}{dt}\underline{\psi}_{qdr} \tag{9.10}$$

$$T_e \frac{3}{2}\left(\frac{p}{2}\right)\text{Re}\left(j\underline{\psi}_{qds}\right) = \frac{3}{2}\left(\frac{p}{2}\right)\text{Re}\left(j\underline{\psi}_{qdr} \cdot \underline{i}_{qdr}\right) \tag{9.11}$$

$$T_e = \frac{3}{2}\left(\frac{p}{2}\right)\left(\psi_{ds}i_{qs} - \psi\right) = \frac{3}{2}\left(\frac{p}{2}\right)\left(\psi_{dr}i_{qr} - \psi_{qr}i_{dr}\right) \tag{9.12}$$

$$\frac{d\omega_r}{dt} = \frac{1}{J}\left(T_e - T_L\right) \tag{9.13}$$

9.5 FIELD-ORIENTED CONTROL

The objective of field-oriented or vector control is to establish and maintain an angular relationship between the stator current space vector and one internal field vector, usually rotor flux or stator flux as suggested by Equation 9.12. This angular relationship may be achieved by regulating the slip of the machine to a particular value that will fix the orientation through feedforward online calculations. In this approach, often called indirect vector control (IVC), rotor flux is aligned with the d-axis. IVC is very popular for industrial drives: it is inherently four-quadrant work until zero speed and is suitable for speed-control loop but very dependent on machine parameters. Under flux-aligned conditions along the d-axis of the synchronous rotating reference frame, the stator current is decoupled into two components: i_{qs} is proportional to machine electrical torque, whereas i_{ds} is proportional to the machine flux, similarly to a dc machine control. Therefore, field orientation implies that online transformation of machine variables takes into account a hypothetical rotating frame equivalent to a dc machine. By real-time transformation back to the three-phase stationary frame, impressed stator currents will impose a dc-like transient response.

Field orientation may also be achieved by online estimation of the field vector position, and usually the stator flux will be aligned with the d-axis. Under this approach, called DVC, the control is less sensitive to parameters. It is dependent only on the stator resistance, which is easier to correct. It is very robust due to real-time tracking of machine parameter variation with temperature and core saturation, but it typically does not work at zero shaft speed. For induction generator applications, DVC seems to be a more effective approach because zero speed operation is not required. There are other ways to control an induction machine based on direct torque control (DTC), as developed by ABB based on the theory developed by Depenbrock,[7,8] where two different loops, corresponding to the magnitudes of the stator flux and torque, directly control the switching of the converter. Although DTC boasts a lot of successful applications in drives, its use for induction generators is not yet fully proven and will not be discussed in this book.

9.5.1 INDIRECT VECTOR CONTROL

In IVC, the unit vector, $\sin(\theta_e)$, and $\cos(\theta_e)$, used for vector rotation of the terminal variables is calculated in a feedforward manner through the use of the internal signals ω_r (rotor speed) and i_{qs}^* (quadrature stator current component proportional to torque). The control principles required to implement IVC in the rotor flux oriented reference frame can be derived from the rotor voltage equation (9.10) and the torque equation (9.12). The magnetizing rotor current is defined in Figure 9.5 as[3]

$$\begin{cases} 0 = R_r i_{dr} + \dfrac{d}{dt}\psi_{dr} - \left(\omega_e - \omega_r\right)\psi_{qr} \\[3mm] 0 = R_r i_{qr} + \dfrac{d}{dt}\psi_{qr} + \left(\omega_e - \omega_r\right)\psi_{dr} \end{cases} \tag{9.14}$$

$$\begin{cases} i_{dr} = \dfrac{1}{L_r}\psi_{dr} - \dfrac{L_m}{L_r} i_{ds} \\[3mm] i_{qr} = \dfrac{1}{L_r}\psi_{qr} - \dfrac{L_m}{L_r} i_{qs} \end{cases} \tag{9.15}$$

Substituting Equation 9.15 into Equation 9.14, where $\omega_{sl} = \omega_e - \omega_r$, yields

$$\begin{cases} \dfrac{d}{dt}\psi_{dr} + \dfrac{R_r}{L_r}\psi_{dr} - \dfrac{L_m}{L_r} R_r i_{ds} - \omega_{sl}\psi_{qr} = 0 \\[3mm] \dfrac{d}{dt}\psi_{qr} + \dfrac{R_r}{L}\psi_{qr} - \dfrac{L_m}{L} R_i i_{qs} + \omega_{sl}\psi_{dr} = 0 \end{cases} \tag{9.16}$$

By imposing alignment of rotor flux with the d-axis of the synchronous rotating reference frame, one can consider that flux projection on the q-axis is zero, and so is its derivative, $\psi_{qr} = d/dt\,\psi_{qr} = 0$. Therefore, the vector will be completely along the d-axis, $\psi_{qdr} = \psi_{dr}$. Substituting such requirements into Equations 9.16 through 9.18 establish the control laws to implement in the synchronous rotating reference frame. The torque equation (Equation 9.19) will need only ψ_{dr} and i_{qs}. By observing such equations, it is possible to see that by maintaining i_{ds} constant the rotor flux will be kept constant and by controlling i_{qs} the machine torque will respond proportionally. Therefore, a current controlled converter, as depicted in Figure 9.8, will cause the induction machine to have a fast transient response under decoupled commands, like a dc machine. The q-axis stator reference i_{qs}^* generates the reference slip frequency ω_{sl} (Equation 9.18), which is added to the instantaneous rotor speed ω_r in order to generate the required angular frequency ω_e, which, after integrating the flux angle θ_e, generates the sine and cosine waveforms for vector rotation:

$$\frac{L_r}{R_r}\frac{d}{dt}\psi_{dr} + \psi_{dr} = L_m i_{ds} \tag{9.17}$$

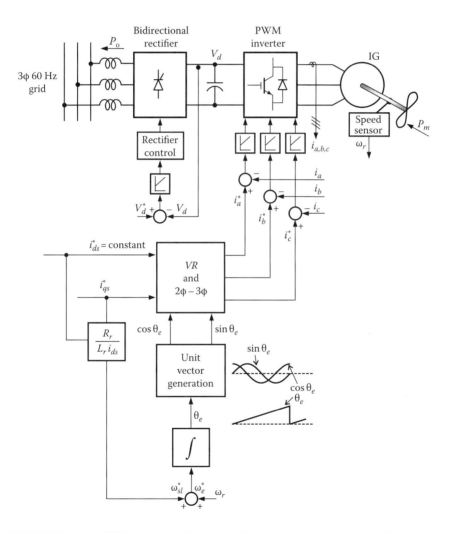

FIGURE 9.8 Basic IVC structure with decoupled stator current input command.

$$\omega_{sl} = \frac{L_m}{\Psi_{dr}} \frac{R_r}{L_r} i_{qs} \tag{9.18}$$

$$T_e = \frac{3}{2}\left(\frac{p}{2}\right)\frac{L_m}{L_r} i_{qs}\Psi_{dr} \tag{9.19}$$

Figure 9.8 shows an IVC-based system with i_{ds}^* and i_{qs}^* command. The power circuit consists of a front-end rectifier connected through a dc link to a three-phase inverter. For generation purposes, the grid-connected rectifier must be operational in four quadrants, delivering incoming machine power to the utility side. By maintaining i_{ds}^* constant, the machine flux will be constant and a positive i_{qs}^* will drive the machine in a motoring mode, whereas for $i_{qs}^* < 0$, the slip frequency ω_{sl} is negative, and the corresponding flux angle θ_e will be such that sine/cosine waveforms will phase-reverse the vector rotation automatically for generating mode.

The block diagram indicates a PWM current control. There are several possibilities when implementing modulators; the reader is referred to Holmes and Lipo.[6] Synchronous current control with feedforward machine counter-EMF is preferable for an optimized operation, as indicated in Figure 9.9, where the reference voltages v_a^*, v_b^*, and v_c^* will drive a voltage PWM controller. Speed and flux outer loops can also be incorporated as outer loops on the basic structure. IVC is a very popular and powerful approach to control induction motors and generators. However, exact machine parameters are needed for improved performance.[7–9]

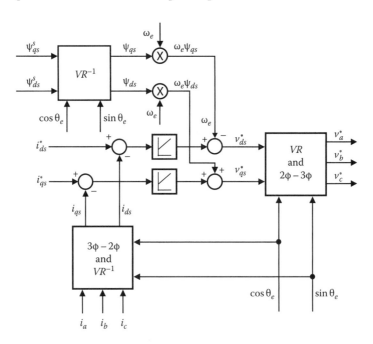

FIGURE 9.9 Synchronous current control for IVC system with flux and speed loops.

9.5.2 DIRECT VECTOR CONTROL

DVC with stator flux orientation is preferred for induction generator systems for two reasons: it is not necessary to use the speed signal in the fabrication of the unit-vector signals, and the estimation is robust against parameter variations for operating speed greater than 5% of base speed. The unit vectors are calculated by the integration of the *d*- and *q*-axis voltages minus the required drop for the correspondent reference frame. The stator flux reference frame has larger flux voltage for extended low-speed operation.

There is a coupling that can be easily compensated for with a feedforward network; the decoupler is in the forward path of the system. Detuning parameters will not affect the steady-state response, and the stator resistance variation can be taken into account with thermocouples installed in the machine slots.

Stator flux orientation also improves transient response with detuned leakage inductance, and there is a limited pull-out torque for a given flux, which improves the stability.

The equations for stator flux orientation are derived from the Blaschke equations, where the *d–q* rotor flux vector (ψ_{dr}, ψ_{qr}) is algebraically replaced by stator flux (ψ_{ds}, ψ_{qs}). Equations 9.20 and 9.21 are the Blaschke rotor equations in the reference frame; the relationship between stator flux and rotor flux is given by Equations 9.22 and 9.23, where $\sigma = 1 - L_m^2/L_s L_r$ is the total leakage factor:

$$\left(1 + sT_r\right)\psi_{dr} = L_m i_{ds} + \omega_{sl} T_r \psi_{qr} \tag{9.20}$$

$$\left(1 + sT_r\right)\psi_{qr} = L_m i_{qs} - \omega_{sl} T_r \psi_{dr} \tag{9.21}$$

$$\psi_{qr} = \frac{L_r}{L_m}\psi_{qs} - \frac{\sigma L_s L_r}{L_m} i_{qs} \tag{9.22}$$

$$\psi_{dr} = \frac{L_r}{L_m}\psi_{ds} - \frac{\sigma L_s L_r}{L_m} i_{ds} \tag{9.23}$$

After substituting Equations 9.22 and 9.23 in Equations 9.20 and 9.21, respectively, and after algebraic manipulations, the Blaschke equations in terms of stator currents become

$$\left(1 + sT_r\right)\psi_{ds} = L_s i_{ds}\left(1 + s\sigma T_r\right) + \omega_{sl} T_r \left(\psi_{qs} - \sigma L_s i_{qs}\right) \tag{9.24}$$

$$\left(1 + sT_r\right)\psi_{qs} = L_s i_{qs}\left(1 + s\sigma T_r\right) - \omega_{sl} T_r \left(\psi_{ds} - \sigma L_s i_{ds}\right) \tag{9.25}$$

In order to attain vector control, the *q* component of the stator flux must be set to zero, and Equations 9.26 and 9.27 follow:

$$\left(1 + sT_r\right)\psi_{ds} = L_s i_{ds}\left(1 + s\sigma T_r\right) - \omega_{sl} T_r \sigma L_s i_{qs} \tag{9.26}$$

$$i_{qs} L_s \left(1 + s\sigma T_r\right) = \omega_{sl} T_r \left(\psi_{ds} - \sigma L_s i_{ds}\right) \tag{9.27}$$

Equations 9.26 and 9.27 indicate coupling between the torque component current i_{qs} and the stator flux ψ_{ds}. The injection of a feedforward decoupling signal i_{dq} in the flux loop eliminates these effects as is indicated in Equation 9.28. The transient nature of i_{dq} is not included in Equation 9.28 since it was found that it does not improve the overall response much.[10,11] The i_{dq} decoupling signal is given by Equation 9.29 obtained from Equation 9.27 and the angular slip frequency ω_{sl} is given in Equation 9.30 obtained from Equation 9.26:

$$\left(1 + sT_r\right)L_s i_{dq} - \omega_{sl}T_r\sigma L_s i_{qs} = 0 \tag{9.28}$$

$$i_{dq} = \frac{\sigma L_s i_{qs}^2}{\psi_{ds} - \sigma L_s i_{ds}} \tag{9.29}$$

$$\omega_{sl} = \frac{i_{qs}L_s\left(1 + \sigma T_r s\right)}{T_r\left(\psi_{ds} - \sigma L_s i_{ds}\right)} \tag{9.30}$$

The control block diagram in stator flux–oriented DVC mode is shown in Figure 9.10. The system uses open-loop torque control with a stator flux control loop. The stator flux remains constant in the constant torque region but can also be programmed in the field weakening region. In addition to Equations 9.29 and 9.30, the key estimation equations can be summarized as follows:[10]

$$\psi_{ds}^s = \int_0^t \left(v_{ds}^S - i_{ds}^s R_s\right)dt \tag{9.31}$$

$$\psi_{qs}^s = \int_0^t \left(v_{qs}^s - i_{qs}^s R_s\right)dt \tag{9.32}$$

$$\psi_s = \left|\sqrt{\left(\psi_{ds}^s\right)^2 + \left(\psi_{qs}^s\right)^2}\right| \tag{9.33}$$

$$\cos\theta_e = \frac{\psi_{ds}^s}{\left|\psi_s\right|} \tag{9.34}$$

$$\sin\theta_e = \frac{\psi_{qs}^s}{\left|\psi_s\right|} \tag{9.35}$$

$$\omega_e = \frac{\left(v_{qs}^s - i_{qs}^s R_s\right)\psi_{ds}^s - \left(v_{ds}^s - i_{ds}^s R_s\right)\psi_{qs}^s}{\left|\psi_s^2\right|} \tag{9.36}$$

The control block diagram of Figure 9.10 shows the principles for controlling the induction generator. The most important parameter to be corrected is the stator

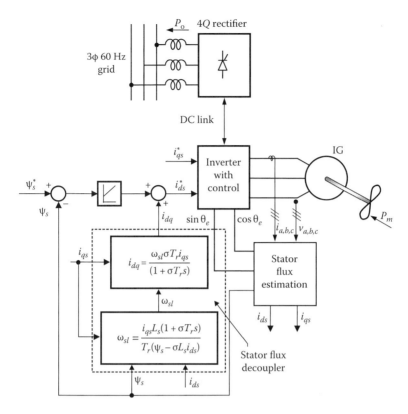

FIGURE 9.10   DVC-based induction generator control with torque and flux command.

resistance, which can be accomplished with a machine thermal model and measurement of terminal voltage and current. Power electronic systems can be either connected in series with the machine, as is discussed in Chapter 10, where a fuzzy-based control optimizes a double-PWM converter system, or they can be used in parallel as static VAR compensation. Figure 9.11 shows the scheme for induction generators. An economic evaluation for each solution should be conducted.

For an induction generator operating in stand-alone mode, a procedure that regulates the output voltage is required, as shown in the control system indicated in Figure 9.11. The dc-link voltage across the capacitor is kept constant and the machine impressed terminal frequency will vary with variable speed. For motoring applications, flux is typically kept constant and only for operating speeds above rated value is a flux-weakening command required. However, induction generators require constant generated voltage. Since the frequency of the generated voltage and current is dependent on the rotor speed, the product of the rotor speed and the flux linkage should remain constant so that the terminal voltage will remain constant, where the minimum rotor speed will correspond to the maximum saturated flux linkage. A look-up table is used to program the right flux level in Figure 9.11. The system starts up with a battery connected to the inverter. Then, as the dc-link voltage is regulated to a higher value across the capacitor, the battery will be turned off by

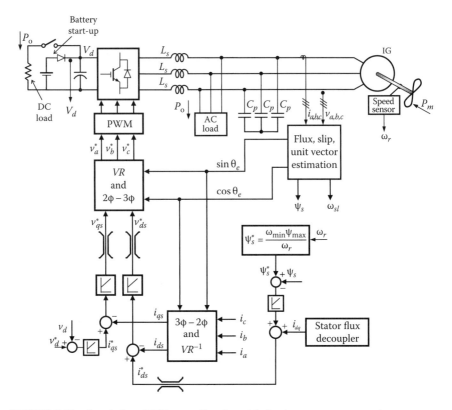

FIGURE 9.11 Stand-alone DVC stator flux–based induction generator control.

the diode and a dc load can be applied across the capacitor. The machine terminals are also capable of supplying power to an auxiliary load at variable frequency.

9.6 SOLVED PROBLEMS

1. Summarize the features of IVC and direct vector control as applied to induction generator systems.

Solution

The objective of field-oriented or vector control is to establish and maintain an angular relationship between the stator current space vector and one internal field vector, usually rotor flux or stator flux. This angular relationship may be achieved by regulating the slip of the machine to a particular value that will fix the orientation through feedforward online calculations. In this approach, often called IVC, rotor flux is aligned with the d-axis. IVC is very popular for industrial drives: it is inherently four-quadrant work until zero speed and is suitable for speed control loop but very dependent on machine parameters. Under flux-aligned conditions along the d-axis of the synchronous rotating reference frame, the stator current is decoupled into two components; i_{qs} is proportional to machine electrical torque whereas i_{ds} is proportional to the machine flux, similarly to a dc machine control. Therefore, field orientation implies that online transformation of machine variables take into account a hypothetical rotating frame equivalent to a dc machine. By real-time transformation back to the three-phase stationary frame, impressed stator currents will impose a dc-like transient response.

Field orientation may also be achieved by online estimation of the field vector position, and usually the stator flux will be aligned with the d-axis. DVC with stator flux orientation is preferred for induction generator systems for several reasons: it is not necessary to use the speed signal in the fabrication of the unit vector signals, and the estimation is robust against parameter variations for operating speed greater than 5% of base speed. The unit vectors are calculated by the integration of the d- and q-axis voltages minus the required drop for the correspondent reference frame. The stator flux reference frame has larger flux voltage for extended low-speed operation. There is a coupling that can be easily compensated for with a feedforward network; the decoupler is in the forward path of the system. Detuning parameters will not affect the steady-state response, and the stator resistance variation can be taken into account with thermocouples installed in the machine slots. Stator flux orientation also improves transient response with detuned leakage inductance, and there is a limited pull-out torque for a given flux, which improves the stability.

2. Do research in the literature on parameter sensitivity for vector control–based systems and write a summary report.

Solution

The choice of reference frame in a DVC has an important impact on the performance of the drive in practical realization. The classical DVC with rotor flux as the reference frame has the parameters in the feedback path. On the other hand, the air gap flux controller and the stator flux controller can be practically realized without any error in the flux position measurement.

In order to avoid flux sensors, the stator flux position is calculated from the stator voltages and currents. In this case, the flux estimation is only parameter dependent on the stator resistance. Adaptation schemes to compensate for temperature drift between the controller parameters and the machine parameters are well known when only stator resistance has to be adapted.

The steady-state analysis showed that direct stator field oriented system always retains maximum torque capability no matter how large the error in the parameters. In contrast, the direct rotor field–oriented system (RFO) is sensitive to the variations in leakage and magnetizing inductances. When detuned, RFO yields reduced torque capability and poorer torque/ampere ratio, or even static instability (pull-out). In case of field weakening, as the machine is operated near its maximum capacity and at reduced flux level, parameter sensitivity becomes more significant. In such a case, stator field–oriented control shows a better performance than the rotor field–oriented control.[12–13] The only problem associated with the stator field orientation is to take care of some feedforwards that can be done easily by simple control strategies. The generation of feedforward signal i_{dq} is dependent on parameters, but being inside the feedback loop this variation can be neglected.

On the other hand, the IVC is generally implemented in the rotor flux–oriented reference frame. In IVC, the unit vector, $\sin(\theta_e)$, and $\cos(\theta_e)$, used for vector rotation of the terminal variables is calculated in a feedforward manner through the use of the internal signals ω_r (rotor speed) and i_{qs}^* (quadrature stator current component proportional to torque). So the control is largely dependent on machine parameters L_m, L_r, R_r in the open loop generation of unit vector. Hence, the effect of parameter variation is more pronounced for IVC-driven induction generator[8,13,14].

9.7 SUGGESTED PROBLEMS

9.1 Do research in the literature on sensorless vector control–based systems and write a summary report.

9.2 Do research in the literature on space vector modulation as applied to vector control–based systems and write a summary report.

9.3 Simulate one of the vector-control–based schemes for induction generators using a block diagram system simulation software.

9.4 Conduct research and reading to understand how multimegawatt induction generators can be controlled by vector control. Assume devices capable of handling high voltages and high currents, maybe not possible to operate with high frequency PWM, and probably connected to a subtransmission line voltage (typically greater than 13 kV). Discuss vector-controlled induction generators using current source inverters, cycloconverters, and multilevel topologies.

REFERENCES

1. Hasse, K., About the dynamics of adjustable-speed drives with converter-fed squirrel-cage induction motor, PhD dissertation in German, Darmstadt Technische Hochschule, Darmstadt, Germany, 1969.
2. Blaschke, F., The principle of field-orientation as applied to the new transvector closed-loop control system for rotating-field machines, *Siemens Rev.*, 34, 217–220, 1972.
3. Kazmierkowski, M.P., Krishnan, R., and Blaabjerg, F., *Control in Power Electronics: Selected Problems*, Academic Press, San Diego, CA, 2002.
4. Murphy, J.M.D. and Turnbull, F.G., *Power Electronic Control of AC Motors*, Oxford University Press, New York, 1988.
5. Trzynadlowski, A.M., *Control of Induction Motors*, Academic Press, San Diego, CA, 2000.
6. Holmes, D.G. and Lipo, T.A., *Pulse Width Modulation for Power Converters: Principles and Practice*, John Wiley & Sons, Hoboken, NJ, 2003.
7. Novotny, D.W. and Lipo, T.A., *Vector Control and Dynamics of AC Drives*, Oxford University Press, New York, 1996.
8. Bose, B.K., *Modern Power Electronics and AC Drives*, Prentice Hall, New York, 2001.
9. Krishnan, R., *Electric Motor Drives: Modeling, Analysis, and Control*, Prentice Hall, New York, 2001.
10. Xu, X., De Doncker, R.W., and Novotny, D.W., A stator flux oriented induction machine drive, *IEEE/PESC Conference Record*, 11–14 April, 1988, Kyoto, Japan. pp. 870–876.
11. Seyoum, D., Rahman, F., and Grantham, C., Terminal voltage control of a wind turbine driven isolated induction generator using stator oriented field control, *IEEE APEC Conference Record*, Miami Beach, FL, 2003, Vol. 2, pp. 846–852.

12. Seyoum, D., The dynamic analysis and control of a self-excited induction generator driven by a wind turbine, PhD dissertation, School of Electrical Engineering and Telecommunications, The University of New South Wales, Sydney, New South Wales, Australia, 2003.

13. De Doncker, R.W. and Novotny, D.W., The universal field oriented controller, *IEEE Trans. Ind. Appl.*, 30(1), 92–100, January/February 1994, doi: 10.1109/28.273626.

14. Profumo, F., Tenconi, A., and De Doncker, R.W., The universal field oriented (UFO) controller applied to wide speed range induction motor drives, *22nd Annual IEEE Power Electronics Specialists Conference, 1991 (PESC'91 Record)*, Cambridge, MA, June 24–27, 1991, pp. 681–686, doi: 10.1109/PESC.1991.162749.

10 Optimized Control for Induction Generators

10.1 SCOPE OF THIS CHAPTER

Efficient electric generation and viable and effective load control for alternative or renewable energy sources usually require heavy investment and are usually accompanied by an ambiguous decision-making analysis that takes into consideration the intermittence of input power and demand. Therefore, in order to amortize the installation costs in the shortest possible time, optimization of electric power transfer, even for the sake of relatively small incremental gains, is of paramount importance. This chapter presents practical approaches based on hill climbing control (HLC) and fuzzy logic control (FLC) for peak power tracking control of induction generator–based renewable energy systems.

10.2 WHY OPTIMIZE INDUCTION GENERATOR–BASED RENEWABLE ENERGY SYSTEMS?

General approaches to optimization utilize linear, nonlinear, dynamic, and stochastic techniques, while in control theory, the calculus of variations and optimal control theory are important. Operations research is the specific branch of mathematics concerned with techniques for finding optimal solutions to decision-making problems. However, these mathematical methods are, to a large degree, theoretically oriented and thus very distant from the backgrounds of the decision makers and management personnel in industry.

Recently, practical applications of heuristic techniques like hill climbing and fuzzy logic have been able to bridge this gap, thereby creating the possibility of applying practical optimization to industrial applications. One field that has proven to be attractive for optimization is the generation and control of electric power from alternative and renewable energy resources like wind, hydro, and solar energy.[1,2] Installation costs are very high, while the availability of alternative power is by its nature intermittent, which tends to constrain efficiency. It is therefore of vital importance to optimize the efficiency of electric power transfer, even for the sake of relatively small incremental gains, in order to amortize the installation costs within the shortest possible time. The following characteristics of induction-generator-based control motivate the use of heuristic optimization:

- Parameter variation that can be compensated for by design judgment
- Processes that can be modeled linguistically but not mathematically
- A need to improve efficiency as a matter of operator judgment

- Dependence on operator skills and attention
- Process parameters affect one another
- Effects cannot be attained by separate PID control
- A fuzzy controller can be used as an advisor to the human operator
- Data-intensive modeling

There are many mathematical algorithms to solve optimization problems for a local minimum based on gradient projection requiring the evaluation of gradient information with regard to a certain objective function. An uncomplicated but powerful technique known as hill climbing is a type of feedforward chaining technique based on an expected maximum value of a function defined in between two consecutive zero-crossing points and choosing actions to achieve a next state with an evaluation closer to that of the goal. Fuzzy and neurofuzzy techniques by now have become efficient tools for modeling and control applications because of their potential benefits in optimizing cost effectiveness.

10.3 OPTIMIZATION PRINCIPLES: OPTIMIZE BENEFIT OR MINIMIZE EFFORT

Examples of state evaluation functions and hill climbing are plentiful in everyday problems.[3] When someone plans a trip on a map, the roads that go toward the right direction are observed. Then, the ones that reduce the distance at the fastest rate are selected. Other considerations may come into the decision, such as scenery, facilities along the way, and weather conditions. In electrical engineering problems, the effort required or the benefit desired in any practical situation can be expressed as a function of certain decision variables, the *objective* or *cost function*, and optimization is defined as the process of finding the conditions that generate a maximum (or minimum) value of such a function.

Figure 10.1 shows a typical function of this kind. The y-axis can represent any physical variable like power, force, velocity, voltage, current, or resistance, while the x-axis might indicate the amplitude of the physical variable to be maximized—for example, a throttle controlling a water pump system (x-axis). As the throttle opens, the water flow increases but the pressure decreases, and the power (y-axis) initially goes up, reaching a maximum and decreasing as the throttle continues opening. Although finding the maximum of such a simple function seems to be trivial, it is not; the maximum can change under some parameter variations: temperature, density, ageing, part replacement, impedance, nonlinearities like dead-band and time delays, and cross dependence of input and output variables. The typical way of dealing with so many interdependencies would be by utilizing sensors to improve parameter robustness, analytical and experimental preparation of look-up tables, and mathematical feedforward decoupling. Such approaches increase costs and development time due to the extra sensors, more laboratory design phases, and the necessity of precise mathematical models.

A metarule is a rule that describes how other rules should be used and establishes how to build the knowledge framework. For example, there is an optimum pump speed for the corresponding maximum of power flow where minimum input pump effort maximizes output power. The heuristic way of finding the maximum might be

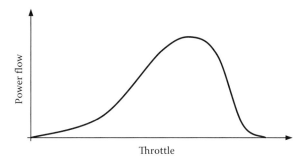

FIGURE 10.1   Objective function for power flow optimization.

based on the following metarule. "If the last change in the input variable x has caused the output variable y to increase, keep moving the input variable in the same direction; if it has caused the output variable to drop, move it in the opposite direction."

10.4   APPLICATION OF HCC FOR INDUCTION GENERATORS

Figure 10.2 shows the systems involved in an induction generator installation. There are several controls related to such systems, classified as mechanical, electromechanical, electroelectronic, and electronic. Mechanical controls use centrifugal weights to close or open the inflow of water for the hydro turbines or to exert pitch control of the blades in wind turbines. Among these are the flywheels used to store transient energy, anemometers to determine the wind intensity, and wind tails and wind vanes for the direction of the winds. They act to control and limit speed. They are slow and rough, yet robust, reliable, and efficient.

Electromechanical controls are more accurate, relatively lighter, and faster since they have less volume and weight than mechanical controls. They use governors, actuators, servomechanisms, electric motors, solenoids, and relays to control the primary energy. Some modern self-excited induction generators also use internal compensatory windings to keep the output voltage constant.

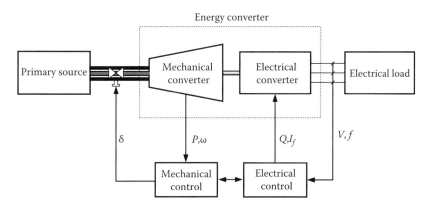

FIGURE 10.2   Transformation of primary energy into electrical energy.

Electroelectronic controls use electronic sensors and power converters to regularize energy and power. They can be quite accurate in adjusting voltage, speed, and frequency.

Electronic controls usually exert their action on the load, which is mechanically coupled to the turbine, and so the load controls them.

The output characteristics of nearly all induction generators resemble each other enough to allow us to describe the mathematical relationships among mechanical power and phase voltage, rotor current, and machine parameters as discussed in Chapter 5. Equation 5.10 is repeated here as

$$P_{mec} = 3I_2^2 \left[\sqrt{\left(\frac{V_{ph}}{I_2}\right)^2 - \left(X_1 + X_2\right)^2} - \left(R_1 + R_2\right) \right] \qquad (10.1)$$

Equation 10.1 is plotted in Figure 10.3 for an induction generator of 20 kW, 220/380 V, 60/35 A, 60 Hz, and four poles. Notice that the power characteristic of the induction generator has a maximum point that in a hydro or wind power plant will be the maximum power to be extracted from nature, which may be different from maximum efficiency. For this, there is no need to know the model of the generator or the turbine, the behavior of the wind, the deterioration of the material due to ageing, or other factors difficult to measure. The right value of the load current is then set up in such a way as to maximize the power extracted from nature. The speed at which the operating point reaches maximum generated power (MGP) is the most important consideration because if the primary energy is not used at the instant it is available from nature, it will be gone.

This speed can be obtained by a step modulation of the load current increment, also called electronic control by the load,[3] depicted in Figure 10.4. The load excess is dissipated by transferring the generator's energy to the public network, battery charge, back pumping of water, irrigation, hydrogen production, heating, and/or

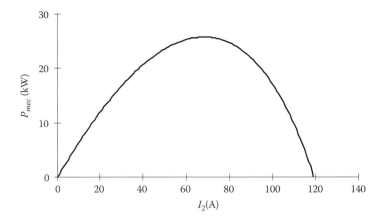

FIGURE 10.3 Mechanical power characteristic of an induction generator with respect to the load.

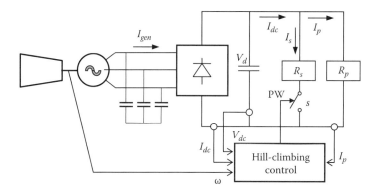

FIGURE 10.4   Induction generator with electronic control by the load.

freezing. Dissipation can also occur by simply burning out the excess of energy in bare resistors to the wind or to the flow of water. The primary load, R_p, is that of the consumer and it has the highest priority. Electronic control by the load for electric micropower plants is the purpose of using HCC. The HCC algorithm will adapt the load current to the available primary energy. There are several methods of using the metarule described in Section 10.3 for induction generators; some of them will be described in Section 10.5. However, the most common techniques to climb to the maximum generated point are fixed step, divided step, adaptive step, and exponential step.[4]

10.5   HCC-BASED MAXIMUM POWER SEARCH

10.5.1   FIXED STEP

The most straightforward HCC method of reaching maximum power peak is the fixed step approach. The logic of this algorithm is based on the output power $P_{dc}(= V_{dc}I_{dc})$ that is measured for each new increment (or decrement) of current ΔI_k being observed per increase (or decrease) of power. If the change in power ΔP_{dc} is positive, with a last variation $+\Delta I_k$ the search for MGP continues in the same direction; if, on the other hand, $+\Delta I_k$ causes a $-\Delta P_{dc}$, the direction of the search is reversed. Figure 10.5 shows a fixed step search where there is some large oscillation about MGP. The fixed step search has the inconvenience of causing oscillations around the MGP, that, if maintained, will cause voltage flickering, electromechanical stresses, and, when there is some resonance present, sustained oscillations. Besides that, if the oscillations have to be limited, the system will be slow at the beginning and during the recovery from surges and transient alterations in the amounts of primary energy.

10.5.2   DIVIDED STEP

Figure 10.6 illustrates the divided step approach, which minimizes the oscillations around MGP. It is similar to the fixed step approach, but, besides having its sign changed, the divided step increment of load current is divided by two after having reached the MGP.

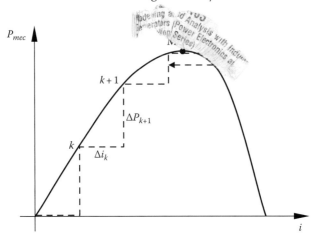

FIGURE 10.5   HCC fixed step.

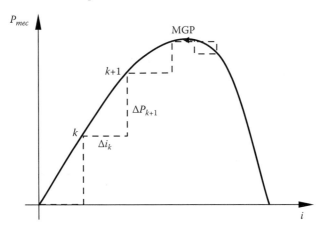

FIGURE 10.6   HCC divided step.

There is a tendency to dissipate the oscillations around that point after the step is applied a certain number of times. The value of the load current is determined by

$$i_{k+1} = i_k + \Delta i_{k+1} \tag{10.2}$$

and every time that MGP is reached, the step changes by

$$\Delta i_{k+1} = \frac{-\Delta i_k}{2} \tag{10.3}$$

When there are variations in the load or in the primary energy, the step resumes its original fixed value and, because of this, each recovery provokes load current oscillations, repeating the inconveniences of the fixed step approach. Another inconvenience of the fixed step and the divided step approaches is that they do not take into account machine inertia, and it is difficult to define after how many applications of the same step, the increment should be changed to a new size. That is, how many steps should be applied before the machine can fully respond to the new situation of the control variable Δi? Many alternatives using fixed and divided steps have been implemented in practice with very poor results with regard to stable operation and load current step size. For a reliable operation, it remains a question of the automatic establishment of a relationship between the inertia of the machine and the application of the current step.

10.5.3 ADAPTIVE STEP

The adaptive step approach is based on an acceleration factor obtained from the tangent to the $P \times I$ curve operating point of the generator. This tangent passes through points k and $k + 1$ in Figure 10.7. The accelerating factor at the point $k + 1$ can be defined as

$$k_{k+1} = \frac{\Delta P_{k+1}}{\Delta i_{k+1}} \left(\frac{I_0}{P_0} \right) \tag{10.4}$$

and the defined instantaneous current for it is $i_{k+1} = i_k + k_{k+1}\Delta i$.

As the operation point approaches the MGP, tangent slope (load voltage) tends to zero, and the current increment tends to zero this way, practically eliminating power oscillations, though sometimes there is a relatively slow oscillation due to the derivative nature of this algorithm. During transient states or under abrupt variations of load or primary power, the algorithm generates an acceleration factor proportional to the tangent slope. Quickly, a new point of stable operation is reached. To accelerate the increment of the generator current with the adaptive step, it should be generated by a large k_{k+1} that decreases as the MGP comes closer. The result is a softer and more stable operation of the load current increase. Two inputs must be carefully selected: the size of the initial step and the rule for the step variation.

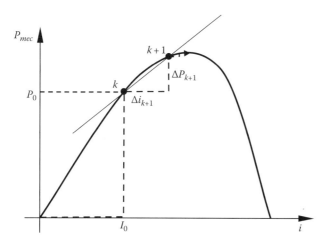

FIGURE 10.7   HCC adaptive step.

10.5.4 EXPONENTIAL STEP

Another possible solution to reach the maximum power peak is the exponential step approach as depicted in Figure 10.8. The exponential step Δi using an exponential incremental approximation gap e^{-ki} to the MGP is given in p.u. as

$$w_i = \frac{\Delta i}{i_{MGP}} = 1 - e^{-ki} \quad \text{for } i = 1, 2, 3, ..., n \tag{10.5}$$

The first iteration ($i = 1$) will give the initial current increment as

$$w_1 = 1 - e^{-k} = W_0 = \frac{\Delta i}{i_{MGP}} \tag{10.6}$$

from which it is possible to obtain

$$k = -\ln\left(1 - W_0\right) \tag{10.7}$$

where W_0 is the minimum step width to guarantee the self-excitation process of the generator. A generic step of current i is given by

$$\Delta i = w_i - w_{i-1} = \left(1 - e^{-ki}\right) - \left[1 - e^{k(i-1)}\right] = e^{-ki}\left(e^k - 1\right) \tag{10.8}$$

From Equation 10.5 for $i = n$, it is possible to determine that

$$n = -\frac{1}{k} \cdot \ln\left(1 - w_n\right) \tag{10.9}$$

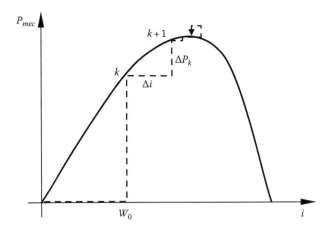

FIGURE 10.8 HCC exponential step.

The MGP exponential approach leads to a very fast transient response and a soft approach to the MGP. For implementation of this method, it is enough to determine the initial increment W_0 and the approaching gap e^{-ki} that determines the iteration number of incremental variations n and W_0, which determines the constant k. In the implementation of this method, two commitments also exist: (1) the transient state is related to the initial increment W_0; and (2) the operating state is related to the approaching gap, e^{-ki}.

In Figure 10.9, k is defined by the step algorithm used to establish the load current (fixed, divided, adaptive, or exponential). W_0 is the initial current step width that is proportional to the period T and it is established according to the secondary load demand or any other acceptable minimum generating level. For fixed step: $k = 1$; divided step: $k_n = 1/2^{n-1}$; adaptive step: $k = (\Delta P/\Delta I)/(\Delta P_o/\Delta I_o)$; exponential step: $k = (1/n)\ell n(1 - w_n)$, where n is nth step aiming at a particular maximum power point.

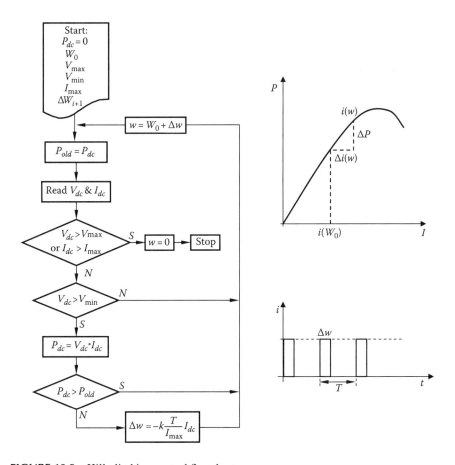

FIGURE 10.9   Hill-climbing control flowchart.

10.5.5 PRACTICAL IMPLEMENTATION OF INCREMENTAL STEPS

The flowchart in Figure 10.9 illustrates the routine sequence used to execute the incremental steps ΔV_c in the control program, which was implemented in LabView®. Small modifications in this program can be made for the fixed, divided, adaptive, and exponential steps as discussed later. In the laboratory tests, the size of each current step was established as a gradual application of a secondary load, R_s, in steps of 2% or so, according to the schematic diagram of Figure 10.4. The primary resistance, R_p, had a fixed value of 90 Ω while the variation of the secondary resistance went from 45 Ω until open circuit.

Because of the application of fixed step, the output power of the generator went according to the curve of Figure 10.10. As can be observed, the gradual application of this step of current went from 425 W, increasing quickly during 12 s, approximately, to stabilize in an oscillatory way around 475 W.

Similarly, to what was done with the fixed step, the application of the divided step resulted in the curve shown in Figure 10.11. Here a similarity was observed in the settling time to reach the set power, only with a difference in the oscillations around the steady-state power, 475 W. The power oscillations, in this case, had a much wider period at similar conditions.

With the adaptive step, it was observed in Figure 10.12 that the initial step was much larger than the subsequent ones, tending to zero at the same time as it approached the steady-state values and that the oscillation period around this point was much larger.

Finally, notice the augmented output power scale for the exponential step in the oscillatory curve shown in Figure 10.13, which presents smaller oscillations along all the evolution of the steady-state power. In spite of this, at the beginning of the power variation, it took a form similar to the other types of steps. The oscillations around the MGP were due to several factors, among which can be mentioned: (1) the

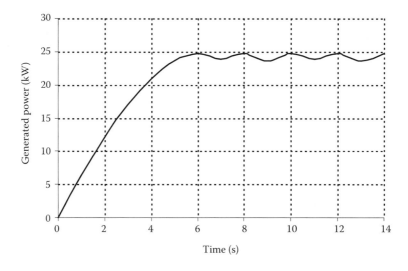

FIGURE 10.10   HCC with fixed step.

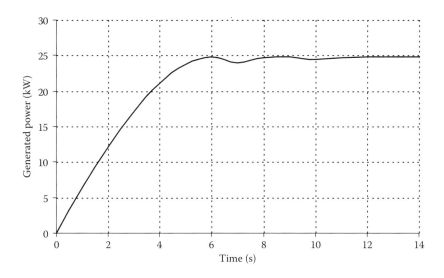

FIGURE 10.11   HCC with divided step.

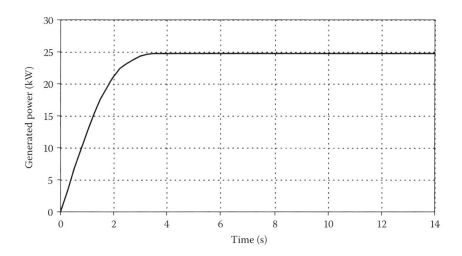

FIGURE 10.12   HCC with adaptive step.

coupling between the turbine and the generator (a rubber belt in these laboratory tests); (2) the inertia and power differences between the turbine and the generator with a small load (19.8 W to 7.5 kW); and (3) the coincidence or near-coincidence of the period of self-oscillation of the generator-turbine set with that of the increment steps used in the control.

In order to better observe the results, no measures were taken to reduce such phenomena during the tests. The major inconvenience of the fixed step and divided step

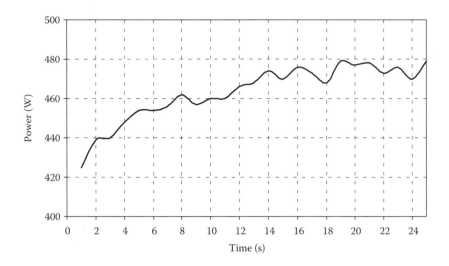

FIGURE 10.13   HCC with exponential step.

approaches is that they do not take into account the inertia of the machine, and it is difficult to define after how many applications of the same step it should be altered. Besides that, this step recovers slowly in load transients or during alterations in the amounts of the primary energy or of the load. With the adaptive and exponential steps, the generator causes an initial large step of load current and then decreases as the point of maximum power gets closer. The result is a softer and stable operation.

10.6   FLC-BASED MAXIMUM POWER SEARCH

FLC is an extension of the methods of load control discussed earlier. Appendix A3 reviews the concepts of FLC required for this chapter. The squirrel-cage induction generator with a variable-speed wind turbine (VSWT) system and double-sided PWM converter with FLC is used to maximize the power output and enhance system performance. All the control algorithms have been validated by simulation and implementation.[5–7] System performance has been evaluated in detail, and an experimental study with a 3.5 kW laboratory drive system was constructed to evaluate performance.

The voltage-fed converter scheme used in this system is shown in Figure 10.14. This work was particularly applied to a vertical-axis wind turbine, but horizontal turbines can be coupled to the shaft as well. A PWM-insulated gate-bipolar transistor (IGBT) rectifies the variable-frequency variable-voltage power from the generator. The rectifier also supplies the excitation needs for the induction generator. The inverter topology is identical to that of the rectifier, and it supplies the generated power at 60 Hz to the utility grid. Salient advantages of the double PWM converter system include

- The line-side power factor is unity with no harmonic current injection (satisfies IEEE 519).
- The cage-type induction generator is rugged, reliable, economical, and universally popular.

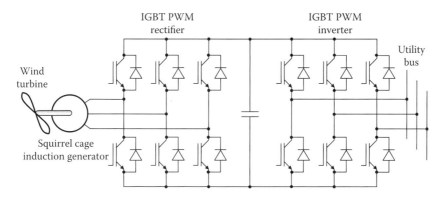

FIGURE 10.14 A double PWM converter–fed wind energy system.

- Induction generator current is sinusoidal.
- There is no harmonic copper loss.
- The rectifier can generate programmable excitation for the machine.
- Continuous power generation from zero to the highest turbine speed is possible.
- Power can flow in either direction, permitting the generator to run as a motor for start-up (required for vertical turbine). Similarly, regenerative braking can quickly stop the turbine.
- Autonomous operation of the system is possible with a start-up capacitor charging the battery.

The wind turbine is characterized by the power coefficient (C_p), which is defined as the ratio of the actual power delivered to the free wind stream power flowing through a similar but uninterrupted area. The tip speed ratio (TSR or λ) is the ratio of turbine speed at the tip of a blade to the free wind stream speed. The power coefficient is not constant, but varies with the wind speed, rotational speed of the turbine, and turbine blade parameters such as angle of attack and pitch angle. C_p is defined as a nonlinear function of the tip-speed ratio as

$$C_p(\lambda) = f(\lambda) \tag{10.10}$$

where

$$\lambda = R\omega/V_w$$

and C_p is the power coefficient

The weighted torque by the power coefficient at the shaft is

$$T_m = \frac{1}{\omega_m}\left[C_p(\lambda)P_m\right] \tag{10.11}$$

The turbine is coupled to the induction generator (IG) through a step-up gear box $(1{:}\eta_{gear})$ so that the IG runs at a higher rotational speed despite the low-speed ω of the wind turbine. Therefore, the aerodynamic torque of a vertical turbine is given by the equation

$$T_m = C_p(\lambda) \cdot \left[0.5 \frac{\lambda \rho_o R^3}{\eta_{gear}} \right] \cdot V_w^2 \tag{10.12}$$

where
    λ is the tip speed ratio (TSR) ($= R\omega/V_w$)
    ρ_o is the air density
    R is the turbine radius
    η_{gear} is the speed-up gear ratio
    V_w is the wind speed
    ω is the turbine angular speed

Figure 10.15 shows a set of constant power lines (dashed) superimposed on the family of curves indicating the region of maximum power delivery for each wind speed. This means that, for a particular wind speed, the turbine speed (or the TSR) is to be varied to get the maximum output power, and this point deviates from the maximum torque point, as indicated. Since the torque–speed characteristics of the wind generation system are analogous to those of a motor-blower system (except that the turbine runs in reverse direction), the torque follows the square-law characteristics and the output power follows the cube law, as indicated earlier. This means that,

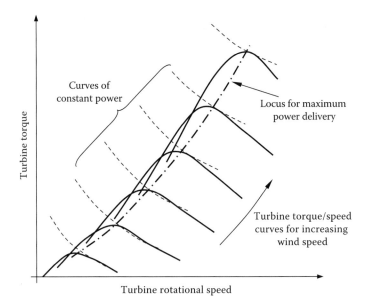

FIGURE 10.15 Family of turbine torque characteristics for variable wind velocities.

under reduced-speed light-load steady-state conditions, generator efficiency can be improved by programming its magnetic flux.

The static torque as given by Equation 10.12 is the average torque production, which must be added to the intrinsic torque pulsation. An actual turbine is quite complex to model since the moving surface of the turbine slips with respect to the wind and there are also extra components of wind in the x, y, and z directions. The blade that moves with the wind during one-half revolution moves against the wind during the other half revolution, and some turbulence occurs behind the moving surface. There are various inertial modes in the system such as rotor blades, hub, gearbox, shaft, generator, and damping caused by the wind itself and the oil in the gearbox. The most important source of torque pulsation taken into account for the present model is that of the rotor blades passing by the tower. The oscillatory torque of the turbine is more dominant at the first, second, and fourth harmonics of the fundamental turbine angular velocity (ω) and is given by Equation 10.13:

$$T_{osc} = T_m \left[A\cos(\omega) + B\cos(2\omega) + C\cos(4\omega) \right] \tag{10.13}$$

In order to maintain the necessary optimum tip-speed ratio, that is, the quotient of wind velocity to the generator speed that extracts the maximum power, an experimental measurement can be conducted to build a look-up table for the actual turbine tip speed ratio profile, and with wind velocity measurements, to change the turbine speed setup. This procedure requires an experimental characterization for each turbine and an anemometer should be used to online track the best operating point. Changing from one optimum point to the next is a necessary stepwise procedure, and resonance must be avoided. The step size is usually not optimum in this approach. So, an FLC can be suggested to cope with these uncertainties.

10.6.1 Fuzzy Control of Induction Generators

The reasons to use an induction generator fuzzy control in wind energy systems are

- To change the generator speed adaptively, so as to track the power point as the wind velocity changes (without wind velocity measurement)
- To reduce the generator rotor flux, boosting the induction generator efficiency when the optimum generator speed setup is attained (in steady state)
- To have robust speed control against turbine torque pulsation, wind gusts, and vortices

Figure 10.16 shows the control block diagram of an induction generator with a double PWM converter–fed wind energy system that uses the power circuit of Figure 10.14. The induction generator and inverter output currents are sinusoidal due to the high frequency of the pulse-width modulation and current control as shown in the figure. The induction generator absorbs lagging reactive current, but the current is always maintained in phase at the line side; that is, the line power factor is unity. The rectifier uses indirect vector control in the inner current control loop, whereas the direct vector control method is used for the inverter current controller.[8,9]

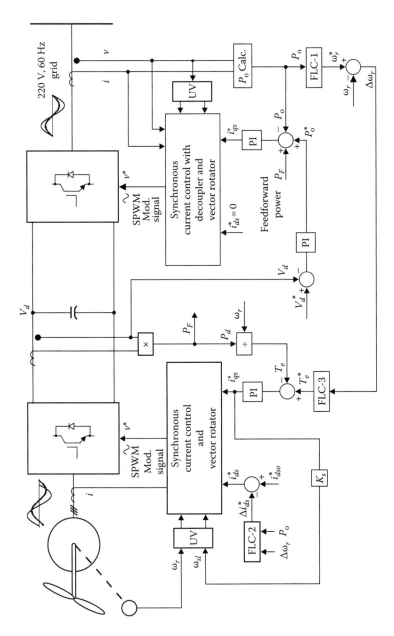

FIGURE 10.16 Fuzzy-controlled double PWM converter for wind energy system.

Vector control permits fast transient response of the system as discussed in Chapter 9. A fuzzy logic–based vector control is used to enhance three characteristics in this system: (1) the search for the best generator speed (FLC-1) command to track the maximum power of the wind, (2) the search for the best flux intensity in the machine (FLC-2) to optimize inverter and induction generator losses, and (3) robust control of the speed loop (FLC-3) to overcome possible shaft resonances due to wind gusts and vortex. Fuzzy controllers are described in detail in Simões et al.[5,6]

For a particular wind velocity V_w, there is an optimum setting of the rotor speed ω_r^*. The speed loop control generates the torque component of the induction generator current to balance the developed torque with the load torque. The variable-voltage variable-frequency power from the supersynchronous induction generator is rectified and pumped to the dc link. The dc-link voltage controller regulates the line power P_o (i.e., the line active current) so that the link voltage always remains constant. A feedforward power signal from the induction generator output to the dc voltage loop prevents transient fluctuation of the link voltage.

There is a local inductance L_s connected between the line-side inverter output and the three-phase utility bus. Such inductance is very important for stable operation of the synchronous current controller and it is selected in such a way that the maximum modulation index at which the line-side inverter operates is as close to one as is permitted by the minimum pulse or notch width capability of the device. Under these conditions, the control varies over a wide range, making the control less sensitive to errors in the controller gains and compensation. The inductance value is such that the slope of the PI output is smaller than the slope of the triangular carrier of the SPWM.[9]

The system can be satisfactorily controlled for start-up and regenerative braking shutdown modes besides the usual generating mode of operation. The flowchart in Figure 10.17 shows the procedure for the initial start-up of a wind turbine. Both inverters are initially disabled during charging of the dc-link capacitor; that is, all the gate pulses are off. The capacitor is charged from the diodes on the line-side inverter with the peak value of the line voltage. To prevent damage to diodes, there is a series resistance for the capacitor initial charging R_s in the dc link. Such resistance is bypassed by an electromagnetic relay when the bus voltage reaches something like 95% of the line voltage. When the capacitor has been charged, the dc-link voltage loop control is exercised, the pulse gates are enabled, and the control loops are activated to establish a smaller voltage than the bus peak value (typically, 75% of the peak value), therefore successfully operating the line-side inverter in PWM mode.

With the dc-link bus voltage fixed, the induction generator can be excited with i_{ds}, as the rated flux is established in the induction generator. Next, a speed reference is commanded to rotate the turbine with the minimum turbine speed required to catch some power from wind (the fuzzy speed controller FLC-3 is always working). The flow of wind imposes a regenerative torque in the induction generator. Of course, the power generation is not optimum yet, but the slip frequency becomes negative and the power starts to flow from the turbine toward the line side. As the power starts to flow to the line side, the dc-link voltage loop control can be gradually commanded to a higher value than the peak value of the line-side voltage (typically, 75% higher).

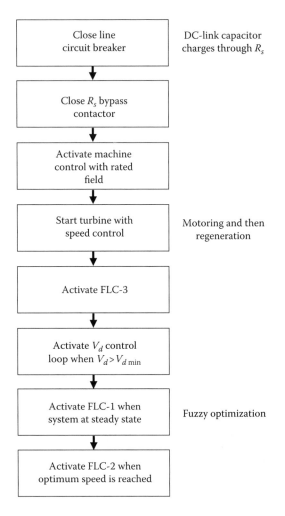

FIGURE 10.17 Initialization of control system.

After the dc-link voltage is established in the new higher value, the system is ready to be controlled by fuzzy controllers FLC-1 and FLC-2.

An IGBT PWM-bridge rectifier that also supplies lagging excitation current to the induction generator rectifies the variable-frequency variable-voltage power generated by the induction generator. The dc-link power is inverted to 60 Hz, 220 V ac through an IGBT PWM inverter and fed to the utility grid at unity power factor, as indicated. The line-power factor can also be programmed to lead or lag by static VAR compensation, if desired. The generator speed is controlled by indirect vector control with torque control and synchronous current control in the inner loops. The induction generator flux is controlled in open loop by control of the i_{ds} current, but in normal condition, the rotor flux is set to the rated value for fast transient response. The line-side converter is also vector-controlled using direct vector control and synchronous current control in the inner loops. The output power P_o is controlled to control the dc-link voltage V_d. Since increase of P_o causes decrease of V_d, the voltage loop error polarity has been inverted.

Tight regulation of V_d within a small tolerance band requires a feedforward power injection in the power loop, as indicated. The insertion of the filter inductance L_s creates some coupling effect, which is eliminated by a de-coupler in the synchronous current control loops. The power can be controlled to flow easily in either direction. A vertical wind turbine requires start-up motoring torque. As the speed develops, the induction generator goes into generating mode. Regenerative braking shuts down the machine.

For the system shown in Figure 10.16, the machine-side inverter uses indirect vector control (IVC). With the control of vector currents i_{ds} and i_{qs}, the rotor flux is aligned with the d-current i_{ds}. With vector control there is no danger of an instability problem; a four-quadrant operation, including zero speed, is possible. The drive has a dc-machine-like transient response: there is no direct frequency control, and the frequency is controlled through the unit vector generated by addition of the induction generator speed ω_{rt} with a feedforward slip-frequency ω_{sl}. The stator angular frequency ω_e is integrated and the flux angle is fed to a sine/cosine table for inverse vector rotation of the machine currents. The PI current controller generates the quadrature reference voltages v_{qs}^* and v_{ds}^*, and a PWM modulator that vector-rotates such voltages with the flux angle in order to generate the gate pulses for the IGBT. The vector rotation, two-phase to three-phase, and three-phase to two-phase transformations are standard and performed as in the literature on vector control.[8,9]

The line-side inverter uses direct vector control (DVC), with a very fast transient response for controlling the dc-link voltage and power flow. The unit vector is generated from the line-side voltages as indicated by the phasor diagram of Figure 10.18, so the vector current I_s is in phase with the vector voltage V_s. Since the reactive current i_{ds} is kept to zero, the line-side current is inverse-vector-rotated, and a synchronous current controller with a decoupling network places the three-phase inverter currents in phase with the three-phase voltages, permitting unity power factor operation. If there is an available inverter power rating, the line-side power factor can be programmable for leading or lagging as required by the utility power system. The equations used in the line-side DVC strategy are available in the literature.[10–12]

10.6.2 Description of Fuzzy Controllers

The optimum fuzzy logic operating control of an induction generator requires three different FLC. The controller here designated as FLC-1 tracks the maximum wind speed. The controller FLC-2 establishes the optimum magnetic flux for the induction generator. FLC-3 is the optimum generator's torque-speed controller.

10.6.2.1 Speed Control with the Fuzzy Logic Controller FLC-1

The product of torque and speed is turbine power, and it equals the line power (assuming a steady-state lossless system). The curves in Figure 10.19 show the line power P_o over a family of rotor speed ω_r in terms of wind velocities. For a particular

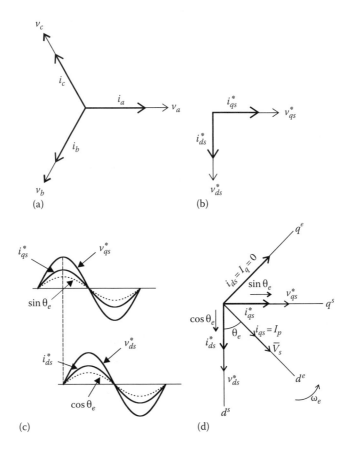

FIGURE 10.18 Phasor diagram for line-side DVC: (a) three-phase line phasors, (b) two-phase line phasors, (c) signal voltage and current waves, and (d) signals in d^s–q^s and d^e–q^e frames.

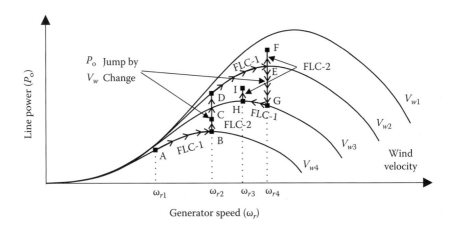

FIGURE 10.19 Fuzzy control FLC-1 and FLC-2 operation showing maximization of line power.

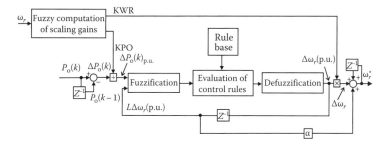

FIGURE 10.20 Block diagram of fuzzy control FLC-1.

value of wind velocity, the function of the fuzzy controller FLC-1 is to seek the generator speed until the system settles down at the maximum output power condition. At a
wind velocity of V_{w4}, the output power will be at A if the generator speed is ω_{r1}. The
FLC-1 will alter the speed in steps until it reaches the speed ω_{r2} where the output power
is maximized at B. If the wind velocity increases to V_{w2}, the output power will jump
to D, and then FLC-1 will bring the operating point to E by searching the speed to ω_{r4}.
The profile to decrease the wind velocity to V_{w3} is also indicated. Therefore, the principle of the fuzzy controller is to increment (or decrement) the speed in accordance with
the corresponding increment (or decrement) of the estimated output power P_o. After
reaching the optimum generator speed, which maximizes the turbine aerodynamic
performance, the machine rotor flux i_{ds} is reduced from the rated value, to reduce the
core loss, thereby further increasing the machine-converter system efficiency.

If ΔP_o is positive with the last positive $\Delta\omega_r$, the search is continued in the same
direction. If, on the other hand, $+\Delta\omega_r$ causes $-\Delta P_o$, the direction of search is reversed.
The variables ΔP_o (variation of power), $\Delta\omega_r$ (variation of speed), and $L\Delta\omega_r$ (last
variation of speed) are described by membership functions given in Figure 10.20 and
the inference is given in the rule in Table 10.1.

In the implementation of fuzzy control, the input variables are fuzzified in accordance with the membership functions indicated in Figure 10.21. The valid control

TABLE 10.1

Rule Table for FLC-1

ΔP_o \\ $\Delta\omega_{r(last)}$	P	ZE	N
PVB	PVB	PVB	NVB
PBIG	PBIG	PVB	NBIG
PMED	PMED	PBIG	NMED
PSMA	PSMA	PMED	NSMA
ZE	ZE	ZE	ZE
NSMA	NSMA	NMED	PSMA
NMED	NMED	NBIG	PMED
NBIG	NBIG	NVB	PBIG
NVB	NVB	NVB	PVB

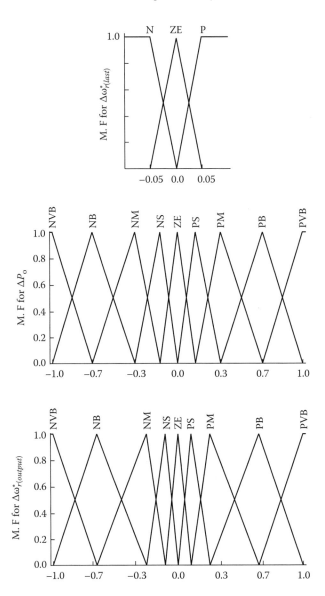

FIGURE 10.21 Membership function for FLC-1.

rules are evaluated and combined through rule Table 10.1, and finally the output is back defuzzified to convert it to a crisp value. The output $\Delta\omega_r$ is added by some amount of $L\Delta\omega_r$ in order to avoid local minima due to wind vortex and torque ripple. The controller operates on a per-unit basis so that the response is insensitive to system variables and the algorithm is universal to any system.

The scale factors K_{PO} and $K\omega_r$ as shown are functions of the generator speed, so that the control becomes somewhat insensitive to speed variation. The scale factor

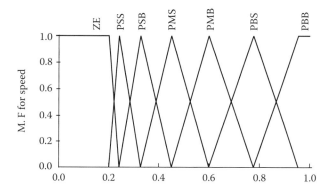

FIGURE 10.22 Membership function for computation of scaling gains.

TABLE 10.2

Rule Table for Computation of Scaling Gains

ω_r	K_{PO}	K_{ω_r}
PSS	40	25
PSB	210	40
PMS	300	40
PMB	375	50
PBS	470	50
PBB	540	60

is generated by fuzzy computation. The speed is first evaluated into seven fuzzy sets as shown in membership functions indicated in Figure 10.22; the scale factors K_{PO} and $K\omega_r$ are generated according to the rule in Table 10.2. From this explanation, the advantages of fuzzy control are obvious. It provides adaptive step size in the search that leads to fast convergence, and the controller can accept inaccurate and noisy signals. The FLC-1 operation does not need any wind velocity information, and its real-time-based search is insensitive to system parameter variation.

10.6.2.2 Flux Intensity Control with the Fuzzy Logic Controller FLC-2

Since most of the time the generator is running at light load, the induction-generator rotor flux i_{ds} can be reduced from the rated value to reduce the core loss and thereby increase the machine-converter system efficiency. The principle of online search-based flux programming control by a second fuzzy controller FLC-2 is explained in Figure 10.23. At a certain wind velocity V_w and at the corresponding optimum speed V_w established by FLC-1, which operates at rated flux (ψ_R), the rotor flux (ψ_R) is reduced by decreasing the magnetizing current i_{ds}. This causes an increasing torque current i_{qs} by the speed loop for the same developed torque. As the flux is decreased,

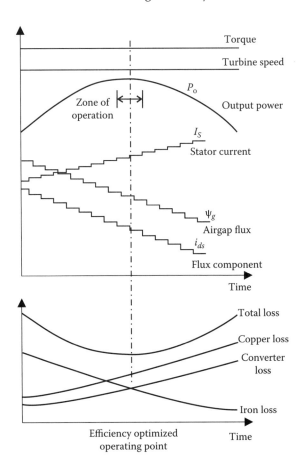

FIGURE 10.23 Fuzzy control FLC-2 operation showing optimization of flux level.

the induction generator iron loss decreases with the concurrent increase of copper loss. However, the total system (converters and machine) loss decreases, resulting in an increase of the total generated power P_o. The search is continued until the system settles down at the maximum power point A, as indicated in Figure 10.23. Any attempt to search beyond point A will force the controller to return to the maximum power point.

The principle of fuzzy logic controller FLC-2 is somewhat similar to that of FLC-1 and is explained in Figure 10.24. The system output power $P_o(k)$ is sampled and compared with the previous value to determine the increment ΔP_o. In addition, the last excitation current decrement $L\Delta i_{ds}$ is reviewed. The membership functions for variation of power ΔP_o, last variation of i_{ds}, $L\Delta i_{ds}$ and change in i_{ds}, Δi_{ds}, are given in Figure 10.25. On these bases, the decrement step of i_{ds} is generated from fuzzy rules given in rule Table 10.3 through fuzzy inference and defuzzification.

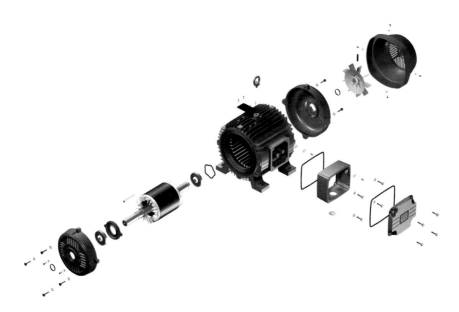

FIGURE 6.2   Exploded view of induction generator major parts. (Courtesy of WEG Motores, Brazil.)

FIGURE 6.4   Finished squirrel cage for large power induction motor. (Courtesy of Equacional Motores, Brazil.)

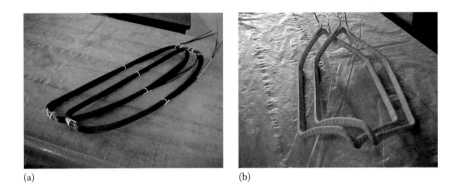

(a)   (b)

FIGURE 6.6 Stator coils: (a) random-wound coil and (b) form-wound coil.

FIGURE 6.13 Medium-sized eight-pole induction machine, with form wound stator coils. (Courtesy of Equacional Motores, Brazil.)

FIGURE 6.14 Squirrel cage rotor for a large induction machine showing the brazed bars to the end rings. (Courtesy of Equacional Motores, Brazil.)

FIGURE 6.15 Stator and rotor punchings for a large induction machine. (Courtesy of Equacional Motores, Brazil.)

FIGURE 6.16 Stator core of a medium-sized induction machine prepared for winding. (Courtesy of Equacional Motores, Brazil.)

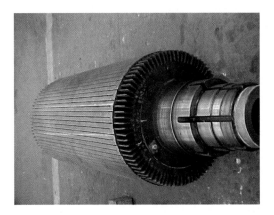

FIGURE 6.17 Medium-sized squirrel cage rotor winding. (Courtesy of Equacional Motores, Brazil.)

FIGURE 6.18 Random stator winding of a small induction machine. (Courtesy of Equacional Motores, Brazil.)

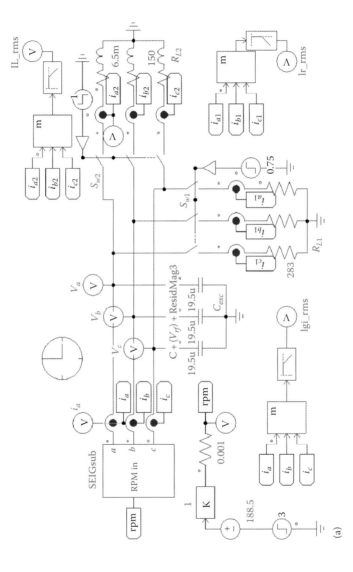

FIGURE 12.31 (a) PSIM SEIG circuit model with linear loads switched at distinct instants.

(Continued)

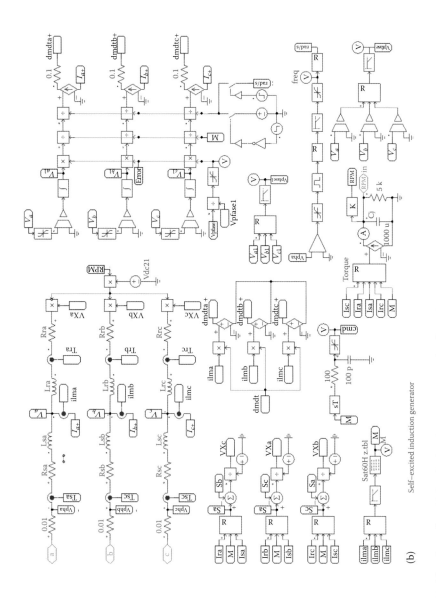

FIGURE 12.31 (*Continued*) (b) SEIGsub subsystem, expansion from (a).

(b)

FIGURE 13.4 CHUJEON-A, a prototype of wave power. (From Jeon, B.-H. et al., Autonomous control system of an induction generator for wave energy device, Ocean Development System Lab., Korea Research Institute of Ships and Ocean Engineering, KORDI, Daejeon, Korea, *Proceedings of The Fifth ISOPE Pacific/Asia Offshore Mechanics Symposium*, Daejeon, Korea, November 2002, pp. 17–20.)

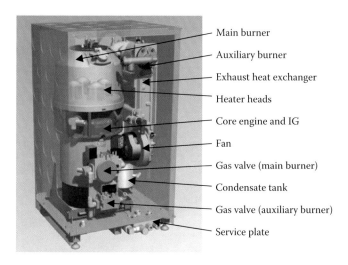

— Main burner
— Auxiliary burner
— Exhaust heat exchanger
— Heater heads
— Core engine and IG
— Fan
— Gas valve (main burner)
— Condensate tank
— Gas valve (auxiliary burner)
— Service plate

FIGURE 13.5 The AC WhisperGen™ gas-fired model. (Courtesy of WhisperGen Ltd., Christchurch, New Zealand.)

FIGURE 13.6 Pima-Maricopa Indian Reservation SunDish. (Extracted from Scott, S.J., Holcomb, F.H., and Josefik, N.M., Distributed electrical power generation, Summary of alternative available technologies, ERDC/CERL SR-03-18, US Corps of Engineers, Engineering Research and Development Center, September 2003.)

FIGURE 13.9 Shaft between the induction machine and the turbine.

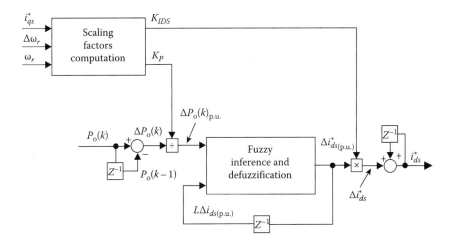

FIGURE 10.24   Block diagram of fuzzy control FLC-2.

The adjustable gains K_P and K_{IDS}, which convert the actual variables to per-unit variables, are given by the respective expressions:

$$K_P = a\omega_r + b \tag{10.14}$$

$$K_{IDS} = c_1\omega_r - c_2T + c_3 \tag{10.15}$$

where a, b, c_1, c_2, and c_3 are derived by trial and error. The effect of controller FLC-2 is to boost the power output.

The FLC-2 controller operation starts when FLC-1 has completed its search at the rated flux condition. If wind velocity changes during or at the end of FLC-2, its operation is abandoned, the rated flux is established, and FLC-1 control is activated.

10.6.2.3   Robust Control of Speed Loop with the Fuzzy Logic Controller FLC-3

Speed loop control is provided by fuzzy controller FLC-3, as indicated in the block diagram of Figure 10.26. As mentioned before, it provides robust speed control against wind vortex and turbine oscillatory torque. The disturbance torque on the induction generator shaft is inversely modulated with the developed torque to attenuate modulation of output power and prevent any possible mechanical resonance effect. In addition, the speed control loop provides a deadbeat-type response when an increment of speed is commanded by FLC-1. Figure 10.26 shows the proportional-integral (PI) type of fuzzy control used in the system. The speed loop error $E\omega_r$ and error change $\Delta E\omega_r$ signals are converted to per-unit signals, processed through fuzzy control in accordance with the membership functions given in Figure 10.27 and the rule in Table 10.4. The output of the fuzzy controller FLC-1 is summed to produce the generator torque reference T_e^*.

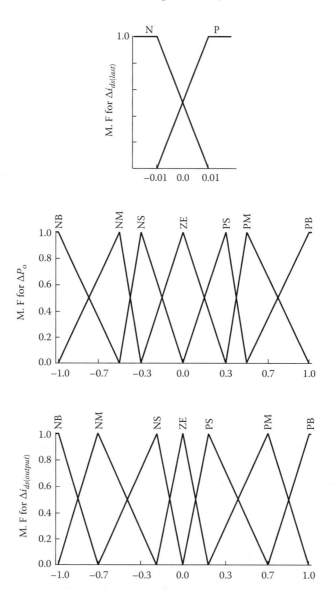

FIGURE 10.25 Membership function for FLC-2.

10.6.3 EXPERIMENTAL EVALUATION OF FUZZY OPTIMIZATION CONTROL

The induction machine used in the experimental setup for this chapter was a standard NEMA Class B type with 220 V, 3.5 hp rating. A 7.5 hp four-quadrant laboratory dynamometer was used to emulate the wind turbine (programmable shaft torque). The system parameters are given in Table 10.5.

While the speed fuzzy controller FLC-3 was always active during system operation, the controllers FLC-1 and FLC-2 operated in sequence at steady (or small

TABLE 10.3

Rule Table for FLC-2

ΔP_o \ $\Delta i_{ds(last)}$	N	P
PB	NM	PM
PM	NS	PS
PS	NS	PS
NS	PS	NS
NM	PM	NM
NB	PB	NB

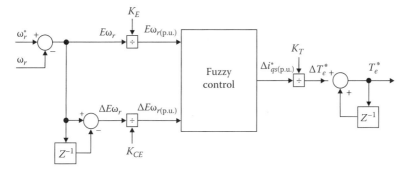

FIGURE 10.26 Block diagram of fuzzy control FLC-3.

turbulence) wind velocity. Besides, a start-up procedure was required for complete activation and a shutdown sequence in case of a fault. There was control coordination for sequencing and the start-up/shutdown procedures.[4–6] For the start-up, the line-side circuit breaker was closed. The dc-link capacitor charged to the peak value of the line voltage through a series resistance, which avoided the inrush charging current. After a delay of 0.5 s, the resistance was bypassed with a relay, and the rated flux was imposed on the induction machine.

The turbine was started with speed control and as the power started to flow, the dc-link voltage rose. The dc-link voltage control was activated when the dc-link voltage was above the limit value and FLC-1 started to search the optimum speed reference ω_r^*. As ω_r^* was altered, the power generation went up, until FLC-1 settled down in the steady-state condition, indicated by a small variation in $\left|\Delta\omega_r^*\right|$ with alternating polarity.

In this condition, the system was transferred to FLC-2 in order to optimize the flux by decreasing i_{ds}^*. During the speed search, any deviation from the expected variation of power indicated that the system was subject to large wind variation and the system was transferred from FLC-1 to nonfuzzy operation, waiting for the transient to vanish. During the flux optimization, the excitation current i_{ds}^* was decreased adaptively by FLC-2. Control was then transferred to optimum operation when the

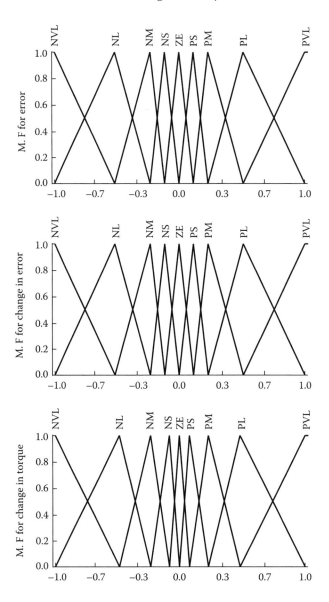

FIGURE 10.27   Membership function for FLC-3.

variation $\left|\Delta i_{ds}^*\right|$ was small with alternating polarity. The search was finished, and the optimum operation state kept the optimum ω_r^* and i_{ds}^*.

The power and variation of power were recorded in order to see if there was any variation in the wind velocity to restart the search. Again, during the operation of FLC-2, any load transient, indicated by variation of torque, transferred the system to the nonfuzzy operation state, waiting for the transient to vanish in order to restart from FLC-1.

TABLE 10.4
Rule Table for FLC-3

CE \ E	NVL	NL	NM	NS	ZE	PS	PM	PL	PVL
NVL					NVL	NL	NM	NS	ZE
NL					NL	NM	NS	ZE	PS
NM				NL	NM	NS	ZE	PS	PM
NS			NL	NM	NS	ZE	PS	PM	PL
ZE		NL	NM	NS	ZE	PS	PM	PL	
PS	NL	NM	NS	ZE	PS	PM	PL		
PM	NM	NS	ZE	PS	PM	PL			
PL	NS	ZE	PS	PM	PL				
PVL	ZE	PS	PM	PL	PVL				

TABLE 10.5
Induction Generator and Turbine Parameters

Induction generator parameters

5 hp	230/460 V	13.4/6.7 A
4 poles	1800 rpm	NEMA Class B
$R_s = 0.370\ \Omega$		$R_r = 0.436\ \Omega$
$L_{ls} = 2.13$ mH	$L_{lr} = 2.13$ mH	$L_m = 62.77$ mH

Turbine parameters

$A = 0.015$	$B = 0.03$	$C = 0.015$
	$\eta_{gear} = 5.7$	

The system had external fault indications from the turbine and inverters, which could indicate dangerous operation during too high wind velocity, or when wind velocity was too low and the power could not be satisfactorily generated. Inverters could indicate a tripping by short-circuit too much current or too-high temperature. If the machine inverter tripped, the line-side inverter could easily shut down the system. On the other hand, if the line-side inverter tripped, the dynamic break in the dc-link bus would keep the bus voltage in a safe range, while the induction generator decelerated to zero speed. Any fault that occurred during the search procedure transferred the system to the shutdown procedure. The fuzzy controllers were disabled, and a deceleration profile was imposed in the speed control mode. The turbine was mechanically yaw-controlled to permit the wind to pass through the blades, and finally the line circuit breaker was disconnected.

Figure 10.28 shows the static characteristics of the wind turbine at different wind velocity. Basically, these are families of curves for turbine output power, turbine torque, and line-side output power as functions of wind velocity and sets of generator speed. For example, if the generator speed remained constant and the wind velocity increased, the turbine power, turbine torque, and line power would increase and then

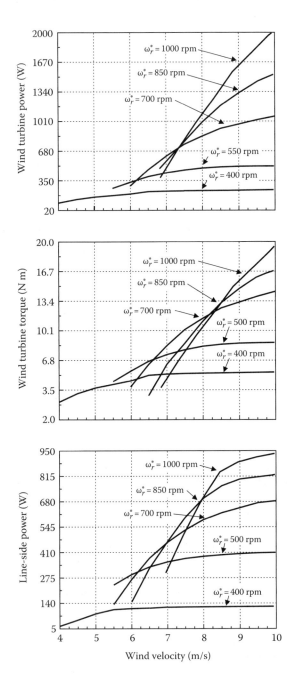

FIGURE 10.28   Static characteristics.

tend to saturate. The slope of increase is higher with higher generator speed. For a fixed wind velocity, as the generator speed increased, the torque and power outputs first increased and then decreased.

The effect of fuzzy controller FLC-1 when the wind velocity slowly increased from 6.125 to 8.750 m/s is shown in Figure 10.29. The generator speed tracked the wind velocity, gradually increasing the line-side power, as shown. As the wind velocity settled down to the steady-state value, the generator speed step size decreased until the optimum power output point oscillated with a small step, as explained before. Note that for a fast transient in wind velocity, the controller FLC-1 remained inoperative. The dc-link voltage is analogous to the level of a low-capacity water tank with input and output water flow pipes, and maintaining its value rigidly constant is extremely difficult. A feedforward power compensator, shown in Figure 10.16, helped reduce the fluctuation. However, as long as the voltage level remained within a tolerance band, the converter system was safe and the performance was satisfactory. The dynamic brake tended to absorb destructive voltage surges in the dc link, as mentioned before.

The fuzzy controller FLC-2 started when the operation of FLC-1 ended. Figure 10.30 shows the performance of FLC-2 at a constant wind velocity of 5.0 m/s and constant generator speed of 520 rpm. The generator excitation current gradually decreased from the initial rated value that increased the line-side power because of improved machine efficiency. The step size of the excitation current gradually decreased as the steady state was approached, and then it oscillated around the steady-state point.

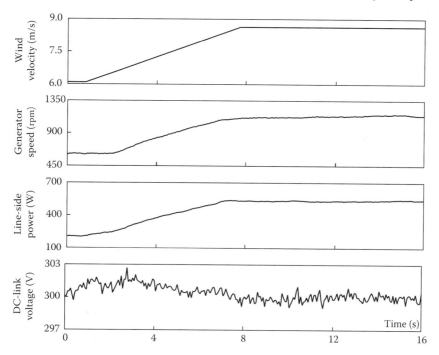

FIGURE 10.29   Aerodynamic optimization with FLC-1.

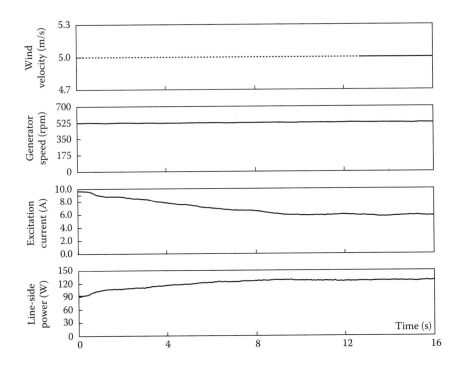

FIGURE 10.30   Flux optimization with FLC-2.

The increment of power was not very large because the induction generator was operating in a light-load condition.

In order to test the robustness of the fuzzy speed controller FLC-3, the oscillatory torque components were added to the dynamometer wind turbine model. Figure 10.31 shows the smooth speed profile (top) at 900 rpm with oscillatory torque (bottom) that swung from 4.35 to 5.65 N m with an average value of 5 N m. The additional pulsating torque introduced by FLC-2 was also highly attenuated by the FLC-3 controller. Figure 10.32 shows an oscillogram of the experimental setup.

Next, the turbine generator system was operated at constant speed (940 rpm), and the wind velocity was varied. At each operating point, the FLC-1 and FLC-2 controllers were operated in sequence and the corresponding boost of power was observed. From these data, the respective efficiency improvement was calculated and plotted earlier. Figure 10.33 indicates that the efficiency gain was significant with FLC-1 control compared with that of FLC-2. The gain due to FLC-1 fell to zero near 0.7 p.u. wind velocity where the generator speed was optimum for that wind velocity, and then rose. The gain due to FLC-2 decreased as the wind velocity increased because of higher generator loading.

Figure 10.34 shows the steady-state performance enhancement of control of the wind turbine at several operating points. After start-up, the system was operating at A. As the speed reference command switched from the host communication interface to the fuzzy controller FLC-1, the reference speed increased, and the optimized point at B (for wind velocity of 5 m/s) was reached. The flux optimization search further

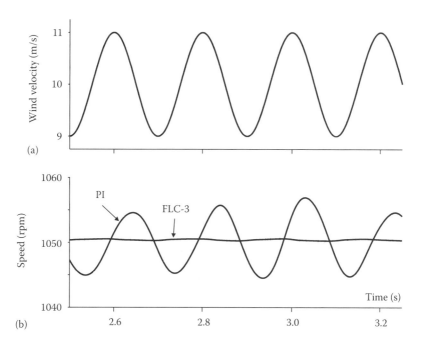

FIGURE 10.31   Comparison of speed controller response by PI and FLC-3 for wind vortex: (a) wind velocity with vortex and (b) FLC and PI responses.

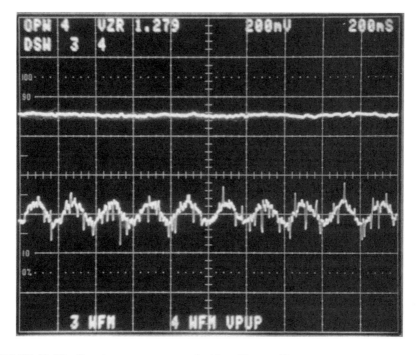

FIGURE 10.32   Speed response compared with turbine oscillatory torque.

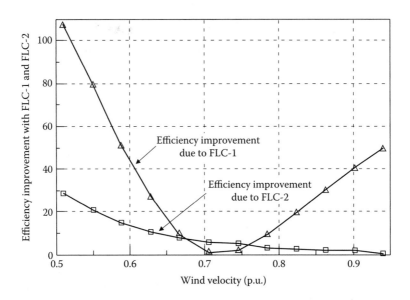

FIGURE 10.33   Wind energy efficiency improvement due to fuzzy control.

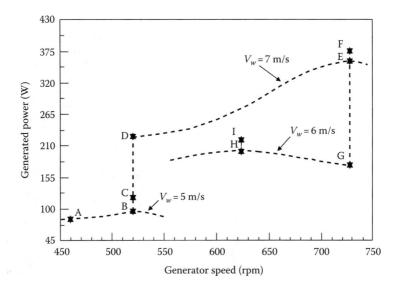

FIGURE 10.34   Fuzzy logic steady-state performance enhancement control of wind turbine at several operating points.

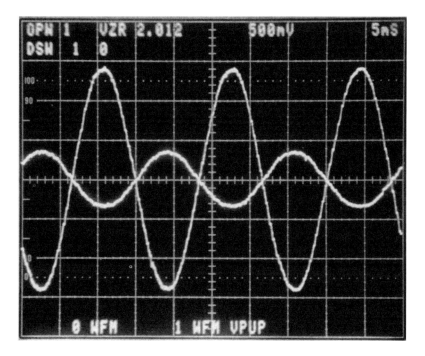

FIGURE 10.35   Line-side voltage and current.

enhanced the power generation by boosting the output power to the point C, as shown in Figure 10.34. The figure also shows the output power level, point D, when the wind velocity stepped from 5 to 7 m/s. The rated flux was established, and the fuzzy controller FLC-1 was enabled to search the optimum point at E. Again, the flux optimization boosted the generated power to F. The path G-H-I indicates the control operation when the wind velocity was stepped down from 7 to 6 m/s. The line-side direct vector control kept $i_{ds} = 0$, maintaining unity power factor all the time. Figure 10.35 shows an oscillogram with a sinusoidal line current and unity power factor. The out-of-phase current wave indicates that the system was in the generation mode.

10.7   CHAPTER SUMMARY

This chapter covered principles of HCC and FLC and applied them to the energy performance enhancement of induction generator–based renewable energy systems. HCC is appropriate to implement electronic control by the load, and having the load follows the incoming input power, whereas FLC is appropriate to adjust the induction generator speed to match the maximum aerodynamic performance, flux programming, and robust speed control with three fuzzy controllers. HCC and FLC have also been applied for optimization of other types of renewable energy sources and performed well.[13–16] The advantages of both hill climbing and fuzzy control are that the control algorithms are universal, fast converging, parameter insensitive, and accepting of noisy and inaccurate signals.

10.8 SOLVED PROBLEMS

1. What are the characteristics of the optimization methods related to renewable sources of energy? Why may efficiency in hill climbing and fuzzy control be irrelevant with respect to the maximum generated power?

Solution

(a) The renewable sources of energy-based electric power are a good case for using heuristic optimization techniques. For such power systems, the installation costs are very high while the availability of power is intermittent. It is therefore of vital importance to optimize the efficiency of electric power transfer, even for the sake of relatively small incremental gains, in order to amortize the installation costs within the shortest possible time. Given next are some important characteristics of such system optimization:

- Parameter variation of the system
- Need for linguistically modeled process
- A need to improve efficiency as a matter of operator judgment
- Dependence on operator skills and attention
- Process parameters are correlated to some extent
- Effects cannot be attained by separate PID
- Data-intensive modeling

(b) Most of the renewable sources of energy (like wind, solar) are nondispatchable by nature. That means we have to use the primary energy when they are available and convert it to electrical energy, otherwise that primary energy is gone. For that reason, the system should be able to generate the maximum power and the control should be based on that. The efficiency is not that important because though the system may be in maximum efficiency, this operating point may differ from the maximum generated power point, which in turn fails to capture the maximum primary energy available at that instant.

2. Give five reasons why optimization in renewable energy systems is important.

Solution

Renewable energy systems are or have

- Nondispatchable.
- High installation costs.
- Intermittent and need of storage devices associated to the optimization methodology.
- Generally, small power sources require different control than conventional huge power plants particularly because of parameter variation and correlation of several different variables.
- The optimization of renewable energy systems brings the maximum power that is available for the resource; anything less than such maximum is not recoverable, that is, the wind or water will flow, the sunlight will not be converted. On the contrast, fossil fuel sources can be shutdown and save their fuel reserves for latter.

10.9 SUGGESTED PROBLEMS

10.1 Describe the tracking methodology used in HCC and discuss each method to provide a basis for choosing one of them for a particular application.

10.2 Describe the tracking methodology used in fuzzy logic control and compare it with that of HCC.

10.3 Explain under what conditions and why HCC adaptive control may be unstable and why the HCC exponential control tends to be the most stable, although limited.

10.4 Using the Fuzzy Logic Toolbox in MATLAB®/Simulink®, design a controller capable of setting the temperature of room in a comfortable zone, suppose you have an electrical machine to control the HVAC.

REFERENCES

1. Toumiya, T., Suzuki, T., and Kamano, T., Output control method by adaptive control to wind power generating system, *Proceedings of IECON'91*, Vol. 3, November 1991, pp. 2296–2301.
2. Datta, R. and Ranganathan, V.T., A method of tracking the peak power points for a variable speed wind energy conversion system, *IEEE Trans. Energy Convers.*, 18(1), 163–168, March 2003.
3. Farret, F.A., Pfitscher, L.L., and Bernardon, D.P., A heuristic algorithm for sensorless power maximization applied to small asynchronous wind turbogenerators, *Proceedings of IEEE Industrial Electronics Symposium*, Vol. 1, December 2000, pp. 179–184.
4. Farret, F.A., Gomes, J.R., and Padilha, A.S., Comparison of the hill climbing methods used in micro powerplants, *IEEE Industry Applications Conference*, Porto Algre, RS-Brazil, November 2000, pp. 756–760.
5. Simões, M.G., Bose, B.K., and Spiegel, R.J., Fuzzy logic based intelligent control of a variable speed cage-machine wind generation system, *IEEE Trans. Power Electron.*, 12, 87–95, January 1997.
6. Simões, M.G., Bose, B.K., and Spiegel, R.J., Design and performance evaluation of a fuzzy-logic-based variable-speed wind generation system, *IEEE Trans. Ind. Appl.*, 33, 956–965, July/August 1997.
7. Bose, B.K. and Simões, M.G., Fuzzy logic based intelligent control of a variable speed cage machine wind generation system, *Environmental Protection Agency EPA/600/SR-97/010*, published by National Technical Information Service (Report # PB97-144851), March 1997.
8. Bose, B.K., *Modern Power Electronics and AC Drives,* Prentice Hall, New York, 2001.
9. Novotny, D.W. and Lipo, T.A., *Vector Control and Dynamics of AC Drives*, Oxford University Press, New York, 1996.
10. Sukegawa, T., Kamiyama, K., Takahashi, J., Ikimi, T., and Matsutake, M., A multiple PWM GTO line-side converter for unity power factor and reduced harmonics, *Conference Records of IEEE Industry Applications Society Annual Meeting,* Vol. 1, September/October 1991, pp. 279–284.
11. Kazmierkowski, M.P., Krishnan, R., and Blaabjerg, R.F., *Control in Power Electronics: Selected Problems,* Academic Press, San Diego, CA, 2002.
12. Duarte, J.L., Van Zwam, A., Wijnands, C., and Vandenput, A., Reference frames fit for controlling PWM rectifiers, *IEEE Trans. Ind. Electron.*, 46(3), 628–630, June 1999.
13. Simões, M.G. and Franceschetti, N.N., Fuzzy optimization based control of a solar array system, *IEEE Proc. Electr. Power Appl.*, 146(5), 552–558, September 1999.

14. Teulings, W.J.A., Marpinard, J.C., Capel, A., and O'Sullivan, D., A new maximum power point tracking system, *Conference Records of 24th IEEE Power Electronics Specialists Conference*, June 1993, pp. 833–838.
15. Senjyu, T. and Uezato, K., Maximum power point tracker using fuzzy control for photovoltaic arrays, *Proceedings of the IEEE International Conference on Industrial Technology*, December 1994, pp. 143–147.
16. Niimura, T. and Yokoyama, R., Water level control of small-scale hydro-generating units by fuzzy logic, *Proceedings of IEEE International Conference on Systems, Man and Cybernetics*, Vol. 3, October 1995, pp. 2483–2487.

11 Doubly Fed Induction Generators

11.1 SCOPE OF THIS CHAPTER

There are two main types of induction machines: squirrel cage machines, with rotor windings short-circuited, and wound rotor induction machines, with accessible slip rings that can be either short-circuited or connected to an external circuit. Wound rotor machines are used in applications where it is desirable to influence the rotor circuit, keeping the stator supply constant. Before the advent of semiconductors, external resistors were used to control the slip, and, consequently, power was lost in those resistors. By using power electronic converters, it is possible to recover such slip power. Thus, a wound rotor induction generator (WRIG) with power electronic converters, that is, fed on both stator and rotor, is called a doubly-fed induction generator (DFIG).

In this chapter, a theoretical and practical coverage of DFIG systems will provide a foundation for its application to high-power renewable energy systems. Since the control is tied to the rotor, only the slip power is processed where the purpose of the control is to synchronize the rotor current with respect to stator reference. Thus, the smaller the operating slip range, the smaller is the required power converter. The size of the power converter may strongly affect the installation cost of energy ($/kW). The utility side of the power converter of the DFIG can be controlled differently from both sides of the power converter. The power factor can be controlled either by the utility side or by the rotor side of the DFIG. The generator operates in four quadrants around the synchronous speed. Therefore, it can easily operate in motoring mode, for example, for pumped hydro applications.

11.2 FEATURES OF DFIG

The WRIG is not as rugged as the squirrel cage type, but the brushes have little wearing and sparking when compared with dc machines and they are the only acceptable alternatives for alternative energy conversion on the megawatt power range.

The fundamentals of slip recovery systems have been known since the beginning of the twentieth century. Between 1907 and 1913, Kraemer and Scherbius proposed cascaded connections to the rotor of induction machines using rotary machines. The Kraemer system transformed the slip energy back into mechanical energy by a second machine in tandem connection to the shaft, while Scherbius proposed a second induction machine connected to the rotor to send the slip power back to

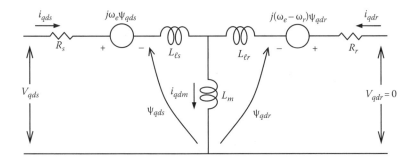

FIGURE 11.1 Complex synchronous dq_s equivalent circuit for wound rotor induction machine.

the line. By using modern power electronic devices, it is possible to recover the slip otherwise dissipated in resistances. Therefore, wound rotor generators use an inverter system connected between the rotor and the grid, while the stator is directly connected to the grid. When the mechanical speed is confined to ±20% around the synchronous speed, the rotor converter is rated only for a portion of the rated stator power, a clear advantage for very-high-power systems.

Figure 11.1 shows a model for a wound rotor induction machine.

Compared with the squirrel cage induction generator (SCIG), the DFIG operating at variable speed has the following advantages:

- The SCIG requires a power converter with full power-processing capability, whereas the DFIG requires a smaller power converter to process the slip power. Thus, the smaller the range of the operating slip, the smaller the required power converter. The size of the power converter may strongly affect the cost of energy ($/kWh).
- The DFIG has a simpler control because the magnetizing current is practically constant regardless of the rotor frequency. The purpose of the control is to synchronize the rotor current with respect to stator reference.
- The utility connection to the DFIG can be controlled from both machine sides, and the power factor can be controlled either by the stator or by the rotor side of the DFIG.
- The DFIG can be commanded by two techniques: either using (1) active and reactive power from the stator side to control the rotor converter or (2) setting up the stator voltage from the rotor side.
- If a DFIG system is capable of operating in four quadrants around the synchronous speed, it can then operate in motor mode, for example, for pumped hydro applications.
- The DFIG has increased power system dynamics and stability, permitting the suppression of power system fluctuations by quickly exchanging energy from the electric system to the machine inertia without decreasing synchronism as happens with synchronous machines. The system can also compensate for quick reactive power needs, and for this reason, it can improve the overall power system dynamics.

11.3   SUB- AND SUPERSYNCHRONOUS MODES

The WRIG is usually fed by the stator and the rotor, which is why it is frequently called a DFIG in the literature. Although WRIG is related more to the machine itself and DFIG embraces the whole system, both acronyms will be used synonymously in this book.

A complex synchronous reference frame equivalent circuit as indicated in Figure 11.1 can represent a DFIG. The main equations related to this model are

$$\underline{v}_{qds} = R_s \underline{i}_{qds} + j\omega_e \underline{\psi}_{qds} + \frac{d}{dt} \underline{\psi}_{qds} \tag{11.1}$$

$$\underline{v}_{qdr} = R_r \underline{i}_{qdr} + js\omega_e \underline{\psi}_{qdr} + \frac{d}{dt} \underline{\psi}_{qdr} \tag{11.2}$$

$$\underline{\psi}_{qds} = L_s \underline{i}_{qds} + L_m \underline{i}_{qdr} \tag{11.3}$$

$$\underline{\psi}_{qdr} = L_r \underline{i}_{qdr} + L_m \underline{i}_{qds} \tag{11.4}$$

$$T_e = \frac{3}{2}\left(\frac{p}{2}\right)\mathrm{Re}\left(j\underline{\psi}_{qds} \cdot \overline{\underline{i}_{qds}}\right) = \frac{3}{2}\left(\frac{p}{2}\right)\mathrm{Re}\left(j\underline{\psi}_{qdr} \cdot \overline{\underline{i}_{qdr}}\right) \tag{11.5}$$

where
   p is the number of poles per phase

   $\overline{\underline{i}_{qds}}$ and $\overline{\underline{i}_{qdr}}$ are the complex conjugates of the stator current and rotor current space vectors

Stator and rotor inductances are defined by $L_s = L_{es} + L_m$ and $L_r = L_{er} + L_m$.

The complex torque in Equation 11.5 can be resolved in the reference frame d_e–q_e, leading to

$$T_e = \frac{3}{2}\left(\frac{p}{2}\right)\left(\psi_{ds}i_{qs} - \psi_{qs}i_{ds}\right) = \frac{3}{2}\left(\frac{p}{2}\right)\left(\psi_{dr}i_{qr} - \psi_{qr}i_{dr}\right) \tag{11.6}$$

In order to find the no-load mechanical speed (ω_{r0}) for given steady-state voltages ($\underline{v}_{qds0}$) and ($\underline{v}_{qdr0}$), one can assume that Equations 11.1 and 11.2 have $\frac{d}{dt}(\cdot) = 0$ (steady state) and $\underline{i}_{qdr} = 0$ (no load)

$$\underline{v}_{qds0} = R_s \underline{i}_{qds0} + j\omega_e L_s \underline{i}_{qds0} \tag{11.7}$$

$$\underline{v}_{qdr0} = js\omega_e L_m \underline{i}_{qds0} \tag{11.8}$$

Combining Equation 11.7 with Equation 11.8,

$$\underline{v}_{qds0} = \left(R_s + j\omega_e L_s \right) \frac{\underline{v}_{qdr0}}{js\omega_e L_m} \tag{11.9}$$

For high-power machines, the stator resistance can be neglected in Equation 11.9. Since $s = 1 - \dfrac{\omega_{r0}}{\omega_e}$, the following relation holds for calculation of the no-load rotational speed in accordance with stator- and rotor-impressed magnitude voltages V_{qdr0} and V_{qds0}

$$\omega_{r0} = \omega_e \left[1 - \frac{L_s}{L_m} \cdot \frac{V_{qdr0}}{V_{qds0}} \right] \tag{11.10}$$

Equation 11.10 needs to be interpreted with care due to some restrictions. The phasor voltages in the reference frame are dc quantities; when positive, they indicate a positive phase sequence (a–b–c) in the three-phase terminals, whereas negative values indicate a negative phase sequence (a–c–b). Therefore, assuming stator voltage to be always positive ($\underline{v}_{qds0}$), the rotor-impressed voltage can be positive ($\underline{v}_{qdr0} > 0$) with the corresponding rotor power ($P_r > 0$) and the rotational speed (ω_{r0}) lower than the electrical synchronous speed (ω_e) with the machine operating in subsynchronous mode with positive slip ($s > 0$). By impressing voltage ($\underline{v}_{qdr0}$), the machine is naturally stable in the subsynchronous region only. Rotor voltage control leads to motoring-only stable operation in subsynchronous mode and to generating-only stable operation in supersynchronous mode. For rotor current control, both motoring and generating forms are stable in sub- and supersynchronous modes.[1–3]

The operating characteristics of the induction generator from a current source are quite different from a voltage source. High-power machines have small stator resistance drop. When under impressed voltage source, the air-gap voltage is close to the stator voltage, but under impressed current, the air-gap voltage and stator voltage vary with shaft loading. Therefore, with current control, it is possible to have a negative rotor power ($P_r > 0$) for positive slip ($s > 0$) by changing the phase sequence of currents in stator and rotor. Equation 11.10 also holds for the supersynchronous mode where $\underline{v}_{qdr0} > 0$ with a consequent $s < 0$. Since the machine is stable with rotor voltage control only, reversing the phase sequence of voltages will impose a generating mode for rotational speed greater than the electrical synchronous speed.

The rotor-injected power P_{ROTOR} coming from a prime mover on the mechanical shaft can be considered as

$$P_{ROTOR} = \left(\underline{v}_{qdr} \frac{3}{2} \overline{\underline{i}_{qdr}} \right) = R_s \underline{i}_{qdr} + js\omega_e \underline{\psi}_{qdr} \cdot \frac{3}{2} \overline{\underline{i}_{qdr}}$$

and therefore, the rotor power can be calculated by $P_{ROTOR} = \dfrac{3}{2}\mathrm{Re}\left(v_{qdr} \cdot \overline{i_{qdr}}\right)$, leading to

$$P_{ROTOR} = \frac{3}{2}R_r i_r^2 + s\omega_e T_e \tag{11.11}$$

Since $\omega_e T_e$ is the air-gap power (P_g) and neglecting the rotor copper loss, the relation in Equation 11.12 holds and Equation 11.13 is the torque equation in steady state:

$$P_r = sP_g \tag{11.12}$$

$$T_e = \frac{P_g}{\omega_e} = 3\left(\frac{p}{2}\right)I_r^2 \frac{R_r}{s\omega_e} \tag{11.13}$$

For high-power machines, the stator resistance is neglected, and the stator terminal power is P_g. Considering that the power flowing out of the machine is negative (generating mode), the induction generator has a power balance in accordance with the torque-slip curve indicated in Figure 11.2. The power distribution for the generator operating at subsynchronous and supersynchronous regions is indicated in the operating region from $0.7\omega_s$ to $1.3\omega_s$. For operation at the subsynchronous region, the slip is positive, and therefore, the rotor circuit receives power from the line, whereas for the supersynchronous region, the slip is negative, and the rotor power supplements extra generating power to the grid.

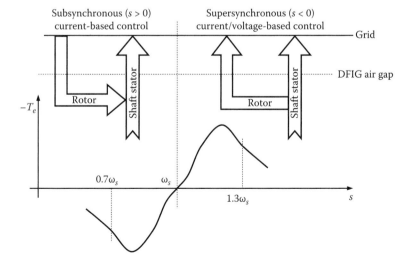

FIGURE 11.2 Torque-slip curve for DFIG in sub- and supersynchronous modes.

Figure 11.3 shows the power balance in a DFIG at subsynchronous generation where $s > 0$ and the power flow into the rotor by a current-controlled inverter. A step-up transformer is usually connected between the low-frequency low-voltage requirements and the grid in order to alleviate the rotor converter ratings. Figure 11.4 shows the supersynchronous generating mode where the mechanical speed is greater than the electrical synchronous speed, so the slip is negative ($s < 0$). The rotor voltages will have their phase sequence reversed; since $P_g < 0$ and $P_r < 0$, the rotor circuit contributes in generating power to the line with improved efficiency. It is important

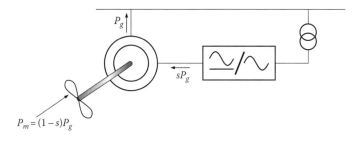

FIGURE 11.3   Subsynchronous generating mode ($s > 0$).

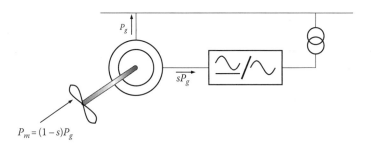

FIGURE 11.4   Supersynchronous generating mode ($s < 0$).

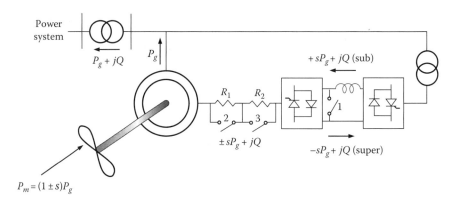

FIGURE 11.5   WRIG with a subsynchronous mode back-to-back double converter.

to note that the shaft incoming power indicates $P_m = (1 + s)P_g$ to show the extra capability of the power conversion, but the slip is actually negative. Thus, very efficient generating systems can be achieved using the supersynchronous region. Because the operating region is limited, the main drawback is the starting-up sequence of the system. One possible way around this is to use auxiliary resistors in the rotor circuit as indicated in Figure 11.5, then drive the machine in motoring mode, and, just after the cut-in speed, plug in the controller, which imposes regenerative operation.

11.4 OPERATION OF DFIG

In order to have a variable speed with improved operation covering the operating region $0.7\omega_s < \omega_r < 1.3\omega_s$, the rotor inverter must be bidirectional as in Figure 11.5 with a smooth transition through $s = 0$, that is, at synchronous speed, the rotor needs dc currents and the system behaves as a pure synchronous generator. In the past, line-commutated and load-commutated inverters were used in the rotor circuit, but close to zero slip, those circuits do not work well. For very-high-power systems, cycloconverters are still a very good choice. For medium-power back-to-back double converters (CSI or VSI), matrix converters are adequate for the scope, if a controlled resistor is introduced at the rotor terminals for starting the system in motoring mode. If the rotor converter injects the reactive power (jQ) in accordance with Equation 11.14, the stator reactive power can be programmed as indicated by Equation 11.15[2]:

$$Q_{ROTOR} = s\omega_e \frac{3}{2}\left(j\underline{\psi}_{qdr} \cdot \overline{\underline{i}_{qdr}} \right) \tag{11.14}$$

$$Q_{STATOR} = \frac{3}{2} L_{ls} \left(i_{qds} \right)^2 - \frac{Q_{ROTOR}}{|s|} \tag{11.15}$$

If reactive power is injected in the rotor, it will be subtracted from the reactive power injected in the stator. Theoretically, leading power factor ($Q < 0$) is possible at the cost of lagging power factor in the rotor. The rotor losses become very high for high slip. In addition, higher slip requires a higher rotor voltage, and leading power factor at the utility side becomes a very difficult scheme to implement. However, decreased VAR from the utility side is possible to implement.

Figure 11.2 showed the operating range of the DFIG with a torque-slip curve for sub- and supersynchronous modes. The actual torque can be calculated from a transient d–q model using Equation 11.6. However, a steady-state approach can be useful in the evaluation of the generator torque response. Figure 11.6a shows the steady-state per-phase equivalent circuit for a squirrel cage induction machine (short circuit in the rotor is considered) where the transformer coupling shows the turns ratio from stator to rotor turns. Figure 11.6b considers the wound rotor induction machine with unity turns ratio where the rotor side is available to apply voltage. Figure 11.6c shows a phasor diagram for the voltages and currents interacting at the T-branch of the induction generator, leading to the rough model indicated in Figure 11.7, that is, an approximate per-phase equivalent circuit for high-power machines where $|(R_s + j\omega_e L_{ls})| \ll \omega_e L_m$; such

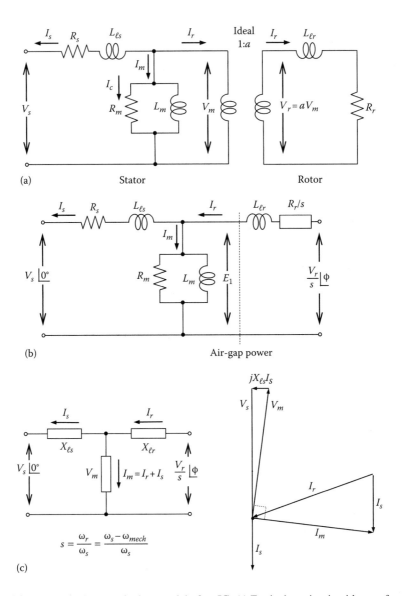

FIGURE 11.6 Per-phase equivalent model of an IG: (a) Equivalent circuit with transformer coupling, (b) equivalent circuit with respect to the stator, and (c) T-branch of an IG and correspondent phasor diagram for rotor current control.

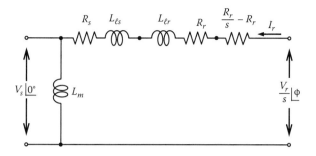

FIGURE 11.7 Approximate high-power equivalent circuit.

a simplified circuit provides performance prediction within 5% of the actual machine. In Figure 11.7, the rotor current I_r is imposed by a current-controlled converter on the rotor side; therefore, the impressed rotor voltage V_r/s on the rotor side has some phase shift ($|\phi$) with respect to the stator side. From this figure, the rotor current (I_r) is related to the stator terminal voltage (V_s) by Equation 11.16. Substituting I_r into Equation 11.13, one gets Equation 11.17, which is a function of slip s for constant frequency (ω_e) and supply voltage (V_s). In order to plot the family of curves, the slip is extended for subsynchronous ($0 < s < 1$) and supersynchronous ($-1 < s < 0$) for fixed synchronous rated stator frequency ($\omega_e = \omega_s$) as indicated in Figure 11.8.

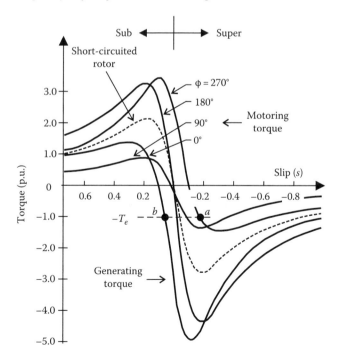

FIGURE 11.8 Effect of injected slip rotor power on the torque-slip characteristics of a DFIG.

$$\bar{I}_r = \cfrac{V_s \lfloor 0° - \cfrac{V_r}{s} \lfloor \phi}{\sqrt{\left(R_s + \cfrac{R_r}{s}\right)^2 + \omega_e^2 \left(L_{ls} + L_{lr}\right)^2} \lfloor \tan^{-1}\left[\cfrac{\omega_e \left(L_{ls} + L_{lr}\right)}{R_s + \cfrac{R_r}{s}}\right]} \tag{11.16}$$

$$T_e = 3\left(\frac{p}{2}\right)\frac{I_r^2\left(\cfrac{R_r}{s} - R_r\right)}{\omega_e} = \left(\frac{3p}{2\omega_e}\right)\frac{\left(\cfrac{R_r}{s} - R_r\right)\left[V_s^2 - 2\cfrac{V_s V_r}{s}\cos\phi + \left(\cfrac{V_r}{s}\right)^2\right]}{\left(R_s + \cfrac{R_r}{s}\right)^2 + \omega_e^2\left(L_{ls} + L_{lr}\right)^2} \tag{11.17}$$

Figure 11.8 shows a family of curves where magnitudes and the phase shift between stator voltage and current will define the position and shape of the torque-slip curve. The rotor side voltage is assumed nearly constant, and the phase shift would be varied to match the required constant air-gap flux. Figure 11.8 shows the torque-slip characteristic for short-circuited rotor windings (passive rotor power flowing in a squirrel cage–type machine). As negative rotor power is extracted from the rotor, the machine increases speed to the supersynchronous region in the curve operating at point "*a*." The machine can have positive rotor power injected, causing a decrease in speed to the subsynchronous region and operates at point "*b*." It is clear that although supersynchronous operation is preferable to provide generation by both stator and rotor, a stable position at the same torque level requires higher slip (*s*), and more losses are incurred at the rotor side. Therefore, it is advisable to implement management of the optimum operating point that minimizes the losses on the supersynchronous mode.

11.5 INTERCONNECTED AND STAND-ALONE OPERATIONS

When directly interconnected with the distribution network, an induction machine must change its speed (increased up to or above the synchronous speed). The absorbed mechanical power at the synchronous rotation is high enough to withstand the mechanical friction and resistance of the air. If the speed is increased just above the synchronous speed, a regenerative action happens, yet without supplying energy to the distribution network. This will happen only when the demagnetizing effect of the current of the rotor is balanced by a stator component capable of supplying its own iron losses and, above that, supplying power to the external load. Figure 11.9 is plotted by using the equations in Section 2.5. Figure 11.9 shows the speed versus torque and output power as well as efficiency versus output power. It must be observed that there is a rotation interval above the synchronous speed in which the efficiency is very low. This effect is caused by the fixed losses related to the low level of generated power and torque.

Some steady-state conditions[3] can be used for designing and predicting the operation performance with wind, diesel, or hydro turbines. Aspects such as high slip

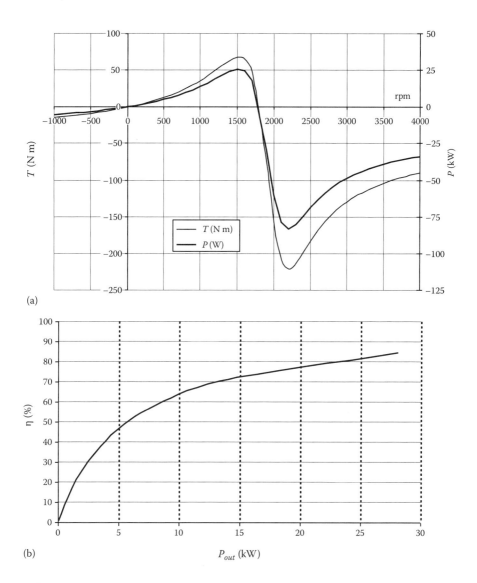

FIGURE 11.9 Typical performance of a four-pole 50 kW induction generator. (a) IG speed versus torque and power characteristics and (b) IG efficiency versus output power characteristic.

factor, torque curve, winding size to support higher saturation current, and increased number of poles can enhance the overall performance of the system.

As predicted by the equivalent model, from an increase of stator voltage, an increase in the output torque and power is obtained. There is an almost direct relationship among these three variables and the increase in the rotor speed (Figure 11.10a). Similar proportional results are obtained when there is an increase in the rotor external source voltage for the power injected through the rotor (Figure 11.10b). Notice this torque versus speed characteristic, there is a shift to the left (lower rotor speeds)

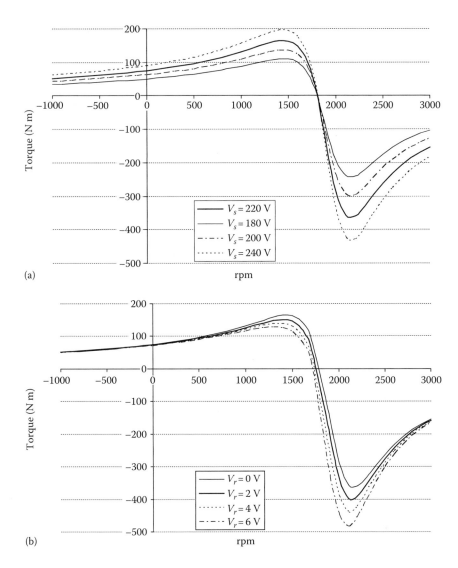

FIGURE 11.10 Control variables in the torque x speed characteristics of the induction generator for several parameters and conditions. (a) Stator voltage, (b) rotor voltage. (*Continued*)

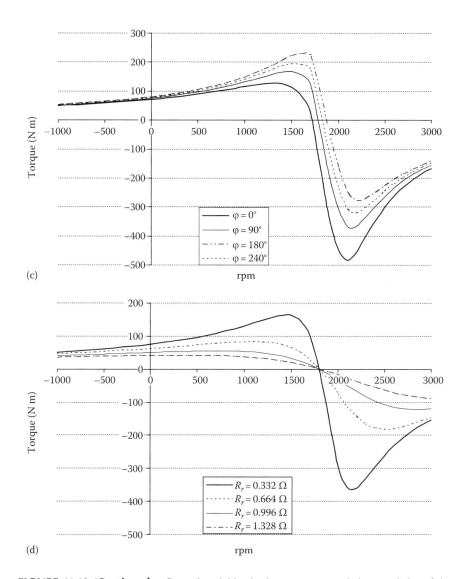

FIGURE 11.10 (*Continued*) Control variables in the torque x speed characteristics of the induction generator for several parameters and conditions. (c) Stator–rotor voltage phase shift, (d) rotor resistance. (*Continued*)

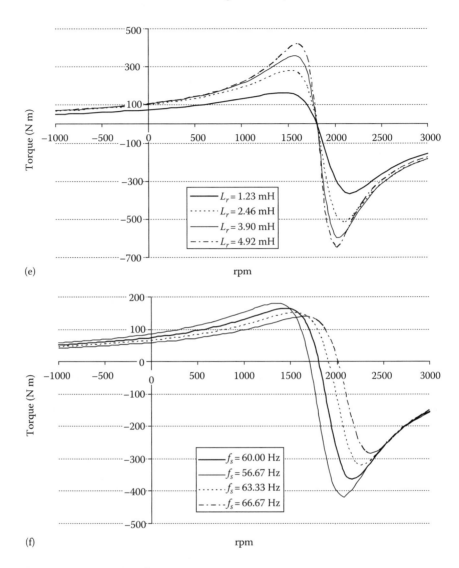

(e)

(f)

FIGURE 11.10 (*Continued*)   Control variables in the torque x speed characteristics of the induction generator for several parameters and conditions. (e) Rotor inductance, (f) frequency of the stator voltage.   (*Continued*)

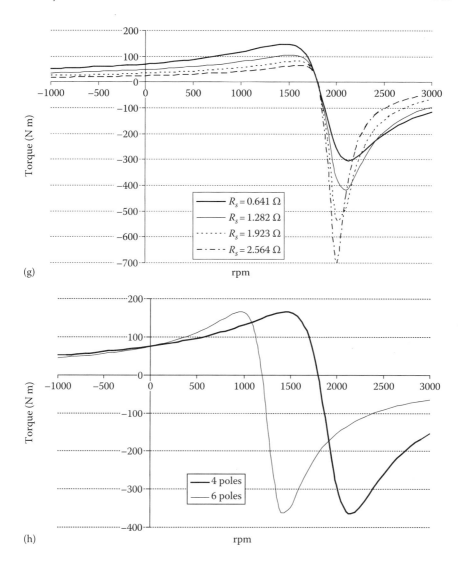

FIGURE 11.10 (*Continued*)   Control variables in the torque *x* speed characteristics of the induction generator for several parameters and conditions. (g) Stator resistance, and (h) number of stator poles.

of the pullout torque caused by changes in the rotor parameters. The pullout torque is defined as the maximum torque the generator is subject to. This is a significant parameter in the sense that the maximum power transfer from the induction generator to the load happens whenever the source impedance is equal to the load impedance, that is,

$$\frac{R_r}{s} = \sqrt{R_s^2 + (X_s + X_r)^2}$$

which when replaced in the torque Equation 11.17 for $V_r = 0$ gives

$$T_e = \left(\frac{3p}{2\omega_e}\right) \frac{V_s^2 \sqrt{R_s^2 + (X_s + X_r)^2}}{\left[R_s + \sqrt{R_s^2 + (X_s + X_r)^2}\right]^2 + \omega_e^2 (L_{ls} + L_{lr})^2} \qquad (11.18)$$

Wider variations are obtained through higher rotor voltages. The shift to the left in the pullout torque becomes much more pronounced if the phase angle ϕ between the stator voltage and rotor voltage is varied (Figure 11.10c).

The slope of the torque curve close to the synchronous speed determines the variation range of the rotor speed during load operation. The steeper the torque is when crossing the synchronous speed point, the stiffer the system becomes and the narrower the speed regulation. This slope increases to the left as the rotor resistance increases in the torque characteristics, yet there is not much effect on the magnitude of the pullout torque (Figure 11.10d). One problem here is that the heat so generated in the rotor needs to be dissipated to the outside, limiting the slip factor in large machines. The stall control used in these large machines has been replaced by a fast pitch control in order to compensate for the rotor speed stiffness caused by the small slip factor.

Rotor reactance can be changed through rotor design, rotor frequency control, or external insertion of modulated values of inductance. The quadratic effects of the rotor reactance on the torque versus speed characteristics are noticeable in Figure 11.10e. Different rotor angular frequencies can only be observed by the fact that the resistance depends on the slip factor. It is interesting to point out here that these changes in X_r have very little effect on the rotor losses, which is not the case with the actual rotor resistance changes. It is important to note also that mismatching the rotor frequency may cause undesirable resonant and antiresonant effect interaction between rotor and stator.

A very effective way of changing the synchronous speed of the induction generator is through stator frequency changes. These changes are exactly proportional to the load frequency and the crossing line over the rotor speed (Figure 11.10f).

One way to change the grid speed ranges of the induction generator is through machine pole changing. The phase circuitry of the stator is designed so that a different configuration of the poles (pole number) can produce quite different synchronous speeds. None of the other variables is comparable in their effects on the pullout torque as is the changing of pole number (Figure 11.10g).

Notice in all graphs except Figure 11.10d that the changes proposed in the stator and rotor parameters have pronounced influence not only on the crossing over of the torque characteristic of the rotor speed axis but also on the magnitude of the corresponding pullout torques.

11.6 FIELD-ORIENTED CONTROL OF DFIG

Several converter topologies can be used to control a DFIG, and many control approaches have been studied and implemented. Nonetheless, it is possible to classify two broad approaches to control the DFIG: (1) a full complete controller where the prime mover active power and the line-side reactive power are the set points for a rotor frame controller for a grid-connected system and (2) a simplified current controller with a line voltage control for stand-alone applications that do not require full active/reactive power management. Both approaches are currently implemented using vector control theory. The converter connected to the rotor side will receive the set points from the vector-controlled-based system. Depending on the topology (cycloconverters, matrix converters, back-to-back double CSI or VSI converters), an inner control for a possible dc link (for the back-to-back double converters) is required, or there will be a natural bidirectional operation (for cycloconverters and matrix converters).

In order to apply field-oriented control (or vector control) for a DFIG, a d–q transformation is required. This is made in two steps: initially stationary three-phase variables (a–b–c) are transformed into stationary two-phase (d^s–q^s) variables and then into rotating two-phase (d^e–q^e) variables. Figure 11.11 shows the vector representation of stationary three-phase variables a, b, c and stationary two-phase variables d^s–q^s. For the sake of simplicity, vector d^s is assumed to be aligned with vector "a." Equation 11.19 shows the transformation matrix. For a three-phase balanced system, the zero-axis component is zero. Therefore, the transformation can be simplified as shown in Equation 11.20. Inverse transformation can also be calculated as shown in Equation 11.21.

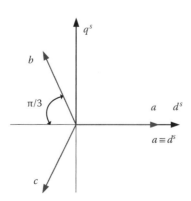

FIGURE 11.11   Vector representation of a–b–c and stationary d^s–q^s frame.

$$\begin{bmatrix} x_d^s & x_q^s & x_0^s \end{bmatrix} = \begin{bmatrix} x_a & x_b & x_c \end{bmatrix} \frac{2}{3} \begin{bmatrix} 1 & 0 & \dfrac{1}{2} \\ -\dfrac{1}{2} & \dfrac{\sqrt{3}}{2} & \dfrac{1}{2} \\ -\dfrac{1}{2} & -\dfrac{\sqrt{3}}{2} & \dfrac{1}{2} \end{bmatrix} \tag{11.19}$$

$$x_d^s = x_a$$

$$x_q^s = \frac{1}{\sqrt{3}}\left(x_b - x_c\right) \tag{11.20}$$

$$x_a = x_d^s$$

$$x_b = -\frac{1}{2}x_d^s + \frac{\sqrt{3}}{2}x_q^s$$

$$x_c = -\frac{1}{2}x_d^s - \frac{\sqrt{3}}{2}x_q^s \tag{11.21}$$

The second transformation is to rotate the axis of the two-phase system at the flux frequency. Therefore, any vector rotating at this frequency will appear as a dc value in such a frame. Vector representation of these two frames is shown in Figure 11.12. The $d^e\!-\!q^e$ frame is also referred to as a reference frame or synchronous frame when it is rotating at synchronous frequency.

Equation 11.22 shows the transformation matrix. These equations can again be simplified to Equation 11.23. Inverse transformation back to stationary frame is shown in Equation 11.24. The superscript e for reference frame quantities will be omitted in future references for convenience. Equations 11.19 through 11.24 can now be used to transform any three-phase quantities to rotating two-phase quantities.

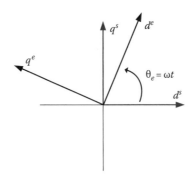

FIGURE 11.12   Vector representation of $d^s\!-\!q^s$ and $d^e\!-\!q^e$ frame.

In this chapter, two rotating frames are used, one rotating at the stator electrical frequency for transforming stator quantities and other rotating at the rotor slip frequency for transforming rotor variables:

$$\begin{bmatrix} x_d^e & x_q^e & x_0^e \end{bmatrix} = \begin{bmatrix} x_d^s & x_q^s & x_0^s \end{bmatrix} \begin{bmatrix} \cos\theta_e & -\sin\theta_e & 0 \\ \sin\theta_e & \cos\theta_e & 0 \\ 0 & 0 & 1 \end{bmatrix} \tag{11.22}$$

$$x_d^e = x_d^s \cos\theta_e + x_q^s \sin\theta_e$$

$$x_q^e = -x_d^s \sin\theta_e + x_q^s \cos\theta_e \tag{11.23}$$

$$x_d^s = x_d^e \cos\theta_e - x_q^e \sin\theta_e$$

$$x_q^s = x_d^e \sin\theta_e + x_q^e \cos\theta_e \tag{11.24}$$

The d–q model of Figure 11.1 has terminals for both the stator and the rotor, and the equations are used to model a WRIG. Motoring convention is used; therefore, the currents are entering on the model as inputs, and real and reactive powers have a negative sign when they are fed into the grid. The subscripts s and r indicate stator and rotor quantities, respectively. Stator and rotor voltage equations obtained from Figure 11.1 are shown in Equation 11.25; ω_e and ω_r are stator electrical frequency and rotor electrical angular speed, respectively:

$$v_{ds} = R_s i_{ds} + \frac{d\psi_{ds}}{dt} - \omega_e \psi_{qs}$$

$$v_{qs} = R_s i_{qs} + \frac{d\psi_{qs}}{dt} + \omega_e \psi_{ds}$$

$$v_{dr} = R_r i_{dr} + \frac{d\psi_{qr}}{dt} - \left(\omega_e - \omega_r\right)\psi_{qr} \tag{11.25}$$

$$v_{qr} = R_r i_{qr} + \frac{d\psi_{qr}}{dt} + \left(\omega_e - \omega_r\right)\psi_{dr}$$

Flux linkages developed in the machine are calculated using the equations shown in Equation 11.26. From these calculated fluxes, by a simple matrix inversion, the currents generated are obtained as shown in Equation 11.27:

$$\psi_{ds} = L_{ls} i_{ds} + L_m \left(i_{ds} + i_{dr}\right)$$

$$\psi_{qs} = L_{ls} i_{qs} + L_m \left(i_{qs} + i_{qr}\right)$$

$$\psi_{dr} = L_{lr} i_{dr} + L_m \left(i_{ds} + i_{dr}\right) \tag{11.26}$$

$$\psi_{qr} = L_{lr} i_{qr} + L_m \left(i_{qs} + i_{qr}\right)$$

$$\begin{bmatrix} i_{ds} \\ i_{qs} \\ i_{dr} \\ i_{qr} \end{bmatrix} = \begin{bmatrix} \psi_{ds} \\ \psi_{qs} \\ \psi_{dr} \\ \psi_{qr} \end{bmatrix} \begin{bmatrix} L_{ls} + L_m & 0 & L_m & 0 \\ 0 & L_{ls} + L_m & 0 & L_m \\ L_m & 0 & L_{lr} + L_m & 0 \\ 0 & L_m & 0 & L_{lr} + L_m \end{bmatrix}^{-1} \tag{11.27}$$

The electrical torque generated in the machine is calculated using Equation 11.28, where p is the number of poles; the mechanical angular velocity of the generator is calculated by Equation 11.29, where ω_m is the mechanical angular speed, J is the system inertia and the mechanical torque input or the wind turbine torque input, and T_m is the prime mover (turbine) torque imposed on the generator shaft:

$$T_e = \frac{3p}{2} \left(\psi_{dr} i_{qr} - \psi_{qr} i_{dr} \right) \tag{11.28}$$

$$\frac{d\omega_m}{dt} = \frac{\left(T_m - T_e \right)}{J} \tag{11.29}$$

Equations for active and reactive powers generated or consumed by the induction generator can be written in terms of stator current components as shown in Equation 11.30. With these equations, modeling of the DFIG is complete:

$$Q_s = \frac{3}{2} \text{Im}\left(\underline{v}_{qds} \cdot \overline{\underline{i}_{qds}} \right) = \frac{3}{2} \left(v_{qs} i_{ds} - v_{ds} i_{qs} \right)$$
$$\tag{11.30}$$
$$P_s = \frac{3}{2} \text{Re}\left(\underline{v}_{qds} \cdot \overline{\underline{i}_{qds}} \right) = \frac{3}{2} \left(v_{qs} i_{qs} + v_{ds} i_{ds} \right)$$

Several vector control methods have been proposed in the literature to control a DFIG.[4] Stator flux–oriented rotor current control is the most common way of controlling a DFIG.[4–8] Wang and Ding and Xu and Wei use an air-gap flux–oriented control method.[8,9] For high-power machines, if the stator resistance can be considered small, stator flux orientation gives orientation with the stator voltage.[10,11] According to Morel et al., pure stator voltage orientation can be achieved without any significant error. Rotor flux–oriented current control methods have also been published.[11]

The scheme presented in this section uses stator flux–oriented control for the rotor-end converter (REC) and grid voltage–oriented control is used to control the front-end converter (FEC).[7] Stator flux orientation is achieved by the online estimation of the stator flux vector position. This type of control, also known as direct vector control (DVC), is less sensitive to machine parameters. It is dependent only on stator resistance, which is easy to correct. It is very robust due to real-time tracking of machine parameter variation with temperature and core saturation but typically does not work at zero shaft speed. This approach is suitable for induction generator control because zero-speed operation is not required.

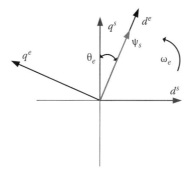

FIGURE 11.13 Stator flux orientation.

11.6.1 ROTOR-END CONVERTER CONTROL

Control on the REC is used to program the active and reactive power generated by the DFIG. It can also be used to achieve harmonic compensation. A stator flux–oriented vector control scheme is used to achieve decoupled control of active and reactive power.[7,12] The same control procedure can also be used to achieve harmonic compensation. The stator flux vector orientation along the rotating reference frame is shown in Figure 11.13. Since the d-axis is aligned along the stator flux vector, the q-axis component of stator flux is zero.

Assuming a small stator resistance voltage drop and since the stator is connected to the grid, the stator flux vector will remain constant and lags the stator voltage vector by 90°. Therefore, due to stator flux orientation, $v_{ds} = 0$ and consequently the stator active and reactive power equations shown in Equation 11.30 can be rewritten as

$$Q_s = \frac{3}{2} v_{qs} i_{ds}$$
$$P_s = \frac{3}{2} v_{qs} i_{qs}$$

(11.31)

Since the control of the induction generator is through rotor currents, stator quantities must be given in terms of rotor quantities. Substituting $\psi_{qs} = 0$ in the stator flux linkage equations shown in Equation 11.26, the d-axis rotor current component is obtained as

$$i_{dr} = -\frac{L_s}{L_m} i_{ds} + \frac{1}{L_m} \psi_{ds}$$

(11.32)

where $L_s = L_{ls} + L_m$.

From Equation 11.32, a positive injection of i_{dr} will result in a lesser value of i_{ds} being drawn from the stator side. That is, the amount of reactive power drawn from the grid can be controlled using i_{dr}. Similarly, the q-axis rotor current component is obtained as

$$i_{qr} = -\frac{L_s}{L_m} i_{qs}$$

(11.33)

From Equation 11.33, the magnitude of i_{qs} is directly proportional to i_{qr}. That is, active power generated by the machine can be controlled using i_{qr}, and the induction generator behaves like a current transformer for active power flow. Injection of positive i_{qr} will result in negative i_{qs} being drawn from grid terminals, that is, active power is being injected into the grid. Substituting Equations 11.32 and 11.33 in the rotor flux linkage equations in Equation 11.26 gives

$$\psi_{dr} = \sigma L_r i_{dr} + \frac{L_m}{L_s}\psi_{ds}$$

$$\psi_{qr} = \sigma L_r i_{qr}$$

(11.34)

where
$$L_r = L_{lr} = L_m$$
$$\sigma = 1 - \frac{L_m^2}{L_r L_s}$$

Using Equations 11.32 through 11.34, the rotor voltage equations shown in Equation 11.25 can be rewritten as

$$v_{dr} = R_r i_{dr} - \left(\omega_e - \omega_r\right)\sigma L_r i_{qr} + \sigma L_r \frac{di_{dr}}{dt} + \frac{L_m}{L_s}\frac{d\psi_{ds}}{dt}$$

$$v_{qr} = R_r i_{qr} + \left(\omega_e - \omega_r\right)\sigma L_r i_{dr} + \left(\omega_e - \omega_r\right)\frac{L_m}{L_s}\psi_{ds} + \sigma L_r \frac{di_{qr}}{dt}$$

(11.35)

From Equation 11.35, it is observed that there is cross coupling between the d- and q-axis due to the presence of rotational emf terms. The current loop dynamics along the d- and q-axis can be made independent by compensating for these terms. However, as the operating slip range of a DFIG is limited, the contributions of these terms are rather small. Also, since the stator flux is practically constant, the transformer emf term, depending on the derivative of flux, can be neglected. Therefore, the governing rotor voltages equations for active and reactive power control can be written as

$$v_{dr}^* = \left(k_{qp} + k_{qi}\,\textstyle\int\right)\left(i_{dr}^* - i_{dr}\right) - \left(\omega_e - \omega_r\right)\sigma L_r i_{qr}$$

$$v_{qr}^* = \left(k_{pp} + k_{pi}\,\textstyle\int\right)\left(i_{qr}^* - i_{qr}\right) + \left(\omega_e - \omega_r\right)\sigma L_r i_{dr} + \left(\omega_e - \omega_r\right)\frac{L_m}{L_s}\psi_{ds}$$

(11.36)

where
k_{qp}, k_{qi} are the proportional and integral gains for d-axis current (reactive power), respectively
k_{pp}, k_{qr} are the proportional and integral gains for q-axis current (active power), respectively
i_{dr}^*, i_{qr}^* are references for reactive and active powers, respectively

These d–q axis reference voltages are converted to an a–b–c frame to generate commands for the rotor-end PWM converter.

11.6.2 HARMONIC COMPENSATION

If a nonlinear load is connected across the stator terminals of the induction generator, it is possible to compensate for the harmonics generated by the load by adding the harmonic current reference to the rotor currents. If i^*_{dLh} and i^*_{qLh} are the d–q axis load currents obtained after filtering out the fundamental from the nonlinear load current, the harmonic reference that needs to be added to the rotor current reference is given by Equation 11.37. These currents are obtained using a band-pass filter that filters out everything except the 5th, 7th, and 11th harmonics. In a three-phase star-connected system, odd harmonic currents that are multiples of three do not exist.

$$i^*_{drh} = -\frac{L_s}{L_m} i_{dLh} + \frac{1}{L_m} \psi_{ds}$$

$$i^*_{qrh} = -\frac{L_s}{L_m} i_{qLh}$$

(11.37)

Therefore, the total rotor current reference is the sum of power reference and harmonic compensation reference given by

$$i^*_{dr} = i^*_{dr-r} + i^*_{drh}$$

$$i^*_{qr} = i^*_{qr-a} + i^*_{qrh}$$

(11.38)

where i^*_{dr-r}, i^*_{qr-a} are, respectively, references for active and reactive power.

Equation 11.39 shows the relation between rotor currents and power demand. Therefore, without making any major changes to the existing DFIG control system, it can be made to operate as an active filter.

$$Q_s = \frac{3}{2} V_{qs} \left(\frac{\psi_{ds} - L_m i_{dr-r}}{L_s} \right)$$

$$P_s = -\frac{3}{2} V_{qs} \frac{L_m}{L_s} i_{qr-a}$$

(11.39)

11.6.3 STATOR FLUX ORIENTATION

Stator flux orientation requires the calculation of magnitude and phase of the stator flux vector without directly measuring the flux. Various methods have been proposed to calculate the stator flux.[13–16] The most common way of calculating stator flux is shown in Equation 11.40, obtained from stator voltage and current measurements:

$$\psi_{ds}^s = \int_0^t \left(v_{ds}^s - R_s i_{ds}^s \right) dt$$

$$\psi_{qs}^s = \int_0^t \left(v_{qs}^s - R_s i_{qs}^s \right) dt$$

(11.40)

Equations 11.41 and 11.42 show the calculations for stator flux vector position and angular speed:

$$\theta_e = \tan^{-1} \frac{\psi_{qs}^s}{\psi_{ds}^s}, \quad \omega_e = \frac{\psi_{ds}^s \frac{\partial}{\partial t}\left(\psi_{qs}^s\right) - \psi_{qs}^s \frac{\partial}{\partial t}\left(\psi_{ds}^s\right)}{\left(\psi_{qs}^s\right)^2 + \left(\psi_{ds}^s\right)^2}$$

(11.41)

$$\cos\theta_e = \frac{\psi_{ds}^s}{\left|\psi_{ds}^s\right|}, \quad \sin\theta_e = \frac{\psi_{qs}^s}{\left|\psi_{ds}^s\right|}$$

(11.42)

This direct integration method to calculate stator flux works very well in a simulation. However, it poses problems in an experimental setup. A few disadvantages of using this method are the sensitivity to stator resistance, drift due to offset in the signals, and incorrect initial value for integration. For generator applications, stator resistance sensitivity can be neglected as the generator usually operates at high speeds. Even if the stator resistance calculation is slightly off, the voltage drop across the resistance is small and variation can be ignored. The other two problems can be compensated for by using a low-pass filter (Equation 11.43) instead of direct integration:

$$\psi_{ds}^s[s] = \frac{1}{s + \omega_c}\left(v_{ds}^s[s] - R_s i_{ds}^s[s]\right)$$

$$\psi_{qs}^s[s] = \frac{1}{s + \omega_c}\left(v_{qs}^s[s] - R_s i_{qs}^s[s]\right)$$

(11.43)

where ω_c is the corner frequency of the filter.

This frequency has to be at least ten times the frequency of the machine. Equation 11.44 shows a recursive filter equation that can be used in a microprocessor-based setup:

$$\psi_{dqs}^s[n] = \frac{1}{\frac{2}{T} + \omega_c}\left[\frac{\left(v_{dqs}^s[n] - R_s i_{dqs}^s[n]\right) + \left(v_{dqs}^s[n-1] - R_s i_{dqs}^s[n-1]\right)}{+ \left(\frac{2}{T} + \omega_c\right)\psi_{dqs}^s[n-1]}\right]$$

(11.44)

where
  T is the sampling time
  ω_c is chosen to be 20 rad/s

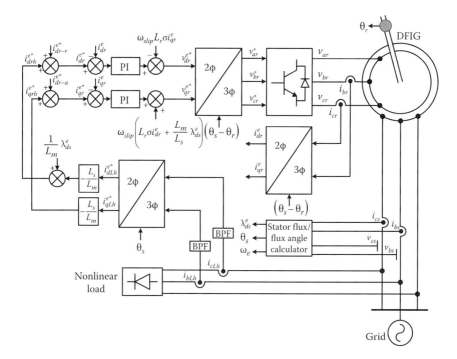

FIGURE 11.14   Vector control scheme for the REC.

Figure 11.14 shows the block diagram for the REC. As shown in this figure, stator flux–oriented vector control requires measurement of stator and rotor currents, stator voltages, and rotor angular position. The harmonic references for active filter operation are obtained from the nonlinear load connected to the grid. If harmonic compensation is not required, then i^*_{drh} and i^*_{qrh} are simply imputed to be zero.

11.6.4   FRONT-END CONVERTER CONTROL

Ideally, when the generator is operating in subsynchronous mode an FEC is not required because power control and harmonic compensation can be accomplished by using just the REC driven by a dc power source. As discussed in Section 11.3, a DFIG generates from both the stator and the rotor when operating above synchronous speed. Therefore, an FEC is used to provide a path for power flow from an REC into the grid in supersynchronous mode as well as power into an REC during subsynchronous mode. It allows bidirectional power flow in the rotor circuit. Another important function of the FEC is to keep the dc-link capacitor voltage constant regardless of direction and magnitude of power flow and to program the dc-link voltage in case a leading power factor on the stator side is imposed. A grid (stator) voltage vector–oriented control scheme is used, allowing decoupled control of the dc-link voltage power and reactive power while allowing bidirectional power flow.[7] The voltage vector is oriented along the d-axis as shown in Figure 11.15.

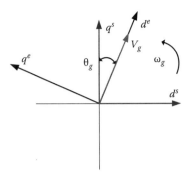

FIGURE 11.15 Voltage vector orientation.

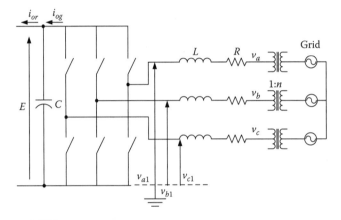

FIGURE 11.16 FEC arrangement.

The control structure of FEC is very similar to REC except that due to d-axis orientation of voltage, active power is related to d-axis current and reactive power is related to q-axis current. Figure 11.16 shows the arrangement of the converter. L and R are the line inductance and resistance, respectively. The converter is connected to the grid through transformers as the dc link is operated at a lower voltage level. The voltage balance equations across the inductors shown in Figure 11.16 are shown in Equation 11.45. After transformation into a d–q reference frame, Equation 11.46 is obtained, where ω_g is the angular speed of the grid voltage vector calculated from the stationary d–q axis grid voltages using Equation 11.47:

$$\begin{bmatrix} v_a \\ v_b \\ v_c \end{bmatrix} = R \begin{bmatrix} i_a \\ i_b \\ i_c \end{bmatrix} + L \frac{d}{dt} \begin{bmatrix} i_a \\ i_b \\ i_c \end{bmatrix} + \begin{bmatrix} v_{a1} \\ v_{b1} \\ v_{c1} \end{bmatrix} \tag{11.45}$$

$$v_d = Ri_d + L\frac{di_d}{dt} - \omega_g Li_q + v_{d1}$$

$$v_q = Ri_q + L\frac{di_q}{dt} + \omega_g Li_d + v_{q1} \tag{11.46}$$

$$\theta_g = \int \omega_g dt = \tan^{-1}\frac{v_q^s}{v_d^s} \quad \text{and} \quad \cos\theta_g = \frac{v_d^s}{\sqrt{\left(v_d^s\right)^2 + \left(v_q^s\right)^2}}, \quad \sin\theta_s = \frac{v_q^s}{\sqrt{\left(v_d^s\right)^2 + \left(v_q^s\right)^2}}$$

$$(11.47)$$

Since the *d*-axis of the reference frame is oriented along the grid voltage vector, the *q*-axis component of the grid voltage is zero. From the active and reactive power equations shown in Equation 11.48, active power is proportional to the *d*-axis component of the current, and reactive power is proportional to the *q*-axis component of the current:

$$P_g = \frac{3}{2} v_d i_d$$

$$(11.48)$$

$$Q_g = -\frac{3}{2} v_d i_q$$

The energy stored in the dc-link electrolytic capacitor is given by $CE^2/2$. The time derivative of this energy must be equal to the sum of instantaneous grid power and the rotor power. With harmonics being neglected due to transistor switching and losses in the system, the following equations are obtained for the dc link shown in Figure 11.16:

$$\frac{1}{2}C\frac{dE^2}{dt} = P_g - P_r \Rightarrow CE\frac{dE}{dt} = P_g - P_r \qquad (11.49)$$

where $P_g = \frac{3}{2}v_d i_d$ and $P_r = \frac{3}{2}\left(v_{dr} i_{dr} + v_{qr} i_{qr}\right)$, from Figure 11.16 $i_{or} = \frac{P_r}{E}$ and $i_{og} = \frac{P_g}{E}$, leading to

$$C\frac{dE}{dt} = i_{og} - i_{or} \qquad (11.50)$$

From Equations 11.49 and 11.50, it can be observed that the dc-link voltage can be controlled using the active power reference current (*d*-axis component of current). Thus, *d*-axis current controls the dc-link voltage, and reactive power is controlled using the *q*-axis current. Equation 11.46 can be rewritten as follows:

$$v_{d1} = -\left(Ri_d + L\frac{di_d}{dt}\right) + \omega_g L i_q + v_d$$

$$(11.51)$$

$$v_{q1} = -\left(Ri_q + L\frac{di_q}{dt}\right) - \omega_g L i_d$$

Therefore, the governing voltage equations for FEC are given by

$$v_{d1}^* = -\left(k_{ep} + k_{ei} \int\right)\left(i_d^* - i_d\right) + \omega_g L i_q + v_d$$

$$v_{q1}^* = \left(k_{qgp} + k_{qgi} \int\right)\left(i_q^* - i_q\right) - \omega_g L i_d \qquad (11.52)$$

where
  v_{d1}^* and v_{q1}^* are the reference values for the FEC
  i_d^* is derived from the dc-link voltage error
  k_{ep}, k_{ei} are the proportional and integral gains for d-axis current (dc-link voltage), respectively
  k_{qgp}, k_{qgi} are the proportional and integral gains for q-axis current (reactive power), respectively

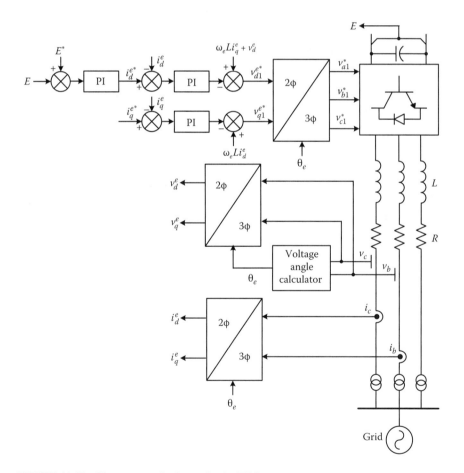

FIGURE 11.17 Vector control scheme for the FEC.

These d–q axis reference voltages are converted to an a–b–c frame to generate commands for the front-end PWM converter. The feed-forward terms shown in the right hand side of Equation 11.52 are the rotational emf terms that appear as cross-coupling terms due to the d–q transformation. These are included as compensation terms to provide proper decoupling between the axes and improve the transient response of the converter. Figure 11.17 shows the block diagram for the FEC.

11.7 ACTIVE–REACTIVE POWER CONTROL FOR A DOUBLY FED INDUCTION GENERATOR

The following control approach will consider that the rotor-side converter receives set points for the rotor currents (i_{ra}^*, i_{rb}^*, i_{rc}^*), for example, with either a hysteresis band inverter or with inner current loop control. The injected generator active and reactive powers are indicated in Equation 11.30. Considering that

$$\bar{i}_{qds} = \frac{1}{L_s}\bar{\Psi}_{qds} - \frac{L_m}{L_s}\bar{i}_{qdr}$$

the following active and reactive power equations hold:

$$P_s = \frac{3}{2}\left[\frac{1}{L_s}\left(v_{qs}\Psi_{qs} + v_{ds}\Psi_{ds}\right) - \frac{L_m}{L_s}\left(v_{qs}i_{qr} + v_{ds}i_{dr}\right)\right] \tag{11.53}$$

$$Q_s = \frac{3}{2}\left[\frac{1}{L_s}\left(v_{qs}\Psi_{ds} - v_{ds}\Psi_{qs}\right) - \frac{L_m}{L_s}\left(v_{qs}i_{dr} - v_{ds}i_{qr}\right)\right] \tag{11.54}$$

By simplifying that stator resistance is considered to be $R_s \approx 0$, there is an alignment of active power with q_e axis and an alignment of reactive power with d_e axis. Therefore, stator flux will also be aligned with d_e axis and $\Psi_{qs} = 0$; consequently

$$\frac{d}{dt}\Psi_{qs} = 0$$

The steady-state values for the following simplified power equations may be used:

$$P_s = -\frac{3}{2}\frac{L_m}{L_s}|\Psi_s|\omega_e i_{qr} \tag{11.55}$$

$$Q_s = \frac{\omega_e}{L_s}\left[|\Psi_s|^2 - L_m i_{qr}|\Psi_s|\right] \tag{11.56}$$

In the coordinate system of the stator voltage, the P_s and Q_s powers are given by

$$P_s = -\frac{3}{2}\frac{L_m}{L_s}V_s i_{qr} \tag{11.57}$$

$$Q_s = \frac{3}{2}\left(\frac{V_s^2}{L_s \omega_e} - \frac{L_m}{L_s}V_s i_{qr}\right) \tag{11.58}$$

Considering the stator flux orientation for the DFIG, the stator current i_{qds} can be considered to have a projection i_{sT} responsible for a hypothetical active torque, which is

$$T_T = \frac{3}{2}\mathrm{Re}\left(j\psi_{qds} \cdot \overline{i_{qds}}\right)$$

under the same reasoning, a reactive torque (i_{sQ}) might be defined as

$$T_Q = \frac{3}{2}\mathrm{Im}\left(j\psi_{qds} \cdot \overline{i_{qds}}\right)$$

The equations for the hypothetical active and reactive torques will help to define the feedback approach. Equations for T_T and T_Q are similar to the electromagnetic torque, but in this case, they are measuring the active and reactive power flows at the stator side and do not need any correction for a pair of poles:

$$T_T = \frac{3}{2}\left(\psi_{ds}i_{qs} - \psi_{qs}i_{ds}\right) \tag{11.59}$$

$$T_Q = \frac{3}{2}\left(\psi_{qs}i_{qs} + \psi_{ds}i_{ds}\right) \tag{11.60}$$

Thus, the current phasor can be defined as $i_{qds} = i_{sT} - j(i_{sQ})$, where i_{sT} is generated from a closed-loop control on T_T and i_{sQ} is generated from a closed-loop control on T_Q. Figure 11.18 shows how the control law for the induction generator can be evaluated for this system; when current control is imposed, only the T-branch of the generator indicated in Figure 11.18 needs to be evaluated. The rotor current i_{qdr} must supply the required stator current i_{qds} subtracted from the magnetizing current i_{qdm}. By assuming that magnetizing current is fully aligned with the reactive stator current component i_{sQ}, the following relations yield

$$i_{qdr}^* = -\left(i_{qds}^* + i_{qdm}^*\right) = -i_{sT} - j(i_m - i_{sQ}) \tag{11.61}$$

$$i_{qr}^* = -i_{sT} \tag{11.62}$$

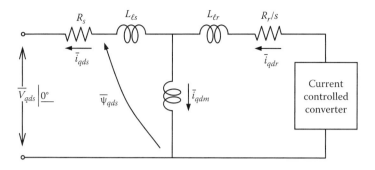

FIGURE 11.18   Distribution of rotor, magnetizing, and stator current.

$$i_{dr}^* = i_m - i_{SQ} \tag{11.63}$$

$$i_m = \frac{|\psi_s|}{L_s} \tag{11.64}$$

Figure 11.19 shows the relationships among space current vectors in respect to stationary, synchronous, and rotor reference frames for supersynchronous operation, that is, the rotor axis leading in respect to d_e axis. The rotor current provides stator current (with active and reactive components) plus the magnetizing current. The rotor current has a locus of constant active power that keeps a prescribed i_{sT} and a locus of constant reactive power that keeps a prescribed i_{sQ}. The calculation of the magnetizing current makes the reactive component very sensitive to stator voltage integrations so that variations in the grid ac voltage and frequency are taken into consideration. However, the saturation of the generator is not considered by Equation 11.64. The diagram of Figure 11.19 can also be drawn for subsynchronous operation; when the rotor axis is lagging the d_e axis, the slip becomes negative, the vector rotation that takes the field coordinates to the rotor frame automatically reverses the phase sequence, and current is injected in the rotor.

Figure 11.20 shows the complete block diagram for the active–reactive-based power control of a DFIG. The three-phase currents and voltages are acquired and converted to a two-phase stationary frame (d_s–q_s). Because the frequency of operation is high, a low-pass filter performs the function of integrator to obtain the stationary stator flux projections (ψ_{qs}, ψ_{ds}) and the stator flux amplitude $|\Psi_s|$. The hypothetical active and reactive torques (T_T, T_Q) are computed and compared to the required values of active and reactive powers (P_s, Q_s) through a division by ω_e. Such error signals (E_T, E_Q) drive the PI controllers to generate the reference values for stator current (i_{sT}^*, i_{sQ}^*). The reactive component is subtracted from i_m, and after limiting blocks, the reference values for rotor currents, i_{qr}^* and i_{dr}^*, are provided in synchronous coordinates. Therefore, vector rotation must consider the angle for

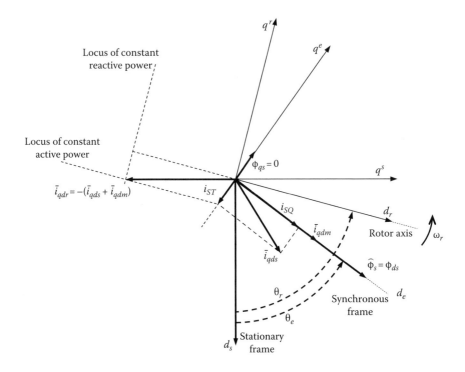

FIGURE 11.19   Diagram of space current vectors for supersynchronous operation.

transformation as the rotor position in field coordinates, that is, subtraction of the mechanical angle θ_r (compensated by the number of poles) from the stator electrical frequency angle θ_e. A two-phase to three-phase transformation brings the reference values for the three-phase rotor currents (i_{ra}^*, i_{rb}^*, i_{rc}^*). More sophisticated techniques to control the inner loops of the bidirectional ac/ac converter are out of the scope of this chapter. It is possible to generate the rotor voltage set points (v_{qr}^*, v_{dr}^*) and through synchronous PI controllers, or space vector or direct torque control, modulate the rotor currents. Although the inner management of such bidirectional power flow of the rotor-side converter is not treated, there are several references on vector-controlled drive systems helpful in designing such controllers.[17,18] In addition, sensorless techniques developed for SCIGs can also be applied for enhanced performance of DFIG.

The active–reactive power control for a DFIG presented in Figure 11.20 requires the full management of power. Active power can be available from the information of the shaft; for example, wind speed can be measured and a look-up table may generate the active power available on the shaft to be converted for the electric grid. Reactive power needs communication with the grid, and command from neighbor power systems must request lagging/leading power factors.

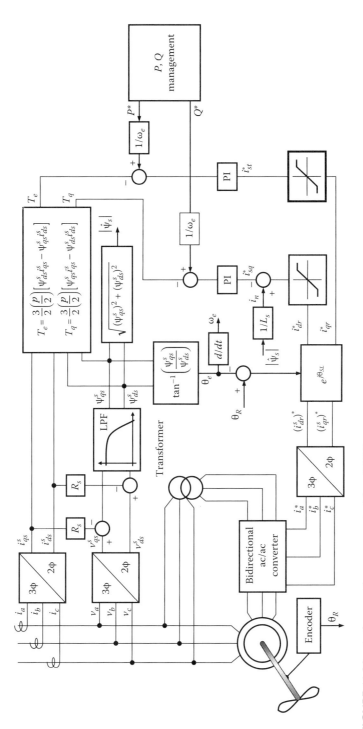

FIGURE 11.20 Active–reactive power control for a DFIG.

11.8   STAND-ALONE DOUBLY FED INDUCTION GENERATOR

Stand-alone applications can be simplified since they do not need as much management, and a simplified current controller with a line voltage control is used. The T-model relating the stator current $\underline{i}_{qds}$, magnetizing current $\underline{i}_{qdm}$, and the rotor current $\underline{i}_{qdr}$ is depicted in Figure 11.18. From Equation 11.3 and considering $L_s = L_{ls} + L_m$, the following d_s–q_s equations hold:

$$\psi_{qs} = L_{ls}i_{qs} + L_m\left(i_{qs} + i_{qr}\right) \tag{11.65}$$

$$\psi_{ds} = L_{ls}i_{ds} + L_m\left(i_{ds} + i_{dr}\right) \tag{11.66}$$

The stator flux will be aligned with the d_s axis and, therefore, $\psi_{qs} = 0$ and, consequently, $\dfrac{d}{dt}\psi_{qs} = 0$, and the following control law of Equation 11.67 must be implemented. Equations 11.68 through 11.70 are valid for these flux aligned conditions:

$$i_{qr}^* = -\frac{\left(L_m + L_{ls}\right)}{L_m}i_{qs} \tag{11.67}$$

$$v_{qs} = R_s i_{qs} + \omega_e \psi_{ds} \tag{11.68}$$

$$v_{ds} = R_s i_{ds} + \frac{d}{dt}\psi_{ds} \tag{11.69}$$

$$\left|\psi_s\right| = \psi_{ds} = L_{ls}i_{ds} + L_m\left(i_{ds} + i_{dr}\right) \tag{11.70}$$

Reference i_{qr}^* commands the injection of active power. The reactive power required for the generator can be commanded by a closed-loop control on the output voltage. The stator voltage is calculated by

$$\left|v_s\right| = \sqrt{\left(v_{ds}\right)^2 + \left(v_{qs}\right)^2}$$

and a PI controller is used to generate the reactive current set point i_{dr}^* as indicated in Figure 11.21. Set points i_{qr}^* and i_{dr}^* are on the synchronous rotating reference d_e–q_e frame, and the inverter needs commands on the rotor frame d_r–q_r as previously considered for the DFIG active–reactive power control system of Figure 11.20. This is done by a subtraction of the mechanical angle θ_r (compensated by the number of poles) from the stator electrical frequency angle θ_e, which is generated internally since this system is in stand-alone mode.

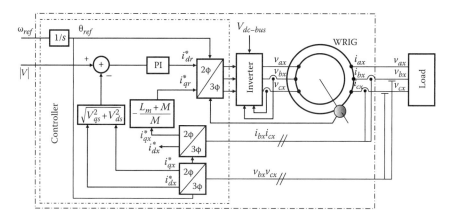

FIGURE 11.21 Stand-alone simplified control of DFIG.

11.9 SOLVED PROBLEMS

1. Explain, using the DFIG power flow fundamentals, why in a very large wind turbine converter the rotor-based converter system is rated a fraction of the total power.

Solution

Typically, the grid-side converter is controlled to have a unity power factor and a constant voltage at the dc link as indicated in the Figure 11.22.

The rotor-side converter is usually controlled to have (1) optimal power extraction from the wind and (2) a specified reactive power at the generator terminal.

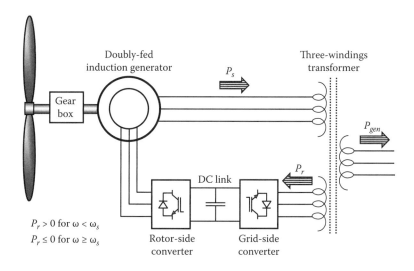

FIGURE 11.22 Wind turbine–powered DFIG with transformer-based utility connection.

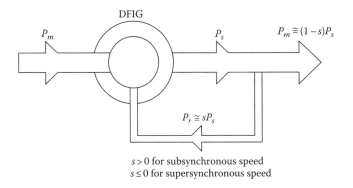

FIGURE 11.23   Wind turbine power flow through the DFIG.

Note that this converter provides sinusoidal three-phase voltages at the slip frequency. Therefore, assuming that the converters are lossless, the net power injected
by the generator to the grid can be modeled. A scheme of the active-power balance
under super- and subsynchronous speed is presented in Figure 11.23, where P_m
corresponds to the mechanical power sent into the machine shaft by the wind turbine. Because typically $R_s \ll X_m$ and $R_r \ll X_m < X_r$, the real part of the equation that
defines the rotor power is given by $P_r = \Re(S_r) \approx s\Re(S_{rs}) = sP_s$.

Note that the imaginary part of such power corresponds to the reactive power
injected to the rotor, supplied by the rotor-side converter, instead of the grid, in
summary, $S_{gen} = S_s - sP_s$, whereas the rotor converter must handle sP_s; that is, for
an operational range around the synchronous speed of $\pm s$, the converter is rated
for a fraction s of the total power. Of course, a start-up and shutdown mechanism
must be incorporated in order to avoid the rotor converter to handle the total
power during start-up and shutdown procedures.

2. Using the transient d–q DFIG model, derive a steady-state d–q model.

Solution

Using Equations 11.25 and 11.26, one can define the slip as $s = \dfrac{(\omega_s - \omega_r)}{\omega_s}$, and con-

sidering that a synchronously rotating frame is used, where all variables become
constant at steady, the machine steady-state equivalent circuit is obtained by setting the differential terms equal to zero. The d–q equations are as follows:

$$V_{qs} = -R_s I_{qs} - X_s I_{ds} + X_m I_{dr}$$

$$V_{ds} = -R_s I_{ds} - X_s I_{qs} - X_m I_{qr}$$

$$\frac{V_{qr}}{s} = \frac{R_r}{s} I_{qr} - X_m I_{ds} + X_r I_{dr}$$

$$\frac{V_{dr}}{s} = \frac{R_r}{s} I_{dr} + X_m I_{qs} - X_r I_{qr}$$

These steady-state equations can support the formulation of an equivalent electrical circuit model. Such exercise is left for the reader to derive.

11.10   SUGGESTED PROBLEMS

11.1 Conduct research to find out applications for DFIGs and write a summary report. Focus your analysis on power rating, economical reasons, control approach, and flexibility of such systems.

11.2 A transformer is frequently used in the connection of the rotor side to the grid. What will be the effect if a lower transformer ratio is selected?

11.3 Why is DFIG not considered for motoring applications for speed range from zero to rated speed?

11.4 Equation 11.17 relates stator voltage, rotor voltage, and torque. Use a spreadsheet program to develop a macro to plot a family of curves in relation to the input parameters. Assume the following machine parameters $R_s = 0.598 \ \Omega$, $R_r = 0.271 \ \Omega$, $L_{ls} = 3$ mH, $L_{lr} = 1.2$ mH, $V_s = 220$ V, $V_r = 10$ V, $p = 4$, and $\omega_e = 377$ rad/s.

11.5 Study the article by Da Silva et al.[19] Propose a modern and contemporary circuit and control techniques to achieve a similar system performance for a DFIG wind- or hydro-based energy system.

REFERENCES

1. Boldea, I. and Nasar, S.A., *Electric Drives: CD-ROM Interactive*, CRC Press, Boca Raton, FL, 1998.
2. Azaza, H. and Masmoudi, A., On the dynamic and steady state performances of a vector controlled DFM drive, *IEEE Int. Conf. Syst. Man Cybern.*, 6, 6–12, 2002.
3. Cadirici, I. and Ermis, M., Double-output induction generator operating at subsynchronous and supersynchronous speeds: Steady-state performance optimization and wind-energy recovery, *IEE Proc. B Electr. Power Appl.*, 139(5), 429–442, 1992.
4. Hopfensperger, B., Atkinson, D.J., and Lakin, R.A., Stator-flux-oriented control of a doubly-fed induction machine with and without position encoder, *IEE Proc. Electr. Power Appl.*, 147(4), 241–250, 2000.
5. Leonhard, W., *Control of Electrical Drives*, 2nd edn., Springer-Verlag, Heidelberg, Germany, 1996.
6. Bose, B.K., *Modern Power Electronics and AC Drives*, Prentice Hall, New York, 2001.
7. Pena, R., Clare, J.C., and Asher, G.M., Doubly-fed induction generator using back-to-back PWM converters and its application to variable speed wind energy generation, *IEE Proc. Electr. Power Appl.*, 143(3), 231–241, May 1996.
8. Wang, S. and Ding, Y., Stability analysis of field oriented doubly-fed induction machine drive based on computer simulation, *Electr. Mach. Power Syst.*, 21(1), 11–24, 1993.
9. Xu, L. and Wei, C., Torque and reactive power control of a doubly fed induction machine by position sensorless scheme, *IEEE Trans. Ind. Appl.*, 31(3), 636–642, 1995.

10. Datta, R. and Ranganathan, V.T., Decoupled control of active and reactive power for a grid-connected doubly fed wound rotor induction machine without position sensors, *The 34th IEEE Industry Applications Conference* (Cat. No.99 CH36370), Hyatt Regency, Phoenix, Arizona, October 1999, pp. 2623–2628.

11. Morel, L., Godfroid, H., Mirzaian, A., and Kaumann, J.M., Double-fed induction machine: Converter optimisation and field oriented control without position sensor, *IEE Proc. Electr. Power Appl.*, 145(4), 360–368, 1998.

12. Abolhasanni, M.T., Toliyat, H.A., and Enjeti, P., Stator flux oriented control of an integrated alternator-active filter for wind power applications, *Proceedings of the International Electric Machines and Drives Conference (IEMDC'03)*, Madison, WI, June 1–4, 2003, pp. 461–467.

13. AlakÄula, M. and Carlsson, A., An induction machine servo with one current controller and an improved flux observer, *IEEE IECON*, 3, 1991–1996, 1993.

14. Hu, J. and Wu, B., New integration algorithms for estimating motor flux over a wide speed range, *IEEE PESC*, 2, 1075–1081, 1997.

15. Jansen, P.L., Lorentz, R.D., and Novotny, D.W., Observer-based direct field orientation: Analysis and comparison of alternative methods, *IEEE Trans. Ind. Appl.*, 30(2), 945–953, 1994.

16. Brown, G.M., Szabados, B., Hoolbloom, G.J., and Poloujadoff, M.E., High-power cycloconverter drive for double-fed induction motors, *IEEE Trans. Ind. Electron.*, 39(3), 230–240, 1992.

17. Palle, B., Godoy Simões, M., and Farret, F.A., Dynamic interaction of an integrated doubly-fed induction generator and a fuel cell connected to grid, *36th IEEE Power Electronics Specialists Conference (PESC)*, Recife, Brazil, June 12–16, 2005, pp. 185–190.

18. Palle, B., Dynamic integration of a doubly fed induction generator with a fuel cell connected to grid, MSc thesis, Colorado School of Mines, Golden, CO, April 2006.

19. Da Silva, L.B., Nakashima, K., Torres, G.L., Da Silva, V.F., Olivier, G., April, G.E., Improving performance of slip-recovery drive: An approach using fuzzy techniques, Industry Applications Society Annual Meeting, 1991., *Conference Record of the 1991 IEEE*, 1, pp. 285–290, September 28, 1991–October 4, 1991.

12 Simulation Tools for Induction Generators

12.1 SCOPE OF THIS CHAPTER

Modern control of induction machines is a very challenging area of electrical engineering, especially when control is based on detailed machine physical and mathematical models. The correct transient state representation of an induction generator is a very complex matter because of the various electromagnetic and electromechanical phenomena involved in a detailed equivalent circuit. This complexity is mostly related to the dependence of the parameters on the frequency, transient mechanical and electromagnetic states, and degree of accuracy of the approximations included in the transient model.

Distributed stray inductances and capacitances in the machine windings and core, imbricated variable magnetic path, and the exact core geometry representations are some of the major difficulties that make this one of the most complex fields in electrical engineering. Nevertheless, the exact representation of so many details can lead to important findings in terms of reduction of electrical and mechanical stresses and to an improvement of the electric field distribution. Most of these details have been resolved in daily practice and experience accumulated over the years. For these reasons, many simplifications are usually made throughout the kinds of practical representations in Chapters 2 through 5. Industrial simulations of such scalar and vector fields use finite elements and are beyond the scope of this book. For most practical purposes, an extremely simplified model is sufficient to represent the main variations of mechanical speed, frequency, torque, power, voltage, and current. Peaks of torque, current, and voltage approximate the real ones encountered in practice, though electromagnetic interference and shaft oscillations are still not quite clear.

This chapter discusses the application of computer tools to the induction generator. The design of a small wind power plant and an appreciation of overall machine behavior are proposed as applications for MATLAB®, PSpice®, the Pascal computer language, the C computer language, Excel®, and PSIM®. Each of these computer tools has features more or less suitable for applications in the field of renewable sources of energy that lead to precious increments of efficiency and durability of equipment under operating conditions. Furthermore, electrical and mechanical stresses can be minimized with knowledge of the causes of torque, voltage, and current peaks or the voltage and current intensity of switching surges.

12.2 DESIGN FUNDAMENTALS OF SMALL POWER PLANTS

When designing a small power plant, the first thing to be done is to determine the energy potential of the primary source and the potential load. During project planning and development, several factors are taken into consideration[1,2]:

- Primary energy resources and siting
- Financial resources
- Local primary energy analysis
- Consideration of the consumer and environment
- Project planning and analysis
- Project development
- Economical operation of the power plant
- Expansion planning

The higher the final costs of the whole installation are expected to be, the more accurate the studies leading up to it need to be. In large power plants (power plants of more than 10 MW), minute details of the natural behavior of the primary source, the distance to the main consumers, and the quality of the energy and voltage levels of transmission and distribution will be very relevant and decisive issues because they will be closely related to every watt-hour generated.

On the other hand, micropower plants (power plants of less than 100 kW) are generally installed at distant sites where there is no other viable alternative for energy supply. Therefore, most of those restrictions will not apply or the load will be so sophisticated and self-regulated that quality of energy and other constraints may be made irrelevant by the filtering, storing, and regulating capabilities of the load itself as long as the average energy is available. Typically, in the present state of the art, an induction generator does not supply more than 1 MW of energy, and so it should be considered only for micro and mini power plants (power plants up to 500 kW). Particularly, the self-excited induction generator is definitely only suitable for micropower plants since it cannot supply viably more than 20 kW without using advanced power electronics and digital controls of speed and voltage.

Useful indicators of the primary energy potential are

- Data already available for surrounding areas
- Topographical evidence (mountains, valleys, sea surface)
- Other visual site phenomena (like those suggested by Beaufort tables)
- Social and cultural identification
- Measurement and recording of the primary source intensity

As this book is mostly about induction generators, our main interest is directed to the types of primary sources related to rotating primary movers such as turbine, internal combustion engines (ICE), and microturbines. Worldwide, the best-known primary sources of energy for rotating conversion are hydro and wind power.

12.3 SIMPLIFIED DESIGN OF SMALL WIND POWER PLANTS

This section will present a brief discussion to motivate an appreciation of the induction generator models introduced in other chapters. A typical application of an induction generator in a small wind power plant with variable speed will be studied. Obviously, there is a minimum amount of available wind energy required to have a viable small power plant. This minimum is based on the cubic wind power density of the site since the wind power is proportional to the cube of the wind speed. In practice, most of these studies base their final decision on sites where the speed of the wind is at least 3 m/s (11 mph); below that, it is almost impossible to have a viable power plant.

Several factors affect wind speed, some natural, some caused by civilization (cities, wildfires, roads, forest devastation). Since our source of energy changes as the cube of the wind speed changes, the output power of the generator is very sensitive to these changes. A synchronous source of conversion is recommended. The main factors affecting these data usually considered are weather systems, local terrain geography, and height above the surrounding area. Such factors because periodic changes in the wind speed are hard to predict accurately. Some studies recommend 10 years or more of statistical data, which for mini and micropower plants could be well beyond the budget. In that case, a year's data from a nearby site can be used to estimate the long-term annual wind speed. During 1 year, most variations in wind speed happen; minor modifications can be introduced. Statistically, such wind speed variations can be described by a probability function known as Rayleigh distribution. The formulas usually used for this purpose are given in Table 12.1.

TABLE 12.1

Usual Formulas for Wind Speed, Root Mean Cubic Power, and Energy Density

Average Speed	Root Mean Cubic Speed	Cubic Power Density	Cubic Energy Density
$V_{av} = \dfrac{1}{n}\sum_{i=1}^{n} v_i$	$V_{rmc} = 3\sqrt{\dfrac{1}{n}\sum_{i=1}^{n} v_i^3}$	$P_{rmc} = \dfrac{C_p}{2n}\sum_{i=1}^{n} \rho_i v_i^3$	$E_{yw} = \dfrac{8760 C_p}{2n}\sum_{i=1}^{n} \rho_i v_i^3$
m/s	m/s	W/m²	Wh/m²/year
		$\rho_o = 1.225$ kg/m³	

v_i is the wind speed sample at the instant i in m/s
V_{av} is the average wind speed in m/s
V_{rmc} is the root mean cubic wind speed in m/s
P_{rmc} is the root mean cubic power in W/m²

E_{yw} is the root mean cubic yearly energy in Wh/m²/year
ρ_i is the air density at the instant i in kg/m³
n is the number of samples

The output power in the turbine shaft including all losses and geometrical details can be approximated by

$$\frac{P}{A} = 0.25 \, V_{rmc}^3$$

where
P is the electrical generator output power
A is the area swept by the turbine blades in m²

It is important to observe at this point that the normal speed of high winds is not high enough to deliver sufficient speed directly to the electrical generator shaft to produce electrical energy. The required speeds are related to the rated synchronous speeds (3600 rpm, 1800 rpm, 900 rpm, and 450 rpm, respectively, for commercial generators of two, four, six, and eight poles), which are very high for the usual wind speed patterns (equivalent to less than 600 rpm). Therefore, it is always necessary to use a speed multiplier with inherent losses.

Once the turbine size is selected for its output power, the next step is to determine the electrical generator size. Though there is no universal standard procedure to obtain the rated speed of a turbine, a good starting point is to use the European manufacturers' recommendation called specific rated capacity (SRC), the ratio of the generator electrical capacity to the rotor swept area. If this recommendation is followed, then it is easy to determine the generator size once the wind turbine is selected.

We can relate the blade tip speed to the wind speed known as $TSR = \lambda$ by

$$\lambda = \frac{v}{V_w} = \frac{\omega R}{V_w}$$

where
v is the blade tangent tip speed
V_w is the instant wind speed
R is the length of each blade
ω is the shaft angular speed

Using the blade-tip tangent speed, we obtain the angular speed. From basic electromechanical theory, we can relate the turbine-generator common shaft power to the electrical generator torque as

$$P_{mec} = kT_{electrical}\omega$$

where k is the speed ratio of the mechanical multiplier and its inherent losses.

The load on the electrical generator should match the rotor speed of the turbine using a good control strategy, since the maximum torque of the turbine always happens before the maximum power with respect to the speed increase. The electrical generator selected should be one that always runs at the turbine maximum power.

In order to maximize the turbine output power, a very important definition in wind turbine design is the power coefficient or the rotor efficiency expressed by Betz as

$$C_p = \frac{\left(1 - \frac{v_2^2}{v_1^2}\right)\left(1 + \frac{v_2}{v_1}\right)}{2}$$

where v_1 and v_2 are, respectively, the wind speed immediately before and soon after reaching the rotor blades.

The maximum theoretical value of C_p is $16/27 = 0.5926$. So, the aim of any wind turbine control system should be to reach this maximum value, which is strongly dependent on the wind speed. This suggests that for maximum output power at any time the rotor speed must be adjusted in such a way as to have the correct v_2/v_1 ratio. Any modern control system operates at variable rotor speed.

After the generator has been selected, the power system components must be sized on the peak capacity of the electrical generator. The control system is in charge of keeping these limits within the right boundaries, which usually follows the bands given in Figure 12.1.

The components of a power system are circuit breakers, lightning arresters, switch gears, transformers, and transmission lines; they are dealt with in many books of power systems design. In addition, a protection system is always embedded in any generating systems.

An important matter to consider is the steady-state matching of both generator-load and turbine-generator connections. In addition, transient analysis must be performed. In steady-state operation, the maximum load should not ever go much beyond the nominal electrical ratings of the generator for the sake of its physical integrity. In electromechanical transient states, the high-generated currents will create intense

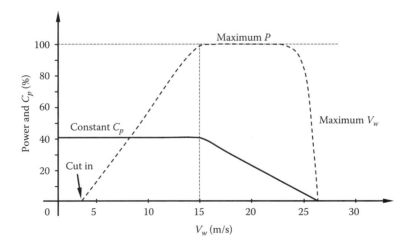

FIGURE 12.1 Speed control bands for wind turbines.

magnetic forces and intense vibrations on the generator windings against the metallic machine core and bearings. As a result, mechanical and electrical stresses are produced over the machine's useful life. Furthermore, electromagnetic noise is also generated, which can have devastating effects on sophisticated control and communication circuits.

Maximum power and torque speeds are not usually the same, as can be observed in Figure 12.2; power speed is higher than torque speed. So the electrical load of the generator should always try to match the maximum power of the turbine. As wind speed changes randomly, the maximum power point will be changing all the time. For this reason, a well-designed speed control should allow for changes in rotor speed so as to keep a maximum value of the power coefficient, C_p.

For injection of energy into the public network, it is always expected that the generator load remains constant or within a controllable range, since the public network is considered to be a machine of infinite size and almost constant voltage and frequency. In stand-alone operation, it must be ensured that the load size would not represent an impact to the generator rotation.[2–5]

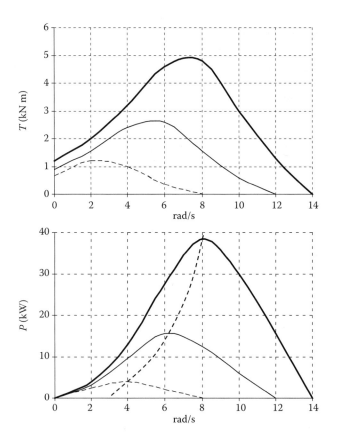

FIGURE 12.2 Torque and power characteristics of a small wind turbine.

12.4 SIMULATION OF SELF-EXCITED INDUCTION GENERATORS IN PSPICE

This example of simulation of the induction generator under transient state uses an approach based on the development of an equivalent triple-T circuit referred to a three-axis stationary reference frame as presented by Szczesny and Ronkowski,[3] adapted by Marra and Pomilio,[4] and called the $\alpha\beta\gamma$ induction machine model (see Figure 12.3). This model has proved to be accurate and quite handy for simulating the aggregation of three terminal induction machines with another device connected to its stator terminals, as, for example, a three-phase inverter. The $\alpha\beta\gamma$ transformation does not affect the stator parameters and likewise the parameters of any other devices connected to the stator.

The circuit parameters of the $\alpha\beta\gamma$ model are derived from the no-load and blocked rotor tests, in the same way they can be obtained for the single-phase equivalent T circuit. As the $\alpha\beta\gamma$ induction machine model provides an immediate circuit representation, it is especially suitable for carrying out simulations assisted by circuit symbolic interface simulation programs, such as PSpice and PSIM. Furthermore, the quality of the representations of line stator currents and voltages obtained from these simulations has to be the same as those measured in the physical machine. An outstanding advantage of the $\alpha\beta\gamma$ model, when compared with the two-axis conventional models, is the ability to simulate unbalanced conditions by simply changing parameter values.

The following assumptions are made in the formulation of the $\alpha\beta\gamma$ model:

- Rotor variables are referred to the stator.
- All magnetic and mechanical losses are ignored.
- The machine is three-wire fed and three phase.
- The magnetomotive force (mmf) has a sinusoidal pattern distributed along the stator (harmonic free).

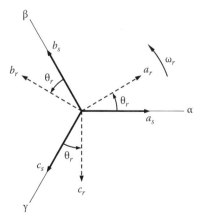

FIGURE 12.3 $\alpha\beta\gamma$ stationary reference frame.

The third of these assumptions imposes the following constraints to the stator and rotor currents:

$$i_{as} + i_{bs} + i_{cs} = 0 \qquad (12.1)$$

$$i'_{ar} + i'_{br} + i'_{cr} = 0 \qquad (12.2)$$

The induction machine model in the $\alpha\beta\gamma$ stationary reference frame is derived from the projection of the stator and rotor quantities in the $\alpha\beta\gamma$ axis, respectively, positioned in the magnetic axis of the stator a, b, and c windings of a virtual two-pole machine,[3,4] as shown in Figure 12.3.

Symbols ω_r and ω_{mr} stand for, respectively, the rotor electrical and mechanical speed in rad/s in relation to the reference axis; p is the number of poles; and θ_r is the rotor angular position in electrical radians:

$$\omega_r = \frac{p}{2}\omega_{mr} \qquad (12.3)$$

The cage rotor induction machine model written in the abc reference frame with machine variables leads to rotor flux and stator flux dependence on the rotor position θ_r as shown in the equations that follow[3,4]:

$$\left[v_{abcs}^a \right] = R_s \left[i_{abcs}^a \right] + \frac{d\left[\lambda_{abcs}^a \right]}{dt} \qquad (12.4)$$

$$0 = R'_r \left[i_{abcr}^{\prime a} \right] + \frac{d\left[\lambda_{abcr}^{\prime a} \right]}{dt} \qquad (12.5)$$

$$\left[\lambda_{abcs}^a \right] = \left(L_{ls} + M \right)\left[i_{abcs}^a \right] + \left[L_{sr} \right]\left[i_{abcr}^{\prime a} \right] \qquad (12.6)$$

$$\left[\lambda_{abcr}^{\prime a} \right] = \left(L'_{lr} + M \right)\left[i_{abcr}^{\prime a} \right] + \left[L_{sr} \right]^T \left[i_{abcs}^a \right] \qquad (12.7)$$

$$\left[L_{sr} \right] = L_{ms} \begin{bmatrix} \cos\theta_r & \cos\left(\theta_r + \frac{2\pi}{3}\right) & \cos\left(\theta_r - \frac{2\pi}{3}\right) \\ \cos\left(\theta_r - \frac{2\pi}{3}\right) & \cos\theta_r & \cos\left(\theta_r + \frac{2\pi}{3}\right) \\ \cos\left(\theta_r + \frac{2\pi}{3}\right) & \cos\left(\theta_r - \frac{2\pi}{3}\right) & \cos\theta_r \end{bmatrix} \qquad (12.8)$$

$$M = \frac{3}{2} L_{ms} \qquad (12.9)$$

where
  $[v]$ is the phase voltage vector
  $[i]$ is the line current vector
  $[\lambda]$ is the flux linkage vector
  L_{sr} is the stator and rotor mutual inductance matrix
  R_r and R_s are the rotor and stator winding resistances, respectively
  M represents the magnetizing saturated inductance (in other sections of this book it may also be denoted as L_m)

The transformation from reference frame *abc* to αβγ for the induction machine model is defined as

$$\left[f_{\alpha\beta\gamma} \right] = \left[K_{\alpha}^{a} \right] \cdot \left[f_{abc} \right] \tag{12.10}$$

where

[$f_{\alpha\beta\gamma}$] and [f_{abc}] are the matrix coefficients for the αβγ stationary and *abc* reference frames, respectively

$\left[K_{\alpha}^{a} \right]$ is the *abc* to αβγ reference frame transformation matrix

Matrix $\left[K_{\alpha}^{a} \right]$ may be divided into a matrix for stator transformation $\left[K_{\alpha s}^{a} \right]$ and another one for rotor transformation $\left[K_{\alpha r}^{a} \right]$.[3,4,7] Matrixes $\left[K_{\alpha s}^{a} \right]$ and $\left[K_{\alpha r}^{a} \right]$ are defined as follows:

$$\left[K_{\alpha s}^{a} \right] = diag \begin{bmatrix} 1 & 1 & 1 \end{bmatrix} \tag{12.11}$$

$$\left[K_{\alpha r}^{a} \right] = \frac{2}{3} \begin{bmatrix} \cos\theta_r + \frac{1}{2} & \cos\left(\theta_r + \frac{2\pi}{3}\right) + \frac{1}{2} & \cos\left(\theta_r - \frac{2\pi}{3}\right) + \frac{1}{2} \\ \cos\left(\theta_r - \frac{2\pi}{3}\right) + \frac{1}{2} & \cos\theta_r + \frac{1}{2} & \cos\left(\theta_r + \frac{2\pi}{3}\right) + \frac{1}{2} \\ \cos\left(\theta_r + \frac{2\pi}{3}\right) + \frac{1}{2} & \cos\left(\theta_r - \frac{2\pi}{3}\right) + \frac{1}{2} & \cos\theta_r + \frac{1}{2} \end{bmatrix} \tag{12.12}$$

Application of Equation 12.10 to the machine model described by Equations 12.4 through 12.9, taking Equations 12.1 and 12.2 into account, yields

$$\left[v_{abcs} \right] = R_s \left[i_{abcs} \right] + \frac{d\left[\lambda_{abcs} \right]}{dt} \tag{12.13}$$

$$0 + R_r \left[i'_{\alpha\beta\gamma r} \right] + \frac{d\left[\lambda'_{\alpha\beta\gamma r} \right]}{dt} + \frac{\omega_r}{\sqrt{3}} \left[\lambda'_{\alpha x} \right] \tag{12.14}$$

$$\left[\lambda_{abcs} \right] = \left(L_{ls} + M \right) \left[i_{abcs} \right] + M \left[\lambda'_{\alpha\beta\gamma r} \right] \tag{12.15}$$

$$\left[\lambda'_{\alpha\beta\gamma r} \right] = \left(L_{lr} + M \right) \left[i'_{\alpha\beta\gamma r} \right] + M \left[i_{abcs} \right] \tag{12.16}$$

$$\left[\lambda'_{\alpha x} \right] = \begin{bmatrix} \left(\lambda'_{\beta r} - \lambda'_{\gamma r} \right) \\ \left(\lambda'_{\gamma r} - \lambda'_{\alpha r} \right) \\ \left(\lambda'_{\alpha r} - \lambda'_{\beta r} \right) \end{bmatrix} \tag{12.17}$$

The notation for stator variables (voltage, current, and flux) are kept as originally used for the *abc* reference frame apart from the omission of the superscript *a*. The αβγ*r* subscript indicates rotor variables referred to the αβγ stationary frame. The [λ$_{ax}$] vector comes out because of the *abc*-to-αβγ frame transformation.

The mechanical system equation is

$$T_e = \left(\frac{2}{p}\right) J \frac{d\omega_r}{dt} + \left(\frac{2}{p}\right) B_m \omega_r + T_L \qquad (12.18)$$

where

J is the rotor inertia in kg m^2
B_m is the rotational friction constant in kg m^2/s
T_e is the electromagnetic torque
T_L is the load torque in N m

Torque T_e may be defined as

$$T_e = \sqrt{3}\left(\frac{p}{2}\right) M \left(i_{as} i_{\gamma r} - i_{cs} i_{\alpha r}\right) \qquad (12.19)$$

The αβγ stationary frame induction machine model is governed by Equations 12.13 through 12.19.

The magnetizing current of the induction machine [i_{abcm}] is the sum of the stator and rotor currents given as

$$\left[i_{abcm}\right] = \left[i_{abcs}\right] + \left[i_{\alpha\beta\gamma r}\right] \qquad (12.20)$$

The saturation of the air-gap magnetizing inductance M is taken into account, considering that the machine is magnetically and electrically symmetrical. Hence, the saturation, caused by the air-gap flux, affects all windings equally and is independent of the direction of the flux at any instant and dependent only on its magnitude.[4] The magnetizing current space vector, as defined in Equation 12.21, is the quantity that determines the saturation degree; that is, it defines the value of M from

$$\bar{I}_m = \frac{2}{3}\left(i_{am} + i_{bm} e^{j\frac{2\pi}{3}} + i_{cm} e^{-j\frac{2\pi}{3}}\right) \qquad (12.21)$$

When considering saturation, the derivatives present in the second member of Equations 12.13 and 12.14 yield

$$\frac{d\left[\lambda_{abcs}\right]}{dt} = L_{ls} \frac{d\left[i_{abcs}\right]}{dt} + M \frac{d\left[i_{abcm}\right]}{dt} + \left[i_{abcm}\right] \frac{dM}{dt} \qquad (12.22)$$

$$\frac{d\left[\lambda'_{\alpha\beta\gamma r}\right]}{dt} = L'_{lr} \frac{d\left[i'_{\alpha\beta\gamma r}\right]}{dt} + M \frac{d\left[i_{abcm}\right]}{dt} + \left[i_{abcm}\right] \frac{dM}{dt} \qquad (12.23)$$

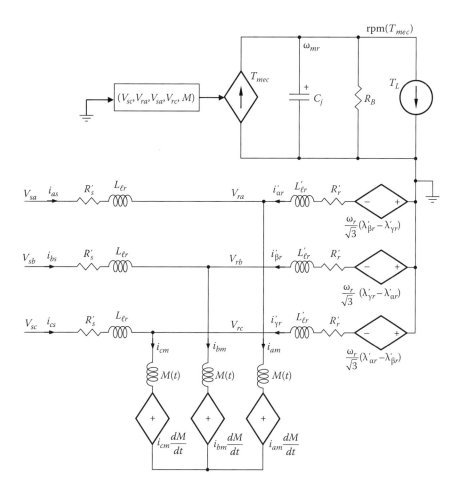

FIGURE 12.4 Equivalent triple-T circuit for the $\alpha\beta\gamma$ machine model.

Notice in Equations 12.22 and 12.23 that the leakage inductances are assumed constant and the saturation inclusion leads to a new element in the magnetizing branch of the equivalent T circuit, which is proportional to the derivative of M with time.

The aforementioned theory is fundamental to the $\alpha\beta\gamma$ machine model represented by the equivalent triple-T circuit presented in Figure 12.4. The mechanical system Equation 12.18 is represented by an equivalent circuit as shown in Figure 12.5, referring to the same grounding point as the circuit in Figure 12.4. In this circuit, T_e is obtained from Equation 12.19, C_j represents J, R_B represents the inverse of B, and T_L is the mechanical load torque.

The $\alpha\beta\gamma$ model described by Equations 12.13 through 12.19 can be simulated by the implementation of the circuits in Figures 12.4 and 12.5 in PSpice graphical language. As can be seen in that figure, the stator currents, rotor currents, and magnetizing currents are sampled with a current-controlled voltage source (H device).

The value of M is determined by the output of a PSpice®ETABLE block whose entry is the magnitude of the magnetizing current space vector. This ETABLE block

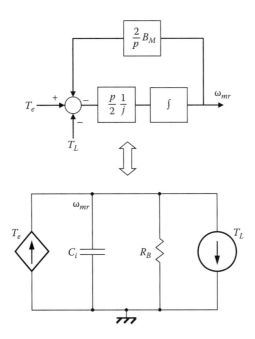

FIGURE 12.5 Mechanical system equivalent circuit.

comprises values of M as a function of the magnitude of the magnetizing current space vector, assessed from the machine-magnetizing curve, experimentally obtained. The output of the ETABLE block is the input of the three variable inductance ZX blocks, which are the inductances M of each phase of the magnetizing branches. The output of the ETABLE block is also an entry to a DIFFER block to determine the value of dM/dt. The output value of the DIFFER block is then multiplied by each one of the magnetizing currents and used to control three voltage-controlled voltage sources, which represent the sources in the magnetizing branch of Figure 12.4.

The voltage sources present in the rotor circuit, relative to the $\left[\lambda'_{ax}\right]$ vector, are calculated using ABM blocks whose entries are the sampled rotor and stator currents M and the rotor electrical speed ω_r. The value of ω_r is obtained from the mechanical system circuit by the multiplication of ω_{mr} and $p/2$. The electromagnetic torque T_e at the mechanical system equivalent circuit in Figure 12.5 is calculated with an ABMI block whose entries are the rotor and stator currents and M.

In this example, the machine represented in the simulation is a four-pole, ½-hp induction machine whose parameters and magnetizing curves were obtained from experimental tests. The machine self-excitation as an induction generator was simulated using a Y-connected 37 µF capacitor bank and sinusoidal sources, which were disconnected at 15 ms, to represent the residual flux effects. The parameters of the induction machine were $r_s = 4.4\ \Omega$, $r_r = 5.02\ \Omega$, $L_{\ell s} = L'_{\ell r} = 15.6$ mH, and $J = 6.10^{-4}$ kg m². The self-excitation was carried out at $\omega_{mr} = 2000$ rpm. The line voltage across the machine terminals obtained from simulation is presented in Figure 12.6 and similar voltage was experimentally obtained as shown

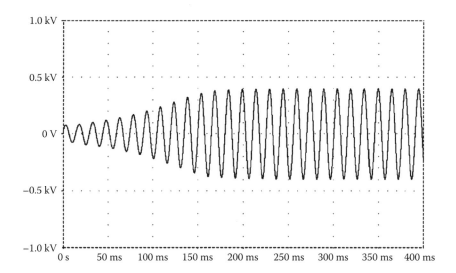

FIGURE 12.6 IG self-excitation line voltage obtained by simulation.

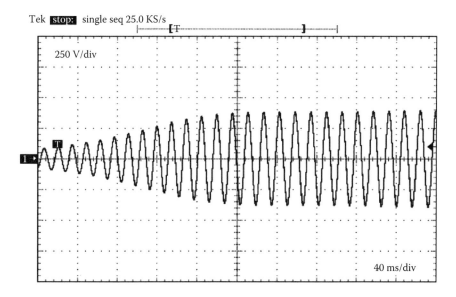

FIGURE 12.7 IG self-excitation line voltage obtained experimentally.

in Figure 12.7. The satisfactory agreement between simulation and experimental results attests to the model reliability.

Although the αβγ model is suitable for representing the average saturation effects, the assumption that the machine remains symmetrical when saturated does not allow for representation of the harmonic phenomena in currents and voltages or effects caused by machine asymmetries.

12.5   SIMULATION OF SELF-EXCITED INDUCTION GENERATORS IN PASCAL

Pascal is popular with students, researchers, and teachers because it is a simple, easily understood language, though this simplicity makes it less flexible than other more powerful languages like C and C++. For study of less complex models, it can be a quite acceptable and fast computer language for engineering problems where high speed and machine interactions are not the main issue. Iterative numerical processes and online fast control schemes are the main features of these computer languages.

Two high-level computer languages commonly used in engineering, Pascal and C, can be of extreme value in simulating the transient phenomenon of self-excitation.

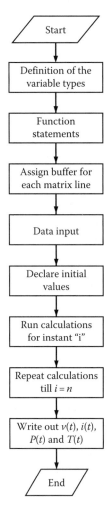

FIGURE 12.8   Pascal flowchart for high-level language simulation of the transient calculations of a SEIG.

In a few minutes, details of the power, torque, voltage, and current transient states can be easily simulated and plotted for several conditions of load and speed of the prime mover. The flowchart in Figure 12.8 can be used to develop a fast-dedicated computer program in either of these languages. The Pascal code of this program for a stand-alone self-excited induction generator is given in Appendix C.

As Pascal differs from C only in the form of statements, the computer program and the output results are left to be discussed in Section 12.10.

12.6   SIMULATION OF STEADY-STATE OPERATION OF AN INDUCTION GENERATOR USING MICROSOFT EXCEL

To understand this section better, read Chapter 4 first. That chapter says that in many applications of the induction generator it is necessary to establish relationships among its parameters and static characteristics in such a way that the designer is able to know in advance if a specified generator is going to be suitable or not for a particular application.[6–8] That is the case, for example, when an induction motor is considered to work as a self-excited induction generator. The self-exciting capacitor bank has to be calculated and sized for extreme conditions of load, and it has to respond to real-world conditions. Preliminary tests can be done based on an Excel spreadsheet, as displayed in the block diagram shown in Figure 12.9 with the theory discussed in Chapter 4.

Table 12.2 displays a fitted Microsoft Excel spreadsheet with all the necessary data for the steady-state calculation of an induction generator in steady-state operation. Table 12.3 defines each variable for the calculations in Table 12.2. An example of the results of these calculations is given in Figure 12.10 where the terminal voltage of the induction generator is plotted versus its output power to the load.

Another example of the use of an Excel spreadsheet is the calculation of the averages and values shown in Table 12.1. This very helpful tool saves a lot of time on tedious performance calculation for every new estimation of either wind or hydro power plant performance. The statistical averages of wind speed distribution or water flow estimation and the discrete integration processes involved are a straightforward task for such spreadsheets. The main limitation of most spreadsheets is that they cannot perform iterative processes unless helped by a lengthy user data transfer.

12.7   SIMULATION OF VECTOR-CONTROLLED SCHEMES USING MATLAB®/SIMULINK®

Simulink is a subprogram in the MATLAB environment. It uses a graphical user interface and enables a system to be simulated by block diagrams and equations. For all of the programs in this section, the versions that are used are as follows:

- MATLAB 6.1.0.450 (R12.1)
- MATLAB Toolbox 6.1
- Simulink 4.1
- DSP Blockset 4.1

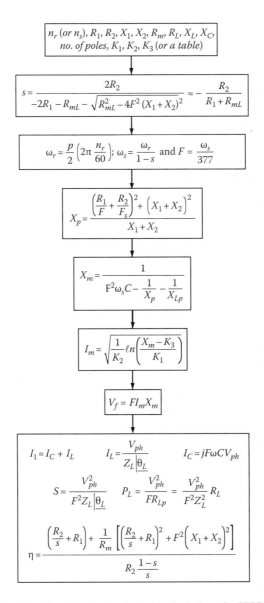

FIGURE 12.9 Excel flowchart for the steady-state calculation of a SEIG.

The DSP blockset is used to run the Butterworth filters in the system. If it is not available in this blockset, then the Butterworth filters can be replaced by transfer functions in the s-domain that represent such low-pass filters.

Shown in Figure 12.11 is a typical block diagram of one of the models that can be run in Simulink. The system contains all of the equations necessary for an induction machine as described in Chapter 9. The reader is also referred to the literature on vector control, as referenced in Chapter 9. Subsystem #1 in

TABLE 12.2

Example of IG Data Fitting in Excel Spreadsheet for Steady-State Calculations

A	B	C	D	E	F	G	H	I	J	K	L	M	N	O
P_L	V_f	I_m	X_m	R_{mL}	R_p	X_p	s	F	w_s	n_r	w_r	R_1	R_2	X_1
3.573516	267.2788	25.16681	10.62521	918.3069	862,016.9	322,807.1	−0.00044	0.999537	376.8254	1800	376.9911	0.575	0.404	1.3074
71.38949	267.0765	25.1253	10.63878	490.4483	246,036.3	92,149.94	−0.00082	0.999154	376.6812	1800	376.9911	0.575	0.404	1.3074
355.1414	266.1861	24.93403	10.70182	165.5928	28,125.34	10,541	−0.00243	0.997551	376.0768	1800	376.9911	0.575	0.404	1.3074
1387.328	262.2742	23.9844	11.02729	47.53132	2,345.553	881.2441	−0.0084	0.991649	373.8515	1800	376.9911	0.575	0.404	1.3074
1717.418	260.7511	23.58812	11.16917	38.40436	1,538.384	578.4512	−0.01036	0.989719	373.1239	1800	376.9911	0.575	0.404	1.3074
2247.662	257.9207	22.84026	11.44641	29.09348	890.3077	335.214	−0.01362	0.986543	371.9265	1800	376.9911	0.575	0.404	1.3074
2393.875	257.0348	22.60647	11.53557	27.20874	780.6757	294.0474	−0.01454	0.985644	371.5879	1800	376.9911	0.575	0.404	1.3074
2473.964	256.5259	22.47268	11.58711	26.26352	728.4258	274.4247	−0.01505	0.985147	371.4004	1800	376.9911	0.575	0.404	1.3074
2559.254	255.9637	22.32548	11.64426	25.31638	677.9009	255.4478	−0.0156	0.984613	371.1991	1800	376.9911	0.575	0.404	1.3074
2650.198	255.339	22.16283	11.70794	24.36733	629.1119	237.1208	−0.0162	0.984038	370.9822	1800	376.9911	0.575	0.404	1.3074
2747.279	254.6407	21.98229	11.77925	23.41635	582.0695	219.4476	−0.01684	0.983416	370.7479	1800	376.9911	0.575	0.404	1.3074
2851.005	253.8546	21.78089	11.85958	22.46344	536.7848	202.4325	−0.01754	0.982743	370.4942	1800	376.9911	0.575	0.404	1.3074

(*Continued*)

TABLE 12.2 (Continued)

Example of IG Data Fitting in Excel Spreadsheet for Steady-State Calculations

P	Q	R	S	T	U	V	W	X	Y	Z	AA	AB	AC	AD
X_2	$X_1 + X_2$	R_m	X_C	$\text{Cos}(\Theta_l)$	$[Z_l]$	R_{lp}	X_{lp}	P	K_1	K_2	K_3	SQRT(R)	η	f_s
1.3074	2.6148	962.5	10.61499	0.99999	20,000	20,000.2	4,472.147	4	11.88	−0.0022	7.68	918.292	4.585827	59.97363
1.3074	2.6148	962.5	10.61906	0.99999	1,000	1,000.01	223,607.4	4	11.88	−0.0022	7.68	490.4204	48.94519	59.95067
1.3074	2.6148	962.5	10.63613	0.99999	200	200.002	44,721.47	4	11.88	−0.0022	7.68	165.5102	82.30469	59.85448
1.3074	2.6148	962.5	10.69944	0.99999	50	50.0005	11,180.37	4	11.88	−0.0022	7.68	47.24275	93.12882	59.50031
1.3074	2.6148	962.5	10.7203	0.99999	40	40.0004	8,944.294	4	11.88	−0.0022	7.68	38.04663	93.60563	59.38451
1.3074	2.6148	962.5	10.75481	0.99999	30	30.0003	6,708.221	4	11.88	−0.0022	7.68	28.61961	93.79725	59.19395
1.3074	2.6148	962.5	10.76461	0.99999	28	28.00028	6,261.006	4	11.88	−0.0022	7.68	26.70144	93.77368	59.14005
1.3074	2.6148	962.5	10.77005	0.99999	27	27.00027	6,037.399	4	11.88	−0.0022	7.68	25.73759	93.75043	59.11021
1.3074	2.6148	962.5	10.77589	0.99999	26	26.00026	5,813.791	4	11.88	−0.0022	7.68	24.77035	93.71838	59.07817
1.3074	2.6148	962.5	10.78219	0.99999	25	25.00025	5,590.184	4	11.88	−0.0022	7.68	23.79953	93.67648	59.04365
1.3074	2.6148	962.5	10.789	0.99999	24	24.00024	5,366.577	4	11.88	−0.0022	7.68	22.82491	93.62354	59.00637
1.3074	2.6148	962.5	10.79639	0.99999	23	23.00023	5,142.969	4	11.88	−0.0022	7.68	21.84622	93.55815	58.96598

Note: Rated values of the induction generator: power = 3.0 kW; phase voltage = 220 V; f = 60 Hz.

TABLE 12.3
Excel Formula Definitions for Table 12.2 Spreadsheet Cells

A	B	C	D	E
$P_L = B2 * B2/(I2 * V2)$	$V_f = I2 * C2 * D2$	$I_m = SQRT((1/Z2) * LN((D2 - AA2)/Y2))$	$X_m = 1/((I2 * I2/S2) - (1/G2) - (1/W2))$	$R_{mL} = R2 * V2/(R2 + V2)$

F	G	H	I	J
$R_p = ((M2 + (N2/H2)) * (M2 + (N2/H2))/(I2 * I2) + Q2 * Q2)/(M2 + (N2/I2))$	$X_p = ((M2 + (N2/H2)) * (M2 + (N2/H2))/ (I2 * I2) + Q2 * Q2)/Q2$	$s = -N2/(M2 + E2)$	$F = J2/377$	$w_s = L2/(1 - H2)$

K	L	M	N	O
n_r 1800	$w_r = PI() * X2 * K2/60$	R_1 0.575	R_2^* 0.404	X_1 1.3074

P	Q	R	S	T
X_2	$X_1 + X_2$	R_m	X_C	COS(QL)

U	V	W	X	Y
[Z_L] 20,000	$R_{Lp} = U2/T2$	$X_{Lp} = U2/(SQRT(1 - T2 * T2))$	P 4	$K_1 = Y2$

Z	AA	AB	AC	AD
K_2 -0.002202	K_3 7.68	SQRT(R) = SQRT(E2 * E2 - 4 * Q2 * Q2)	η = 100 * H2 * ((M2 + (N2/H2)) + ((M2 + (N2/H2)) * (M2 + (N2/H2)) + (I2 * (O2 + P2)) * (I2 * (O2 + P2))/ R2)/(N2 * (1 - H2))	$f_s = J2/(2 * PI())$

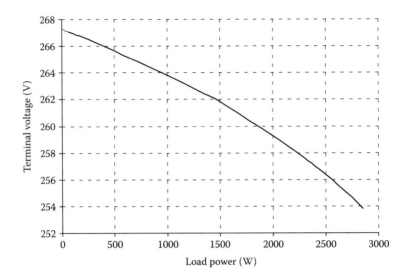

FIGURE 12.10   Example of the voltage regulation Excel plot of an induction generator.

Figure 12.11 is the controller for the model; it changes based on which model is running. Subsystem #2 in Figure 12.11 is the three-phase inverter block, which changes the reference voltages from the controller to the phase voltages through a PWM converter. Subsystem #3 in Figure 12.11 is the induction machine block, which contains all of the equations necessary for an induction machine. The block takes in the phase voltages and calculates speed, currents, fluxes, and developed torque. Subsystem #4 in Figure 12.11 takes the outputs of the model where the matrices are sent to the MATLAB workspace as the value listed. The time index ("tout") is also output to the workspace. Subsystem #5 in Figure 12.11 is a filtering block that filters both the input voltages and output currents with an eighth-order low-pass Butterworth filter.

12.7.1   INPUTS

Before any of the models can be simulated, the *consts.m* file must be run to initialize the variables within the machine and also the controller constants. Table 12.4 displays the inputs to the model, which can be changed; all of the values are included in *consts.m* except for the repeating sequence.

12.7.2   OUTPUTS

In terms of the outputs of the model, every time the model is run it creates the matrices given in Table 12.5.

There are several scopes inside of the blocks to show instantaneous variables such as i_{qs} and i_{ds} and the reference voltages v_{qs}^* and v_{ds}^*. To obtain any other values within

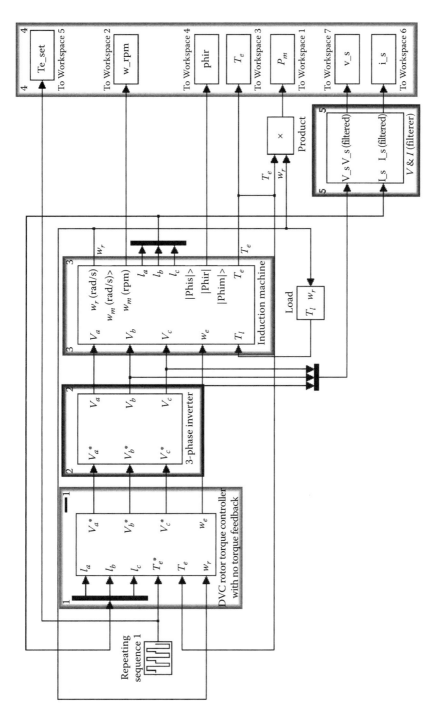

FIGURE 12.11 Simulink simulation diagram for vector controlled schemes.

TABLE 12.4
Inputs for Simulink-Based Vector Control Scheme

Repeating sequence	It applies to all block diagrams and is the reference torque the machine is trying to follow.
J	This is the equivalent inertia on the mechanical shaft assume 0.05.
Kw	This value is the constant to determine rotational losses in the machine when running. $T_L = K_w\omega^2$, usually 0.00047502.
freqt	This value is the frequency that the PWM operates at, usually at 2000 Hz.
Vdc	This is the value of the DC bus, usually at 300 V.
Tstop	This value is how long the simulation runs, usually 10 s.
t_J_large	After this time value, the inertia of the machine will be changed to a very large value so the speed of the machine does not change with torque, usually at 4 s.
fc*	All these values, fci_dvc, fcv, etc., are used to control the filtering in the output, controllers, and the machine. Lower filtering values can create smoother signals, but could also create instabilities in the system.

TABLE 12.5
Output for Simulink-Based Vector Control Scheme

Te_set	Torque command, which derives directly from the repeating sequence in N m.
w_rpm	Mechanical speed of the shaft in rpm.
phir	Rotor flux within the machine in Wb.
T_e	Torque developed by the machine in N m.
P_m	Mechanical power created by the machine in W.
v_s	V_a, V_b, and V_c, filtered to 1000 Hz in V.
i_s	I_a, I_b, and I_c, filtered to 1000 Hz in A.

the machine or the controllers, the reader can add either a scope connected to that point or to a workspace block. To run the model, follow these steps:

1. Change the MATLAB directory to where the files are.
2. Type *consts* in the MATLAB command window. MATLAB will display:
   a. The current revision
   b. "Constants have been initialized"
3. Open one of the models to be run.
4. Click on the icon for simulation at the top of the screen. The time will be displayed at the bottom of the screen and will go to Tstop.

After the simulation runs to the end of the time, all of the output variables will be available to use in the MATLAB workspace. For example, to display the currents on a graph with time, use the following command: *plot(tout,i_s)*, which will plot all three currents in a figure.

All of the simulations run with the following sequence. A low (1 N m) reference torque is applied for 2 s (0–2 s). This allows the fluxes within the machine to stabilize. A high (5 N m) reference torque is applied for 4 s (2–6 s). While this is happening, the machine is running as a motor and the speed will increase until the rotational losses are equal to the applied torque. A very high inertia (10,000 J) is applied at time 4 s. This makes the speed of the machine constant so that power can be generated with a negative torque. Without this change, when a negative torque is applied, the machine will reverse rotation and spin in the other direction. A high negative (–5 N m) reference torque is applied for 4 s (6–10 s). Since the speed is locked due to the high inertia, the motor is now a generator generating power.

The next three sections outline how to run the models for the specific controllers and give typical outputs for these models.

12.7.3   INDIRECT VECTOR CONTROL

Files needed:

- consts.m
- ivc_rotor_model_torque_no_feed.mdl
- ivc_rotor_model_torque_w_feed.mdl

In the indirect-vector torque-control system, both of the IVC rotor models find the electrical speed from the speed of the shaft and the calculated slip angle. The controller imposes the restraint that the rotor flux, Phir, is aligned with the ds-axis. Since I_{qs} is directly proportional to torque, the current I_{qs} is controlled to follow the reference torque applied to the block.

The model without feedback uses no measurement from torque to control the torque of the motor, relying on the proportionality of I_{qs} and torque. The torque response of the system is shown in Figure 12.12, and the torque follows the reference torque, except that when the torque is negative, there is a dc offset. The speed response of the machine is shown in Figure 12.13 and in Figure 12.14 is the power to the machine. As can be seen from those figures, from time 0 to 6 s, the machine is in motoring mode, and from 6 to 10 s, the motor is generating power due to a positive speed and a negative torque.[8,9]

The feedback model assumes that there is a torque sensor on the shaft and therefore the reference i_{qs} is based on the difference between the reference torque and the actual torque, commanded through a PI controller. This provides a significantly better response on the torque, as shown in Figures 12.15 through 12.17. Except for the better torque response, this system is identical to the system without feedback.

12.7.4   DIRECT VECTOR CONTROL WITH ROTOR FLUX

Files needed:

- consts.m
- dvc_rotor_model_torque_no_feed.mdl
- dvc_rotor_model_torque_w_feed.mdl

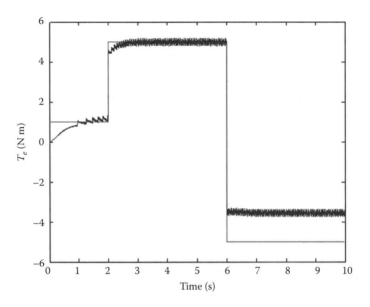

FIGURE 12.12 Torque response for indirect vector control.

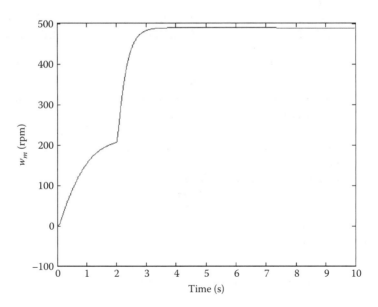

FIGURE 12.13 Speed response for indirect vector control.

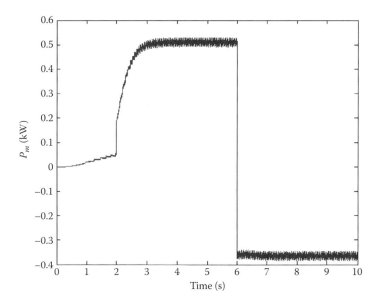

FIGURE 12.14 Power response for indirect vector control.

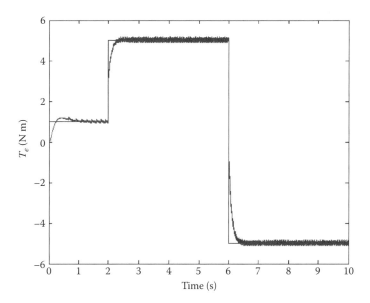

FIGURE 12.15 Closed-loop torque response for indirect vector control.

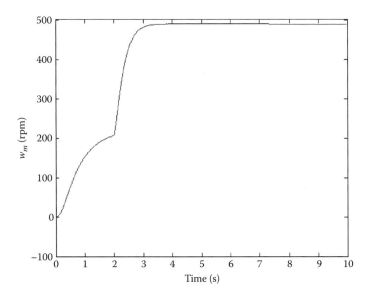

FIGURE 12.16   Closed-loop speed response for indirect vector control.

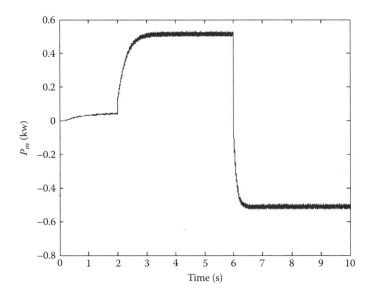

FIGURE 12.17   Closed-loop power response for indirect vector control.

In the direct-vector-control system with rotor flux, both of the DVC rotor models find the electrical speed from value of the rotor flux on the stationary axis. The controller imposes the restraint that the rotor flux is aligned with the ds-axis. Since I_{qs} is directly proportional to torque, the current reference I_{qs} is proportional to the reference torque applied to the block.

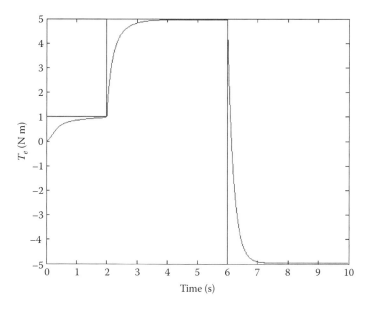

FIGURE 12.18 Torque response for direct vector control with rotor flux.

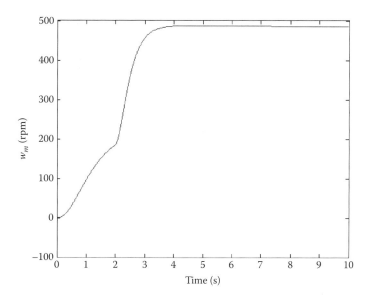

FIGURE 12.19 Speed response for direct vector control with rotor flux.

The model without feedback uses no measurement from torque to control the torque of the motor, relying on the proportionality of I_{qs} and torque. The torque response of the system is shown in Figure 12.18, and the torque follows the reference torque, with a slow response. The speed response of the machine is shown in Figure 12.19, and in Figure 12.20 is the power to the machine. As can be seen from

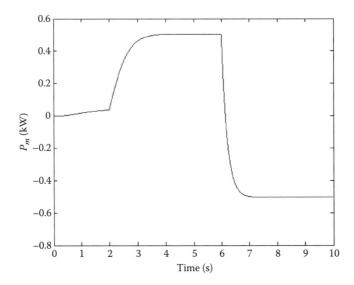

FIGURE 12.20 Power response for direct vector control with rotor flux.

those figures, from time 0 to 6 s, the machine is in motoring mode, and from 6 to 10 s, the motor is generating power due to a positive speed and a negative torque.

The feedback model assumes that there is a torque sensor on the shaft and therefore the reference I_{qs} is based on the difference between the reference torque and the actual torque, commanded through a PI controller. This provides significantly better response on the torque, as shown in Figures 12.21 through 12.23. Except for the better torque response, this system is identical to the system without feedback.

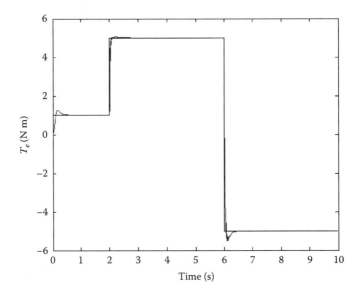

FIGURE 12.21 Closed-loop torque response for direct vector control with rotor flux.

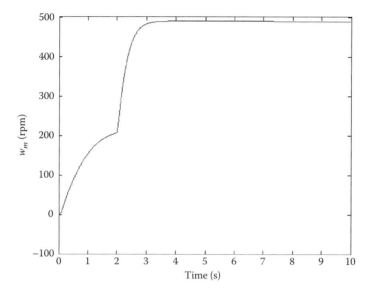

FIGURE 12.22 Closed-loop speed response for direct vector control with rotor flux.

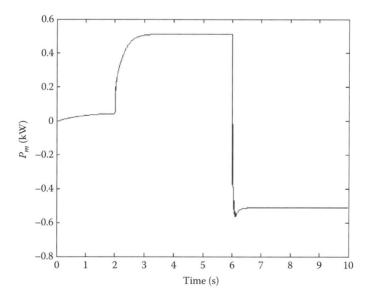

FIGURE 12.23 Closed-loop power response for direct vector control with rotor flux.

12.7.5 DIRECT VECTOR CONTROL WITH STATOR FLUX

Files needed:

- consts.m
- dvc_stator_model_torque_no_feed.mdl
- dvc_stator_model_torque_w_feed.mdl

In direct vector control with stator flux, both of the DVC stator models get the electrical speed directly from the currents and voltages in the machine. The controller imposes the restraint that the stator flux in the machine is aligned with the ds-axis. Since the torque produced by the machine is proportional to the current I_{qs}, this current is used to control the torque developed by the machine.

The model without feedback uses no measurement from torque to control the torque of the motor, relying on the proportionality of I_{qs} and torque. The torque response of the system is shown in Figure 12.24 and follows the reference torque very closely.

This is the best controller in terms of response, especially considering that there is no feedback from the torque being developed by the machine. The speed response of the machine is shown in Figure 12.25, and in Figure 12.26 is the power to the machine. As can be seen from the figures, from time 0 to 6 s, the machine is in motoring mode, and from 6 to 10 s, the motor is generating power due to a positive speed and a negative torque.

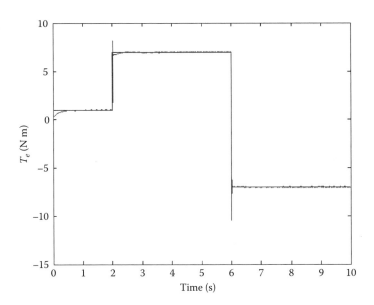

FIGURE 12.24 Torque response for direct vector control with stator flux.

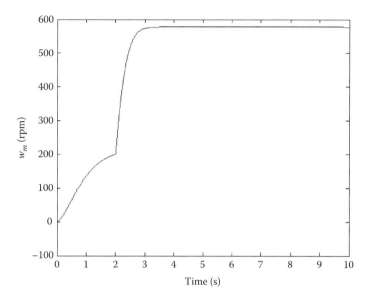

FIGURE 12.25 Speed response for direct vector control with stator flux.

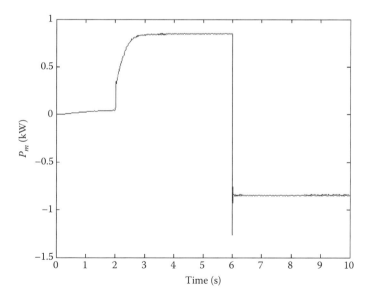

FIGURE 12.26 Power response for direct vector control with stator flux.

The feedback model assumes that there is a torque sensor on the shaft and there-fore the I_{qs} is based on the difference between the reference torque and the actual torque, commanded through a PI controller. Adding this feedback does not make the controller act that much better, as the model without the feedback is very good at following the reference torque. The responses with feedback are shown in Figures 12.27 through 12.29.

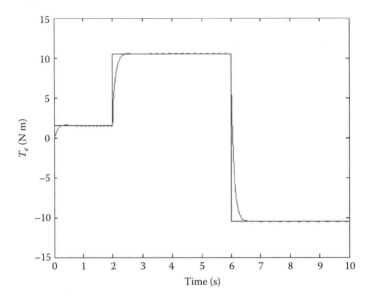

FIGURE 12.27 Closed-loop torque response for direct vector control with stator flux.

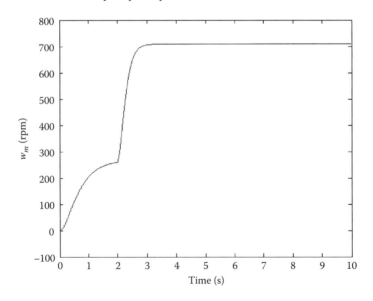

FIGURE 12.28 Closed-loop speed response for direct vector control with stator flux.

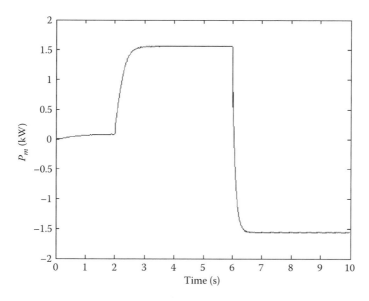

FIGURE 12.29   Closed-loop power response for direct vector control with stator flux.

12.7.6   EVALUATION OF THE MATLAB/SIMULINK PROGRAM

Simulink is a very powerful program that can be used to simulate a wide variety of systems. The graphical user interface is very easy to use, and scopes can be placed at any junction to observe the current state of the system. In addition, the ability of Simulink to output variables to the MATLAB environment is a large plus when dealing with complex systems. Simulink can also interface with hardware and perform calculations in real time to control systems.

However, the graphical user interface imposes problems in terms of simulation time. The models that are shown in this section take 5 min to run a 10 s simulation on a Pentium 3 1.2 GHz processor with 512 MB RAM. When running, the program will consume over 200 MB of RAM; you need a powerful system to run these simulations.

12.8   SIMULATION OF A SELF-EXCITED INDUCTION GENERATOR IN PSIM

PSIM is a commercial program that enables speedy analysis of an induction generator. Readily available tools and electrical and electromechanical components make this program an easy tool to observe phenomena that would otherwise only be observable in a good power-system laboratory. PSIM is an internationally accepted computer program used to demonstrate practical results in many high-level international conferences and journals.

In this section, we present a short introduction to the PSIM block that can be used to simulate the stand-alone operation of a wound-rotor self-excited induction generator (SEIG). Before going through this section, the student should first read Chapter 4.

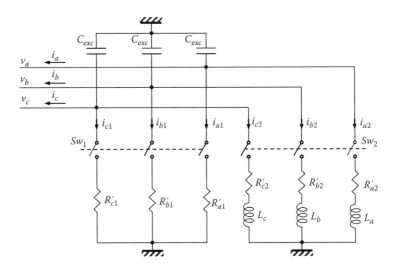

FIGURE 12.30 Load circuit for the self-excited induction generator.

The same model shown in Figure 12.4 is used to simulate the SEIG model in PSIM except by the addition of the load as represented in Figure 12.30. The induction generator parameters in this figure refer to the stator side in the PSIM SEIG block. The mutual inductances are implemented with PSIM computational and mathematical blocks as depicted in Figure 12.31b. The loads can be switched on or switched off after the unloaded generator has completely excited. Computation time in a regular PC is about 25 s for a simulation time of 1.25 s.

- R_s stator resistance, ohms
- $L_{\ell s}$ stator leakage inductance, H
- R_r rotor resistance, ohms
- $L_{\ell r}$ rotor leakage inductance, H
- C_{exc} excitation capacitance, μF
- Sw_1 primary load switch
- Sw_2 secondary load switch
- R_{a1}, R_{b1}, R_{c1} primary load resistances, Ω
- R_{a2}, R_{b2}, R_{c2} secondary load resistances, Ω
- L_{L2} secondary per-phase load inductance, H
- f_{rated} operating frequency, Hz
- V_{rf} equivalent residual flux vector, V
- t_1 instant at which the primary load is connected, s
- t_2 instant at which the secondary load is connected, s

The load parameters are initialized to simulate an open circuit. The induction generator might fail or take a very long time to self-excite if started on a heavy load. A suitable value for self-excitation capacitance should be chosen from the magnetization curves to self-excite the generator up to the rated voltage as discussed in Section 4.3.

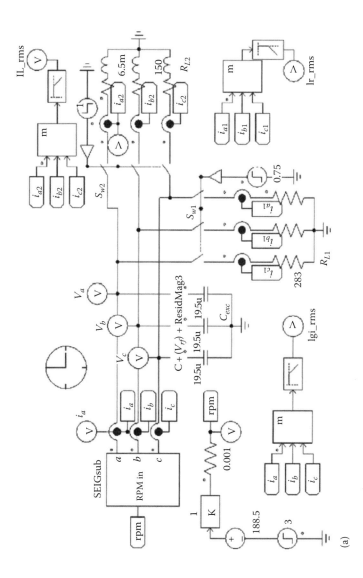

FIGURE 12.31 (a) **(See color insert.)** PSIM SEIG circuit model with linear loads switched at distinct instants.

(Continued)

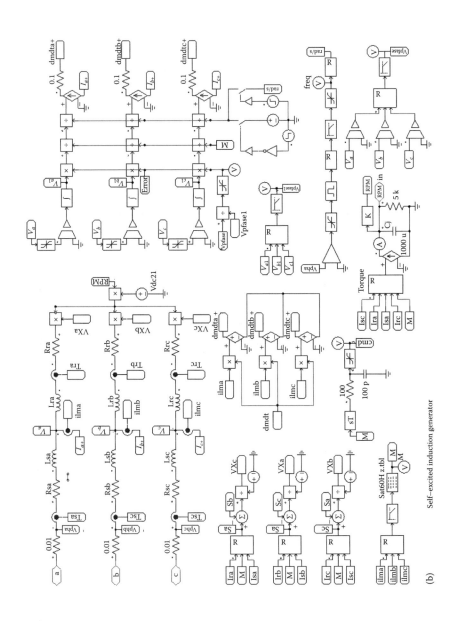

(b)   Self-excited induction generator

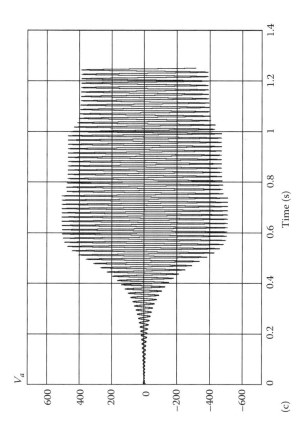

FIGURE 12.31 (*Continued*) (b) **(See color insert.)** SEIGsub subsystem, expansion from (a) (c) voltage build up with the self-excitation response.

TABLE 12.6

Data for the Magnetization Curve

I_m (A)	X_m (H)
0.00	0.715
0.08	0.721
0.12	0.719
0.18	0.715
0.26	0.705
0.35	0.693
0.52	0.667
0.71	0.627
0.85	0.595
0.97	0.570
1.05	0.547
1.23	0.509
1.31	0.497
1.60	0.440
1.96	0.387
2.30	0.346
3.00	0.286
4.00	0.230
6.00	0.153

The nonlinear relationship between the magnetizing current I_m and the magnetizing inductance X_m is expressed in the form of a lookup table and is stored in the file *sat60 Hz.tbl*. The generator torque is calculated in the block *Torque*. The lookup table content has the format and data shown in Table 12.6.

Column 1 is the magnetizing current I_m, in A, and column 2 is the magnetizing inductance L_m, in mH. The PSIM model applies the residual flux voltage V_{rf} in the same way as the MATLAB/C simulation. That is, an initial charge with amplitude V_{rf} is applied to the self-exciting capacitor in phase "a" to simulate the residual magnetism of the machine core. Figure 12.31a and b present the main circuit and its subroutine used to simulate the self-excitation process and the sequential primary and secondary load responses of a stand-alone SEIG as presented in Figure 12.31c whose parameters are shown in Table 12.6.

The primary load is switched on at $t = 0.75$ s after the generator has completely excited. Computation step is 0.00001 s, with the secondary load switched at 1.00 s for a total simulation time of 1.25 s (Table 12.7).

PSIM is faster when compared to a program in MATLAB but usually slower than compiled C. The transient in PSIM has the same overall trend as what is expected in reality, and such model can be very useful in studying and designing alternative energy systems powered by induction generators. The steady-state values in PSIM and in MATLAB (discussed next) are very close.

TABLE 12.7

Parameters Per-Phase Used for the SEIG Simulated in PSIM

SEIG Parameters		Load Parameters	
R_s	4.2 Ω	R_{a1}, R_{b1}, R_{c1}	283 Ω
$L_{\ell s}$	22 mH	R_{a2}, R_{b2}, R_{c2}	150 Ω
R_r	4.32 Ω	L_{L2}	6.5 mH
$L_{\ell r}$	22 mH	Primary load switched at	$t_1 = 0.75$ s
p	4 poles	Secondary load switched at	$t_2 = 1.00$ s
f_{rated}	60 Hz	R	20 Ω
C_{exc}	19.5 μF	L	20 mH
V_{rf}	10 V	Shaft speed	1800 rpm
Calculation step Δt	0.00001 s	Simulation time	1.25 s

12.9 SIMULATION OF A SELF-EXCITED INDUCTION GENERATOR IN MATLAB

This is a brief introduction to the MATLAB program simulating the stand-alone operation of a self-excited induction generator (SEIG). The flowchart in Figure 12.32 describes the systematic simulation procedure. The inputs to the program are the machine parameters, magnetization characteristic, residual magnetism, and self-excitation and load parameters as presented in Table 12.8. A step of 1 ms can produce accurate results without sacrificing execution speed.

An impulse function is used as well to represent the remnant magnetic flux in the core. The load parameters are initialized with large values for load resistance and capacitance to simulate an open circuit or no load condition. On the other hand, the induction generator might fail to self-excite if started on a heavy load. A suitable value for self-excitation capacitance should be chosen from the magnetization curves to self-excite the generator at the rated voltage. The load parameters can be modified after the generator has been completely excited. Mutual inductance is computed at every step using the relation obtained from experimental data. To change the magnetization characteristic, the relationship between mutual inductance and magnetization current has to be entered as shown in Figure 12.30.

The program solves the generator differential equations for the d–q axis voltages and currents using the $c2d$ function in MATLAB. The d–q variables are then converted back to a-b-c variables using the d–q transformation. Table 12.9 gives some SEIG parameters for MATLAB simulation results.

The load is switched on at $t = 2$ s after the generator has completely excited. Computation time is 8 s for a simulation time of 3 s. The prime mover model has not been included in this program. The prime mover speed w_{rl} is assumed to be constant in the program, but any function of time can be assigned to the variable based on load variations or prime mover input. MATLAB is relatively slow compared with similar programs in C or Pascal.

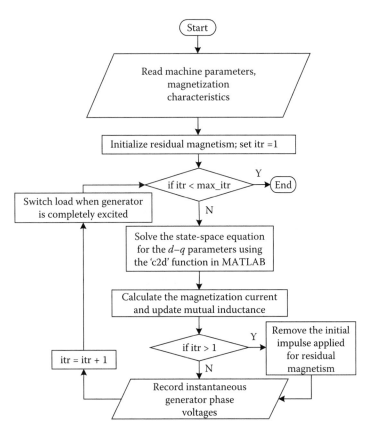

FIGURE 12.32   Flow chart for MATLAB and C simulations.

TABLE 12.8
SEIG Parameters for MATLAB Simulation

Parameters

R_{s1}	Stator resistance, Ohms
X_{s1}	Stator reactance, Ohms
R_{r1}	Rotor resistance, Ohms
X_{r1}	Rotor reactance, Ohms
M_1	Mutual inductance, Ohms
u_1	Residual magnetism vector
C_l	Self-excitation capacitance, F
R	Load resistance, Ohms
L	Load inductance, H
w_{r1}	Prime mover speed, rad/s
Tisim	Simulation time, s
DeltaT	Sampling time for discretization in s

Note: All parameters are referred to the stator side, Figure 12.30.

TABLE 12.9

SEIG Parameters for MATLAB and C Simulations

Simulation Parameters		Open Circuit Load Parameters	
R_{s1}	0.262 Ω	R	1000 Ω
X_{s1}	0.633 Ω	L	100 H
R_{r1}	0.447 Ω	Load switched at	$t = 2$ s
X_{r1}	1.47 Ω	R	20 Ω
M_l	0.25 H	L	0.02 H
u_1	[10 10 0 0] V	w_{r1}	377 rad/s
C_1	180 μF	Tsim3	3 s
$M_1 = 0.423e^{-0.00351 I_m^2} + 0.0236$		DeltaT	0.001 s

12.10 SIMULATION OF A SELF-EXCITED INDUCTION GENERATOR IN C

To better understand these simulations, the student should first read Chapter 4. This is a short introduction to the C program that simulates the stand-alone operation of a self-excited induction generator (SEIG).

C has three modules: the compiler, the assembly, and the debugger. The compiler converts a program statement into object code. When you write a C or C++ program, the filename of your source code usually ends with an extension *.c* or *.cpp*. Examples of that are *counter.c* and *satura.cpp*.

Assembly language is a low-level structured language that allows you to write a computer program as close as possible to the real machine code of the computer. Low level means that its language symbols are very close to the machine's physical connections. Writing assembly language can be a very complicated and laborious task. The advantages of assembly language are speed and direct full control over the hardware circuitry. Assembly language programs can be written separately with an editor and later assembled and linked. The role of a compiler is to convert program code into object code. An object code filename usually ends with an extension *.asm*. Examples of that are *seig.asm* and *charact.asm*.

It is difficult to write a first version of a computer program without making two kind of mistakes: syntax errors and programming errors. Syntax errors occur because of wrong compilation or assembly of the source codes, which can cause abortion of the executable computer program. As a rule, syntax errors are easier to correct than programming errors. Programming errors are mistakes in programming logic. Most of the time, they are not detected in compilation or in code assembly but will show up as unreasonable results. Sometimes these errors are very difficult to detect. The debugger program exists to help the user detect errors in general by giving logical hints or general rules to be followed to shortcut the search for mistakes.

Program simulations in C are developed similarly to MATLAB simulations. The flowchart in Figure 12.32 describes the systematic simulation procedure. Language C statements for simulation of a general SEIG are listed in Appendix B. The inputs to the program are the machine parameters, magnetization characteristic, residual magnetism, self-excitation, and load parameters listed in Table 12.8. Notice that a discretization sampling time of 0.1 ms can produce very accurate results without sacrificing execution speed.

The main difference between the C solution and the MATLAB solution is that in C the generator differential equations for the d–q axis voltages and currents have to be solved using a numerical method like, for example, the Runge–Kutta fourth-order procedure. The d–q variables are then converted back to a–b–c variables using the d–q transformation. Parameters for this simulation are also given in Table 12.9. The following is a discussion of some simulation results.

The load was switched on at $t = 2$ s after the generator had completely excited. Computation time was 2 s for a simulation time of 3 s, as displayed in Figure 12.33. The prime mover was not modeled in this program. The prime mover speed ω_{rl} was assumed constant in the program, but any function of time can be assigned to the variable based on load variations or prime mover input. The C program calculations are also very fast compared to a similar program in MATLAB.

12.11 SOLVED PROBLEM

1. Give good technical and economic reasons for recommending a computer program from among the ones you know to estimate the torque and speed of an induction generator relating it only to its electrical parameters. What would be the best program to recommend as an equivalent circuit for these calculations?

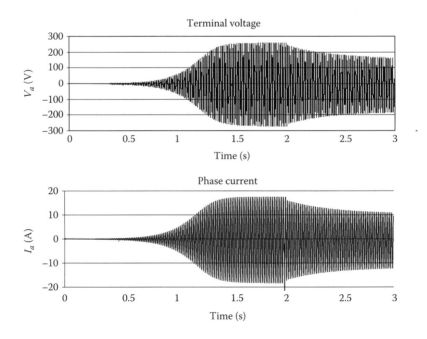

FIGURE 12.33 Self-excitation and load response in C language of a stand-alone SEIG.

Solution

As discussed in this chapter, several computer programs can be used to simulate the induction generator. The choice of a particular program is a trade-off between the computation time and the accuracy of the simulated model. By using Excel spreadsheet, it is very easy to compute torque and speed as shown in the given example. But the results may not be very accurate depending on the simplifications of the used equations. Typically, steady state results can be calculated very easily using a spreadsheet.

If using Simulink, or a similar block-diagram simulator (PSIM has a block-diagram interface, PLECS or SIMPLORER are also other possibilities), both the induction generator and the associated controls can be easily simulated. The simulation speed might be slower than other approach, depending on the step-size and simulation complexity, in addition of the requirement to learn all the building blocks for the system. If the choice is to program in C or Pascal or any other compiled language, the simulation is faster, but if transient analysis is required, some ordinary differential equation function solver must be used. The advantage of a compiled language is to allow computer- or microprocessor-based implementation, customized for their specific operating system.

Therefore, if one wants to observe and study only torque and speed responses of the induction generator without any control, one can use spreadsheet, C, or Pascal program. On the other hand, if the generator and control are to be simulated, the choice could be Simulink for block-diagram implementation or PSIM for electrical circuit implementation.

12.12   SUGGESTED PROBLEMS

12.1 Collect data for a wind power site being considered for the installation of a small power plant. Develop a computer program to estimate the average speed, the root mean cubic speed, the cubic power density, and the cubic energy density. Arrange the data to give the mechanical input torque for an induction generator to be sized for energy conversion of this wind system.

12.2 Do the same as in Problem 2 for a small hydro power plant.

12.3 Develop a program to estimate induction generator performance as an efficient and qualified source of energy for an industrial application.

12.4 Use the MATLAB/Simulink block diagram shown in Figure 12.1 to simulate a vector control for a small hydropower plant of 100 kW so that voltage and frequency control does not exceed a 5% tolerance range.

12.5 Do the same as 12.4, but use PSIM.

12.6 Do the same as 12.4, but use PLECS (another circuit/system simulator).

12.7 Do the same as 12.4, but use SIMPLORER (another circuit/system simulator).

REFERENCES

1. Vosburgh, P.N., *Commercial Applications of Wind Power*, Van Nostrand Reinhold Co., New York, 1983.
2. Patel, M.R., *Wind and Solar Power Systems*, CRC Press, Boca Raton, FL, 1999.

3. Szczesny, R. and Ronkowski, M., A new equivalent circuit approach to simulation of converter induction machine associations, *Proceedings of the European Conference on Power Electronics and Applications (EPE'91),* Firenze, Italy, 1991, pp. 4/356–4/361.

4. Marra, E.G. and Pomilio, J.A., Spice-assisted simulation of the *abg* model of cage rotor induction machines including saturation, *Proceedings of the IEEE-Industry Application Conference, Induscon 2004,* Joinville-SC-Brazil, 2004.

5. Kusko, A., *Emergency/Standby Power Systems,* McGraw-Hill Book Company, New York, 1989.

6. Lawrence, R.R., *Principles of Alternating Current Machinery (Máquinas de Corriente Alterna H.A.S.A.),* Editora Hispano Americana S.A., Buenos Aires, Argentina, 1953.

7. Chapman, S.J., *Electric Machinery Fundamentals,* 3rd edn., McGraw-Hill International Edition, New York, 1999.

8. Krause, C., *Analysis of Electric Machinery,* McGraw-Hill, New York, 1986.

9. Kovács, P.K., *Transient Phenomena in Electrical Machines,* Elsevier, London, U.K., 1984.

13 Applications of Induction Generators in Alternative Sources of Energy

13.1 SCOPE OF THIS CHAPTER

This chapter deals with the most ordinary applications of induction generators. Such applications are usually based on the variable speed features of the induction genera-tor, which enables the control of a wide set of variables such as frequency, speed, output power, slip factor, voltage tolerance, reactive power, and others. In its variable speed operation, an asynchronous link (dc link) is usually interposed between the generator and the load or network to which the power is being delivered. The sim-plest application of the induction generator is its direct connection to a grid, which is sufficient to guarantee the electrical rotation above the synchronous to control a scheduled power flow. Stand-alone applications are also common in cases of low levels of power generation. In all these cases, a well-designed speed control or speed limiter is mandatory for good results.

The primary energy, turbine type, and number of poles in the generator deter-mine the speed, commonly stated in rotations per minute (rpm). Most electrical loads demand that generators must be driven at a speed that will generate a steady power flow at a frequency of 50/60 Hz. The number of poles will define the neces-sary shaft speed of the turbine. For US 60 Hz frequency applications, a two-pole generator demands speeds as high as 3600 rpm, making it too high for practical use with small wind power or hydropower plants. For hydropower systems, the 1800 rpm four-pole generator is the most commonly used. For small wind power systems, the 900 rpm eight-pole is often encountered in field applications. In many cases, it is necessary to use speed multipliers, making the whole system heavier, more main-tenance demanding, and relatively less efficient, so the cost of the generation unit is more or less inversely proportional to the turbine speed and the type of primary energy. The lower the speed, the larger the frame size needs to be for equivalent output power.

Induction generators are generally appropriate for smaller systems. They have the advantage of supplying an almost constant output frequency at variable speeds. The power flow in this case will depend on the slip factor. In addition, they are lighter, rugged, and cheaper than synchronous generators.

As discussed in Chapter 6, the induction generator is a standard three-phase induction motor, wired to operate as a generator. Self-excitation capacitors are used for the voltage building-up process and are popular for smaller systems that generate less than 15 kW. Requirements on constant frequency and voltage are not so demanding. Efficiency of induction generators depends on their size but can be considered approximately 75% at full load and decreases to as low as 60% or even much less at light loads.

There are other factors to consider when selecting a generator for an ac system, such as capacity of the system, types of loads, availability of spare parts, voltage regulation, and cost. Selecting and sizing the generator is very technical, and an energy expert should make the selection during the viability studies. As explained in Chapter 3, if high portions of the loads are likely to be of the inductive type, such as phase-controlled converters, motors, and fluorescent lights, a synchronous generator will be a better choice than an induction generator. Induction generators in stand-alone application mode cannot supply on their own the high starting-up surge power required to achieve the inertial state of motor loads without some special starting procedure.

13.2 VOLTAGE AND FREQUENCY CONTROL OF INDUCTION GENERATORS

From the point of view of the consumer, the voltage and frequency controls of induction generators are fundamental issues since out-of-range settings can damage any appliance connected to the grid. From the electricity company standpoint, accurate controls of voltage and frequency can limit the electrical and mechanical stresses in the power system and deliver good-quality energy. In any case, the drive system needs to transmit power from the turbine to the generator shaft in the required direction, amount, and speed. Controlled either by the source or by the load, an electronic load controller (ELC) can be used to generate electrical power at stable voltage and frequency. Typical drive systems in wind power and microhydropower systems are as follows:

- *Direct coupling*: A direct drive system is one in which the turbine shaft is connected directly to the generator shaft. In many cases, they are a common shaft structure. Direct coupling is obviously efficient, simple, cheap, maintenance free, and light. These drivers are used only for cases where the speed of the generator shaft and the speed of the turbine are compatible.
- *Gearbox coupling or speed multiplier*: Gearboxes are suitable for use with larger machines when belt drives would be a bit unreliable and inefficient. Gearboxes have problems regarding specification, size, weight, alignment, maintenance, efficiency, and cost. These aspects rule them out for small units of wind power or hydropower plants.
- *Coupling with wedge belts and pulleys*: This is the most common choice for microhydropower systems. Belts for this type of system are widely available because they are extensively used in all kinds of small industrial machinery.
- *Coupling with timing belt and sprocket pulley*: These drives are similar to those used in vehicle camshaft drives and use toothed belts and pulleys.

They are efficient and clean-running and are especially worth considering for use in very small system drives (less than 3 kW) where efficiency is critical and aspects of synchronism and weight are taken into account.

* *Clutch coupling*: These are used to provide gradual and smooth coupling for speed-sensitive loads or network connection as in some cases of diesel units to avoid the predictable and pronounced voltage dips, sags, or droops in voltage and frequency.

13.3   APPLICATION OF ELECTRONIC LOAD CONTROLLERS

An ELC is a solid-state electronic device designed to regulate output power of a wind power plant or hydropower system. Maintaining a near-constant load on the turbine generates stable voltage and frequency. The controller compensates for variation in the main load by automatically varying the amount of active power dissipated in a resistive load, generally known as the secondary, ballast, or dump load, in order to keep constant the total load on the generator and turbine.

Wind or water turbines, like petrol or diesel engines, will vary in speed as load is varied, applied, or disconnected. Although not much machinery uses direct shaft power, speed variation will seriously affect the frequency and voltage output from a generator. It could damage the generator by mechanical or electrical stress or/and overloading because of high power demand or over-speeding under light or no-load conditions. Traditionally, complex and costly hydraulic or mechanical speed governors similar to larger hydro systems have been used to regulate the water flow into the turbine or the wind vane direction as the load demand varied.

ELCs have increased flexibility, simplicity, and reliability have been developed for the modern small hydropower–wind power plants. They are usually used with hill climbing control (HCC) in order to get the maximum power from the primary source, which is better than achieving maximum efficiency. A dc–dc converter can work as an ELC and advantageously replace the conventional mechanical controllers to simultaneously regulate the speed and the electrical variables (voltage, either current, or power) of induction generators. Rotation of the turbine generator set is kept almost constant within certain limits through modulation of the active power and using the inherent voltage characteristics of the induction generator.[1–4] The control is based on a current diversion from the main (primary) load to a secondary load (ballast) using a very conventional dc–dc converter connected in parallel to the main load. Changes in the main load current become approximately independent of the generator current changes and vice versa. The excess current may be recovered in a secondary plant with a large time constant such as a heating, cooling, water irrigation, battery charging, electrolysis, or pump-storing scheme. For smoother control of the output voltage level, pump store schemes can be modulated in 1, 2, 4, 8, and so on units of power.[3,4]

The idea behind electronic control by the load is to apply more or less torque on the turbine generator shaft to compensate for the primary energy variations in such a way that speed (or electrical frequency) can be regulated within a narrow range. This shaft loading is electronically controlled by a variable-switched resistor to smoothly

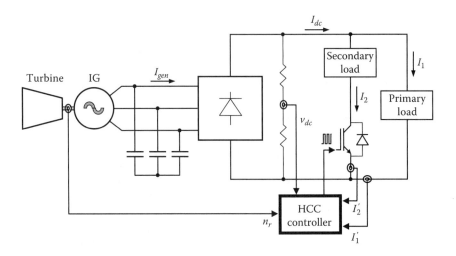

FIGURE 13.1   Electronic control by the load.

adjust the effective resistance across the generator terminals according to the balance between primary energy availability and load demands.[1-5]

A simple and immediate scheme is depicted in Figure 13.1 from which it is possible to understand the switching principle using a dc–dc converter with pulse width modulation (PWM) modulation. Load and primary energy (water, wind) vary randomly so they produce a variable dc current I_{dc}. To cope with this, the switched modulation approach has to distribute the total generated current, I_{dc}, between the primary current I_1 and the secondary current I_2. Current I_2 is controlled by the effective value of the secondary load resistor, which is the current portion diverted from the load current, to control I_{dc} and/or the primary load current itself according to the power demand, as explained in Sections 10.4 through 10.6. All the same, to keep the main load current, I_1, constant during any primary source energy variation, I_2 should be compensated with identical variations but of opposite sign. If both, I_{dc} and I_1, should vary, the dc–dc converter must be able to compensate for them according to its needs:

$$I_{dc} = I_1 + I_2 \tag{13.1}$$

where
  I_{dc} is the average dc link current
  I_1 is the average primary load current
  I_2 represents the average current through the secondary load

The balance between load and source currents in ELCs can be seen as if there were an apparent independence between variations of the average dc link current and the primary load current whose difference has to be drained out by I_2. Therefore, the load current variation is compensated by the effective variation of the current through the variable resistor. In the PWM modulation usually used for these purposes, the duty cycle is defined as

$$D = \frac{W}{T_{sw}}$$ (13.2)

where
  W is the pulse width of the current through the dc–dc converter
  T_{sw} is the switching period.

Converter secondary current is expressed as

$$I_2 = \frac{D \cdot V_{dc}}{R_2}$$ (13.3)

where
  R_2 is the actual value of the secondary load
  V_{dc} is the dc–dc converter output

The resistive effective value may also be expressed as

$$R_{eff} = \frac{R_2}{D}$$ (13.4)

Combining Equations 13.2 through 13.4, the control pulse-width variable W is established to modulate the effective secondary resistor:

$$W = \frac{R_2}{f_{sw}V_{dc}}\left(I_{dc} - I_1\right)$$ (13.5)

From Equations 13.1 through 13.5, the relationship between I_{dc} and $f_{sw} = 1/T$ is given then as

$$I_{dc} = \frac{V_{do}\left(R_2 + R_1 W f_{sw}\right)}{R_1 R_2}$$ (13.6)

In many simplified control algorithms, control of the generator speed may be sufficient to guarantee reasonable control of the generated voltage, rotation, or frequency. In some cases, it is sufficient to keep the load voltage confined within certain limits to guarantee that the generator's speed and power will be reasonably within the operating range. In the example presented in Figure 13.2 for a 7.5 kW induction generator, it is observed that the frequency as much as the terminal voltage and

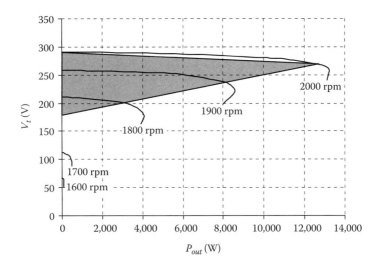

FIGURE 13.2 Effect of the output voltage control on rotor speed and load power. (From Bonert, R. and Hoops, G., *IEEE Proc. B.*, 5, 28–31, 1990.)

output power is directly proportional to the speed of the turbine. The dark area in this figure represents the safe operating area where voltage is limited between 120 and 290 V. However, a good algorithm for the speed controller becomes an essential issue in larger schemes where bulkier inertial rotating masses are involved.

ELCs are suitable for either induction or synchronous generators. In induction generators, the excess generated current is diverted to a ballast load. Voltage control is not required for synchronous generators because they have a built-in automatic voltage regulator. However, the frequency of an induction generator without an ELC would vary as the load and primary source change and, under no-load conditions, could be much higher than the rated frequency. In compensation, ELCs react so fast to load changes that speed changes are not even noticeable unless a very large load is applied.

The major benefit of ELCs is that they replace all mechanical parts of the traditional controls and are reliable, compact, set in modules, and virtually maintenance free. The advent of ELCs has allowed the introduction of simple and efficient multijet turbines for microhydropower and conventional fixed blades for wind power systems that are no longer burdened by the expensive hydraulic governors.

In many cases, ELCs can also be used as a load management system by assigning predetermined prioritized or primary loads (direct consumption) and secondary loads (indirect loads), such as electrolysis, water heating, space heating, battery charging, or other loads. In this way, it can use the available power rather than dumping it into the ballast load or secondary load. It can be used also to connect loads by priority sequence and can thus control loads that total four to five times the actual output of the micropower system.

Depending on the variable being controlled (voltage, power, frequency, rotation, water flux, power coefficient, and efficiency), there exist various specific types of

ELCs on the market that can regulate systems from as small as 500 W to 150 kW. The choice of the controller depends on the type of generator (single-phase or three-phase, dc or ac output, and so on).

Other types of controllers are being introduced to the market based on similar principles that may suit a particular application as the ones based on fuzzy logic approaches. They may control the distribution of electric power to various loads such as lighting, refrigeration, space heating, or water heating in prioritized sequence.

13.4   WIND POWER WITH VARIABLE SPEED

Induction generators and wind energy have a close relationship based on the nature that the wind power varies with the cube of the wind speed. The wind generator efficiency is affected directly by the wind speed, whose rated value in turn depends on rotor design and rotation. The rotation is expressed as a function of the blade tip speed for a given wind speed. Therefore, to maintain a fixed generator speed on a high efficiency level is virtually impossible.[4–7]

The simplest way of driving a wind power turbine is, of course, by not using any special control form. In this case, the speed of the turbine is constant (as it is connected directly to the public grid), which operates at a fixed frequency or under a constant load and primary power. Power generated this way cannot be controlled, and in the case of wind turbines, the wind speed is altered by wind thrusts and in the case of hydro turbines, by the water flow. There are benefits to incorporating some control in a wind power turbine, such as an increase in the energy captured from nature and a reduction in dynamic loads. Furthermore, speed actuators allow flexible adjustment of the operating point, also known as the German concept, and the project's safety margins can be decreased.

Large-scale integration of wind farms involves requirements that are more stringent because the introduction of large amounts of energy in intermittent generation will significantly affect network management mechanisms and have an impact on network stability and security. Therefore, a stable power control is a vital issue for the operation of wind power turbines connected to the grid. There are three different methods of safe power control in wind turbines (WTs): stall regulation, pitch control, and active stall control.[3,7,8]

Most WTs are passive-stall regulated, with the machine rotor blades bolted onto hubs at a fixed attack angle. In these cases, the rotor blade is twisted when moving along its longitudinal axis to ensure that turbulence on the side of the rotor blade not facing the wind is created gradually rather than abruptly when the wind speed reaches its critical value. If stall control is applied, there are no moving parts in the rotor and a complex control system is needed. However, stall control represents a very complex aerodynamic design problem and related design challenges in the structural dynamics of the entire WT to avoid stall-induced vibrations caused by the wind turbulence along the blades. Electronic control by the load and mechanical limiters is often recommended to avoid these critical operating states.[4,5]

Pitch control uses an electronic controller to check the output power several times per second. When the output power becomes too high, it sends an order to the blade pitch mechanism, which immediately turns the hydraulic drive of the rotor blades slightly out of the wind direction, and vice versa when the wind speed drops again. The rotor blades are kept at an optimum angle by the pitch controller. In active pitch control, the blade pitch angle is adjusted continuously based on the measured parameters to generate the desired output power up to the maximum limit, as in conventional regulated pitch control. Electric stepper motors are frequently used for this purpose.

The induction generator is well suited for operation in the four-speed bands to be considered in the operation of WTs (see Figure 13.3). In the first band, which goes from zero to the minimum speed of generation (cut-in), wind speed is usually below 3 m/s. Up to this limit, the power generated just supplies the friction losses. In the second band (optimized constant C_p), the turbine-rated speed is maintained by a system of blade position control with respect to the direction of wind attack (pitch control) or a modulated load (ELC). In the third band, that is, for high-speed winds, the generator capacity limits the speed to a maximum constant output power (constant power). Above this band (at wind speeds around 25 m/s), either the rotor blades are aligned in the direction of the wind or the rotor head is designed to tilt back (furl) in small units, to avoid mechanical damage to the electrical generator (speed limit).

The more commonly used form of speed and power control for wind generators is to control the attack angle of the turbine blades, where the generator is connected directly to the public grid. Another simpler form of control is the use of generators driven by turbines at fixed attack angles of the blades, connected to the public network through a controlled dc link with HCC controllers. Control occurs on the load flow, which in turn acts on the turbine rotation. As the rotor speed must change according to the wind intensity, the speed control of the turbine has to command low speed at low winds and high speed at high winds.

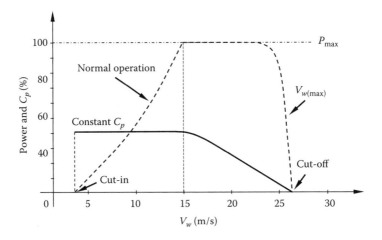

FIGURE 13.3 Speed control range for WTs.

13.5   RUN-OF-RIVER HYDRO GENERATION

Growing interest in water management and sustainable environment toward a sustainable world has awoken new sources of hydro energy.[9–14] Among these are the run-of-river plants to produce electricity using induction generators. Run-of-river hydro plants do not require dams; that is, they are mainly run-of-river with little or no reservoir impoundment. They rely on the natural downward flow of the stream to guide water through pipes to a generating station. The force of the water spins a turbine, which drives an electric generator that causes electromotive forces and thus useful electrical power.

Of the two major types of hydro projects, the environmental impact of run-of-river facilities is considered low compared with facilities that have large storage reservoirs.[9–14] Many Internet sites deal with such small systems.[15–19] A 25 MW project on this basis is located upstream of a 15 m waterfall on the Mamquam River, Canada. Another example is the 11 MW run-of-river hydroelectric facility Furry Creek commissioned in British Columbia, which uses the tremendous force of Furry Creek to generate enough electricity to power 7000 homes annually.

For run-of-river hydro projects, a portion of a river's water is diverted to a channel, pipeline, or pressurized pipeline (penstock) that delivers it to a waterwheel or turbine. The moving water rotates the wheel or turbine, which spins a shaft. The motion of the shaft can be used for mechanical processes, such as pumping water, or it can be used to power an alternator or generator to generate electricity. Conditions to consider with any run-of-river hydro scheme include

- Available and desired power, voltage, and frequency
- Accessibility
- Intake and outfall locations
- Water head
- Water flow
- Availability of local materials
- Pipeline diameter
- Pipeline route
- Purpose of the generation
- Load evolution and prediction
- Powerhouse location
- Possibilities of grid connection
- Ecological issues
- Landscape issues
- Noise containment
- Floating materials

13.6   WAVE AND TIDAL POWERS

Demands for renewable energy power plants have increased during the last few years because of limits including international instability, pollution, reduction of fossil fuels, and the outstanding negative environmental impacts of burning fossil fuels.

Since oceans cover over 70% of the earth's surface and the energy contained in waves and tidal movements is enormous, wave energy devices have lately attracted as much attention as wind energy devices. In many places in the world, the wave and tidal power have been seriously considered in association with the induction generator.[20–25]

Korea Ocean Research and Development Institute (KORDI) has developed a 60 kW prototype of a wave energy device named CHUJEON A (see Figure 13.4).[21] The plant has 18.6 m overall height, 13 m outer diameter, and 10 m inner diameter. The CHUJEON-A has an air chamber inside and is a floating type of wave energy device moored by four-point compliant buoy anchoring on the bottom of the shallow sea.

The main purpose of the CHUJEON-A project is to examine the possibilities for the practical use of wave power plants on the eastern seashore of Korea by gathering and analyzing data related to energy conversion such as wave height, wind speed and direction, rotating speed of the turbine, output power, and so on.

The energy conversion mechanism of CHUJEON-A consists of an air chamber, Wells turbines, and an induction generator (stand-alone). The Wells turbines obtain single-directional torque from the bidirectional flow of air in a narrow duct. This flow

FIGURE 13.4 **(See color insert.)** CHUJEON-A, a prototype of wave power. (From Jeon, B.-H. et al., Autonomous control system of an induction generator for wave energy device, Ocean Development System Lab., Korea Research Institute of Ships and Ocean Engineering, KORDI, Daejeon, Korea, *Proceedings of The Fifth ISOPE Pacific/Asia Offshore Mechanics Symposium*, Daejeon, Korea, November 2002, pp. 17–20.)

of air is caused by pressure variation in the air chamber resulting from phase difference between wave and motion of the floating air chamber. By its variable speed characteristics, the induction generator is very appropriate to convert rotational energy from the naturally irregular and distinct levels of compression or depression of air in the chamber into electrical energy. Therefore, one of the main requirements for CHUJEON-A is to supply a constant voltage at constant frequency to local isolated loads in a wide range of rotor speeds with high efficiency.

Several other prototypes of devices for wave energy harvesting, such as the Wave Swing, are listed in the references.[22–25] Wave Swing is a bottom-standing, completely submersed point absorber, with a linear-direct induction generator to convert the oscillatory motion into electricity.

Another example of wave energy harvest is the AquaBuOY. It is a freely floating heaving-point absorber, reacting against a submersed reaction tube (mass of water). The reaction mass is moving a piston assembly that drives a steel-reinforced elastomeric water pump (hose pump). The hose pump is pumping water on a higher-pressure level. An accumulator is used to smooth the power output, and the pressure head is then discharged onto an impulse turbine to generate electricity. Grid synchronization is achieved using a variable speed drive and step-up transformer to a suitable voltage level.

Another application of the induction generator in wave power is in the Lake William Hovell Hydro Power Project located on the King River, about 25 km from the town of Whitfield in northeastern Victoria, Australia.[14] Construction was completed in 1994, and the station is connected to the outlet works of a 35 m high earth and rockfill dam, which was built in 1971. The reservoir created by the dam has a capacity of 13.5 GL, which is only about 6% of the mean annual runoff at the site. This is insufficient to regulate annual flows to a significant degree. About 80% of the energy is generated from flows that would otherwise pass over the spillway during winter and spring, and the remainder from irrigation flows over the summer and the autumn. The net head is 29 m under maximum flow conditions, and the rated generating capacity is 1500 kW.

Power in the Lake William Hovell Hydro Power Project is generated by a horizontal-shaft Francis turbine directly coupled to an induction generator rotating at 500 rpm. Civil works for the power station comprise a concrete substructure and a steel-clad superstructure. Outgoing power is cabled underground about 50 m to a 22 kV switchyard. Generation is at 6.6 kV, stepped up to 22 kV for mains transmission. During construction, 21 km of line was restrung and remote telemetry and tripping features installed as part of grid interconnection and transmission.[24–25]

Tides are generated by the rotation of the earth within the gravitational fields of the moon and the sun. The relative motions of these bodies that cause the surface of the oceans to be periodically raised and lowered generate tidal power. Therefore, tidal energy is predictable in both its timing and magnitude. It has been estimated that if less than 0.1% of the renewable energy available within the oceans could be converted into electricity, it would satisfy the present world demand for energy more than five times over.

One positive characteristic of tidal energy is found in the high-energy density of ocean currents. Sea water is 832 times as dense as air, providing an 8-knot ocean current with the equivalent kinetic energy of a 390 km/h wind. A second characteristic of the turbine for tidal energy design is that it can optimally capture the kinetic energy of the flowing water. Large units (> 1 MW) can tap this power to satisfy electricity demands in the multiple-gigawatt range. Smaller energy loads can be met by deploying mid-range units (> 100 kW) in off-grid communities, remote industrial sites, and regions with established net metering policies. A third good characteristic of tidal energy is that the power system necessary for its connection to a grid may be common to other ocean energy extraction technologies, PV systems, and wind power generation. These types of energy are very plentiful along all the continental coasts in the world.[26-30]

Tidal power is not a new concept and has been used since at least the eleventh century in Britain and France for milling grain. Tidal energy technology can be considered largely mature with a 240 MW commercial unit operating successfully since 1966 in France at La Rance. Smaller units have been constructed in other countries, namely, Canada (Bay of Fundy, 17.8 MW, 30 GWh/year, 1984), USSR (Kislogubskaya near Murmansk, 0.4 MW, 1968), China (Jiangxia, 3.2 MW, 11 GWh/year, 1980), and others. However, it seems that the energy crisis is not yet having sufficient impact on nature and the environment to justify the already significant development of tidal energy technology that has occurred over the last 40 years, particularly in France, Canada, the former USSR, and China. Presently, only a small number of commercial power plants are in operation.[28-34]

Locations where tidal power could be viably developed are relatively few because a mean tidal range of 5 m or more is needed for the electricity cost to be competitive with conventional power plants. However, based on coastal conditions such as the shape of the ocean floor, water depth, and the shape of bays, in some places, tidal ranges can be as large as 15 m. Additional requirements are a large reservoir and a short and shallow dam closure. A barrage constructed across an estuary is equipped with a series of sluice gates to concentrate on a bank of low-head axial turbines. The most common construction method involves the use of caissons.

The usual technique such as *barrage* technology is to dam a tidally affected estuary or inlet, allowing the incoming tide to enter the inlet unimpeded, then using the impounded water to generate power. The main barriers to the uptake of the technology are environmental concerns and high capital costs. In recent years, these problems have been mitigated considerably by design, involvement of experts and local communities in the identification and installation of new plants, and growing understanding of how to achieve more sustainable energy development. However, there have been very few studies in the academic literature that analyze the process of how a new form of sustainable energy has been changing to become more acceptable to researchers.[34-37]

With respect to the generation of electricity from tides, it is very similar to hydroelectric generation, except that water is able to flow in both directions. This fact must be taken into account in development projects using induction generators. Control and hydropower management may become a very complex matter. Like other hydro schemes, tidal power also has high capital costs due to the large-scale engineering involved, although the operating costs are low. Tide has the extra problem of being in a coastal environment where engineering is likely to be even more costly due to the changeability of the coastal and environmental effects. Several reasons point to system design for tidal device power plants, performance estimates, and economic assessments in many parts of the world.[34–37]

The main advantages of tidal power are that it is a renewable energy and needs no fuel or oil, it is easy to operate and maintain, it is suitable for use in isolated locations, it is nonpolluting and almost silent when running, it can move when site conditions change, it can be designed for local manufacture and maintenance, and it can operate 24 h/day without a full-time attendant. Disadvantages that must be pointed out are its relatively high capital cost, the negative impact on the migratory fish of tidal power generation (which is not difficult to mitigate), and the construction time, which can be several years for larger projects.[25–29] Tidal power plant operation is intermittent with a load factor of 22%–35%. Plant lifetime can be very long (120 years for the barrage structure and 40 years for the equipment). The high capital costs and long construction time have deterred the construction of large tidal schemes, as have perceived impacts from the coastal engineering. Tidal barrages can cause significant modifications to the basin ecosystem. However, they can also bring benefits like flood protection, road crossing, marinas, jobs, and increased tourism.

Apparently, tidal power is much more attractive than wind power since much more power can be generated from the same size of machine. However, tidal power will become commercially viable only after large-scale designs have been proved. Tidal energy is the most predictable of the marine renewable resources, and its variation over the tidal cycles can be predicted with considerable accuracy well into the future.[30] Fluctuations do occur over, time but they can be predicted hours, days, or even years in advance if the stream has been properly studied. Certainly, it is a good opportunity for the use of induction generators.

13.7 STIRLING ENGINE POWER AND COGENERATION

The Stirling engine is an external combustion engine that converts heat from a variety of sources into mechanical energy that can be used to generate electricity. It is powered by the expansion of a gas when heated followed by the compression of the gas when cooled. It contains a fixed amount of gas, which is transferred back and forth between a *cold* and a *hot* end. There are two pistons: the displacer piston and the power piston. The displacer piston moves the gas between the two ends, and the power piston changes the internal volume as the gas expands and contracts.[38,39]

The gases used inside a Stirling engine never leave the engine. There are no exhaust valves that vent high-pressure gases, as in a gasoline or diesel engine, and there are no explosions taking place. Because of this, Stirling engines are very quiet. The Stirling cycle uses an external heat source, which could be anything from gasoline to solar energy to the heat produced by moldering plants. Advantages of Stirling engines are ease of use with solar energy or any other natural heating source, relatively high efficiency, good operation possibilities with partial load, good modularity, and quiet operation. The best working gas in a Stirling engine is hydrogen. Helium works nearly as well as hydrogen, but it is much more expensive. The cheapest alternative is air, but it has properties inferior to the other two gases.

The heating and cooling sources for the Stirling engine are situated outside the engine. The heating source can be fuel burned in a combustion chamber in the presence of air, solar energy with Fresnel heat concentrators, or natural gas. The heating source warms up the Stirling engine and increases pressure on the gas inside the hot chamber. The cold source can be water or the surrounding air to cool it down. Evidently, water is preferable because in this case the engine can also be used for heating purposes. In cases where combustion chambers are used, the exhaust gases on their way out may first warm up the air coming into the combustion chamber and then the water for the heating system.[38,39]

Home power station products already exist all over the world. In the United States, according to the approval requested by the Southern California Edison Company to the Public Utilities Commission of the State of California, the Stirling engine became a new form of solar energy renewable resource.[38] Another push in cogeneration systems comes from a commercial Stirling unit produced by WhisperGen Limited in Christchurch, New Zealand (see Figure 13.5).[39] Two WhisperGen models exist. One delivers ac power and is intended for grid-connected applications. The other delivers dc power and runs on diesel or kerosene and is intended for small isolated systems like boats. The ac gas-fired unit is a four-cylinder double-acting Stirling cycle: the electrical side is modulated up to 1200 W ac at 220–240 V and the thermal side is modulated up to 8 kW via sealed liquid circulation, sound level 63 dBA (LWA), natural gas, and a grid-connected four-pole induction generator. The induction generator is inside the housing (also called rotary or sealed alternator). The alternator/generator description of course depends on whether it is ac or dc related. The whole module has an International Electrotechnical Commission standardized (IEC) plug and socket connections with reduced dimensions ($500 \times 600 \times 850 - w \cdot d \cdot h$).

Another source of heat could well be the sun, as suggested by the multiple solar parabolic set shown in Figure 13.6.[40–44]

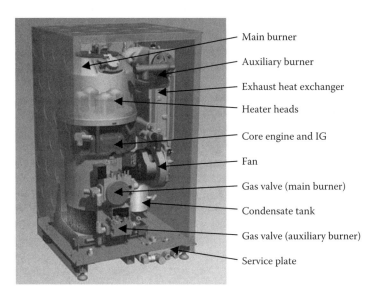

Main burner

Auxiliary burner

Exhaust heat exchanger

Heater heads

Core engine and IG

Fan

Gas valve (main burner)

Condensate tank

Gas valve (auxiliary burner)

Service plate

FIGURE 13.5 (**See color insert.**) The AC WhisperGen™ gas-fired model. (Courtesy of WhisperGen Ltd., Christchurch, New Zealand.)

FIGURE 13.6   (See color insert.) Pima-Maricopa Indian Reservation SunDish. (Extracted from Scott, S.J., Holcomb, F.H., and Josefik, N.M., Distributed electrical power generation, Summary of alternative available technologies, ERDC/CERL SR-03-18, US Corps of Engineers, Engineering Research and Development Center, September 2003.)

13.8   DANISH CONCEPT

The Danish concept refers to the directly grid-connected squirrel cage induction generator, with a gearbox between the turbine rotor and the generator. The generator speed is fixed by the grid frequency and by the number of pole pairs of the generator. Stall effect is used to limit the power output at higher wind speeds. Some induction generators use pole-adjustable winding configurations to enable operation at different synchronous speeds. However, at any given operating point, this Danish turbine has to operate at constant speed. This concept is depicted in Figure 13.7.

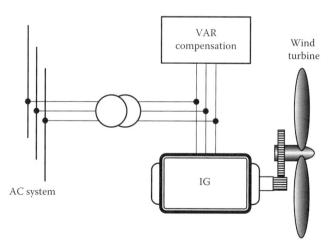

FIGURE 13.7   The fixed speed Danish concept.

Although the Danish concept results in a bulkier configuration, it is advantageous to the relatively low cost of the control mechanism and better suited to the rather pronounced natural variations in wind speed throughout the year. Its disadvantages are the heavy mechanical stress on the gearbox because of sudden interchanges of generators and the need for higher slip factor designs for large generators to allow them to work smoothly. To date, this concept is suitable only for relatively small wind power units. Larger units can use other techniques, such as pole changing. Power electronics may become a very powerful tool to resolve the speed stiffness of the induction generator for existing machines or for new designs, because it allows changes in the rotor resistance through an external control, as suggested in Figure 11.10d, as well as for pole changing through a stator or externally DFIG-driven rotor circuit, as suggested in Figure 11.10.

13.9 DOUBLY FED INDUCTION GENERATOR

It is common sense that an adjustable-speed generator for WTs is necessary when the output power becomes higher than 500 MW. Furthermore, to limit mechanical stresses and power surges in these high-power systems, speed control is necessary. For this purpose, the doubly-fed induction generator (DFIC) system is emerging as having many advantages to reduce cost and has the potential to be built economically at power levels above 1.5 MW for offshore applications and places of highly variable wind speeds.

WTscan operate either at fixed speed or variable speed. For a fixed-speed WT, the generator is directly connected to the electrical grid. For a variable-speed WT, the generator is controlled by power electronic equipment. There are several reasons for using variable-speed operation of WTs, among which are the possibilities to reduce the stresses of the mechanical structure, acoustic noise reduction, and the possibility to control active and reactive power. Encouraged by the facilities presently available in the DFIG as discussed in Chapters 7 and 11, most of the major WT manufactures are developing new larger WTs in the 3–5 MW range. These large WTs are all based on variable-speed operation with pitch control using a direct-driven synchronous generator (without gearbox) or a high-power DFIG. Fixed-speed induction generators with stall control are regarded as unfeasible for these large WTs. Today, DFIGs are commonly used by the WT industry for larger WTs.

According to the WT capability, there is a different level of cost per megawatt installed, efficiency, controllability, and grid performance. WT manufacturers did not consider such impacts of integrating WTs on network power quality levels as a design priority until recently. Since grid requirements are met where possible using additional equipment to compensate for the limited capabilities of WTs, major network augmentations would have made the project financially unattractive in most early cases. These impacts are mostly related to the control of active power control, reactive power, voltage, frequency stability, and fault ride through capability, as displayed in Table 13.1.[3,4]

The capability of an induction generator to have connections made to the rotor (rotating part) as well as to the stator (stationary part) contrasts with the ordinary induction machine where electrical connections are made to the stator only.[33]

TABLE 13.1

Wind Generator Capability Chart

WT Capability	Active Power Control	Reactive Power Control	Voltage Control	Frequency	Fault Ride through Capability	Impact on Power Quality Levels
Fixed-speed machine	No	No	No	No	No	Yes
Two-speed machine	Limited	Limited	No	Limited	No	Yes
Variable slip machine	Limited	Limited	No	Limited	No	Limited
Doubly-fed machine	Yes	Yes	Limited	Yes	Limited	Limited
Synchronous variable-speed machine	Yes	Yes	Yes	Yes	Limited	Limited

Source: Based on Morton, A., Maximizing the penetration of intermittent generation in the SWIS, Econnect Project No. 1465, South West Interconnected System (SWIS), Prepared for Office of Energy, Perth, Western Australia, Australia, July 2005.

Note: "No" means the turbine does not have the capability to do so without adding additional equipment.

As discussed in Chapter 11, the DFIG design provides an attractive trade-off between the cost-competitiveness of a fixed-speed generator and the high performance and controllability of a variable-speed generator and is able to compete aggressively with the control of active power, reactive power, voltage, frequency stability, and fault ride through capability. They are only limited by the size of the power converter connected to the rotor, typically 30% of the nominal rating of the generator.[45]

The strong point of the DFIG is that the line-side converter is able to either absorb or provide real power to the grid, depending on the operating speed of the generator. If the generator is operating below the synchronous speed for the grid frequency and pole number, some amount of real power will flow through the line-side converter to the dc link and then from the machine-side power converter into the rotor circuit. If the turbine is operating above synchronous speed, real power will flow in the opposite direction.[46–49]

To avoid unnecessary grid reinforcements or limitations of wind farms in distribution grids, it is advisable not to use rules of thumb, but rather to carry out load flow analyses, flicker assessment, and so on as outlined in the good practices of power engineering.

The application of dynamic wind farm models as parts of the power-system simulation tools allows detailed studies and development of innovative grid integration techniques. In addition, the importance of accurate modeling may be highlighted by applying alternative simulations of a short circuit event for the illustration of possible outcomes.[50–53]

13.10　PUMP-AS-TURBINE

Perhaps, pump-as-turbine (PAT) is the most spread out possibility for the induction generator since nearly all modern buildings have one or more water pumping systems. For a number of years, there has been wide interest in reverse-engineered conventional pumps that can be used as hydraulic turbines for small amounts of alternative hydroenergy. The action of a centrifugal pump operates like a water turbine when it is run in reverse. Because the pumps are mass-produced in association with induction motors, they are more readily available and less expensive than ordinary turbines. It is estimated that the cost of a PAT is at least 50% less or even lower than that of a comparable turbine. However, for adequate performance, a microhydropower site must have a fairly constant head and flow because PATs have very poor partial-flow efficiency. It is possible to obtain full efficiency from PATs by installing multiple units, with variable sizes or not, where they can be turned on or off depending on the availability of water in the stream. PATs are most efficient in the range of 13–75 m (40–250 ft) of gross head. The higher the head, the less expensive the cost per kilowatt; this is generally the case with all turbines.[53-55]

One of the advantages of using a PAT instead of a conventional turbine is the opportunity to avoid a belt drive. However, in some circumstances, there are advantages of fitting a belt drive to a PAT. The advantages of using a direct drive arrangement are

- Low friction losses and minimized coupling wearing out (saving up to 5% of the output power)
- Less weight and volume
- Lower cost
- Longer bearing life (no lateral forces on bearings)
- Less maintenance services
- Minimized coupling vibrations along generator's life cycle

The use of combined pump motor units is recommended for microhydro schemes that are to be used only for the production of electricity and where the simplest installation possible is required. There are, however, some limitations to using such integral units, as listed in the following. Turbine speed is fixed to the generator speed, thus reducing the range of flow rates when matching the PAT performance to the site conditions:

- Lower efficiency compared to working as a pump for the same water head
- Limited choice of generators available for a particular PAT
- Difficult connection of mechanical loads directly to the PAT

Standard centrifugal pumps are manufactured in a large number of sizes to cover a wide range of heads and flows. Given the right conditions, PATs can be used over the range normally covered by multijet Pelton turbines, cross flow turbines, and small Francis turbines. However, for high-head, low-flow applications, a Pelton turbine is likely to be more efficient than a pump, and no more expensive.

The use of a PAT has greatest advantage in terms of the possibility of reversible use, cost, and simplicity, for sites where the alternative would be either a cross-flow turbine, running at relatively low flow, or a multijet Pelton turbine. For these applications, a cross-flow turbine would normally be very large compared with an equivalent PAT. Very small cross-flow turbines are more expensive to manufacture than larger ones because of the difficulty of fabricating the runner. Therefore, a cross-flow installation would require a large turbine running at slower speed than an equivalent PAT, resulting in the need for a belt drive to power a standard generator. A Pelton turbine for this application would require three or four jets, resulting in a complicated arrangement for the casing and nozzles, although it would be more flexible than a PAT for running with a range of flow rates.

A small Francis turbine could also be used in this range but would be even more expensive than a cross-flow turbine. What dictates the use of a PAT is that it requires a fixed flow rate and is therefore suitable for sites where there is a sufficient supply of water throughout the year. Long-term water storage is not generally an option for a microhydro scheme because of the high cost of constructing a reservoir. Because the selection of a PAT is somewhat difficult, the client should confirm its performance with the designer or pump manufacturer in advance, including the characteristics of the pump and its induction motor because the characteristics of pumps differ according to the manufacturer.[53–57]

13.11 PUMPED-STORAGE PLANTS OR BACK-PUMPING

Pumped storage hydropower comprises an upper storage reservoir and a lower storage reservoir connected by tunnels (Figure 13.8). In general, the greater the difference in elevation between the upper and lower reservoirs, the more economical is the scheme. It is the only proven technology for large-scale energy storage and could be used to store power generated by intermittent generators (such as wind) at times of low demand for use when demand is high.[3]

In the early days, pumped hydroelectric plants used a separate motor and dynamo, mainly because of the low efficiency of dual generators. This increased cost and space for the installations because separate pipes had to be built. A majority of modern plants use an induction generator that can be run backward as an electric motor. The efficiency of the generator has increased, and it is now possible to retrieve more than 80% of the input electrical energy. This leads to significant cost savings. Also, modern generators are of suspended vertical construction. This allows better access to the thrust bearing above the rotor for inspection and repair.[58,59]

An example of this is the 300 MVA doubly-fed cycloconverter-driven induction machine drive for the pump storage station Goldisthal, which was successfully put into operation by the Alston Company (Figure 13.9). The cycloconverter only needs to drive at most one-third of the total power of the machine. The efficiency of the turbine is increased by varying the speed (–10% to 4%). The high-dynamic flux-oriented control (torque step response time <10 ms) can be employed for very fast power control for the stabilization of the grid.[60]

As shown in Figure 13.8, pumped storage projects differ from conventional hydroelectric projects in that they pump water from a lower reservoir to an upper reservoir

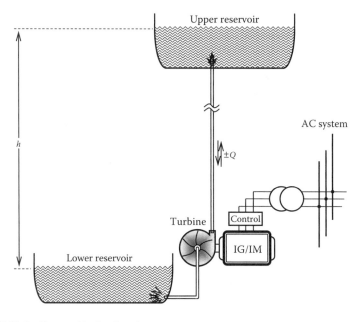

FIGURE 13.8 Pumped hydroelectric storage system.

FIGURE 13.9 **(See color insert.)** Shaft between the induction machine and the turbine.

when demand for electricity is low. The initial potential energy associated with the head is transformed into kinetic energy. One part is the kinetic energy associated with the mass moving with velocity. The other is the pressure part, with the enthalpy given by pressure P over the density of water multiplied by the remaining mass:

$$E_{ini}^{pot} = mgh = \frac{1}{2}m'u^2 + P\Delta V = E^{kin} + \text{enthalpy} \tag{13.7}$$

where ΔV is the volume variation due to the mass difference given by $(m - m')/\rho$.

There are frictional losses, turbulence, and viscous drag, and the turbine has an intrinsic efficiency. For the final conversion of hydropower to electricity, the generator efficiency must also be considered. Therefore, the overall efficiency of pumped hydro systems must consider the ratio of the energy supplied to the consumer and the energy consumed while pumping. The energy used for pumping a volume V of water up to a height h with pumping efficiency η_p is given by Equation 13.8, and the energy supplied to the grid while generating with efficiency η_g is given by Equation 13.9:

$$E_{pump} = \frac{\rho ghV}{\eta_p} \tag{13.8}$$

$$E_{gen} = \rho ghV\eta_g \tag{13.9}$$

During peak demand, water is released from the upper reservoir to the lower reservoir through high-pressure pipelines. Water passes through the reversible turbine and ultimately pools in the lower reservoir. The turbine drives the power induction generator, which creates electricity. Therefore, when releasing energy during peak demand, a pumped hydroelectric storage system works similarly to traditional hydroelectric systems. When production exceeds demand, the control box converts the generator connections into motor connections, allowing the machine to pump and store water back up into the upper reservoir, usually with the early morning surplus.

The amount of electricity that can be stored depends on two factors: net vertical distance through which the water falls, called the effective head h, and the flow rate Q (or the volume of water per second from the reservoir and vice versa). A common false assumption is that head and flow are interchangeable, but in fact simply increasing the volume per second cannot compensate for the lack of head. This is because high-head plants can be quickly adjusted to meet the electrical demand surge. For lower head, the diameter of the pipe would have to be enormous in order to produce the same power except for the run-of-river plants. This would not be economically viable, which is why most plants are of the high-head variety.

The value of hydroelectric storage systems is enhanced by the response speed of the generators. Any of the turbines can be brought to full power in just a few seconds if they are initially spinning in air. Even starting from complete standstill takes only 1 min. Because of the rapid response speed, pumped hydroelectric storage systems are particularly useful as backup in case of sudden changes in demand. Also, partly because of their large-scale and relative simplicity of design, pumped hydroelectric storage systems are among the cheapest in terms of operating costs per unit of energy. The cost of storing energy can be an order of magnitude lower in pumped hydroelectric systems than in storage systems like flywheels and fuel tanks. Unlike hydroelectric dams, pumped hydroelectric systems have little effect on the landscape. They produce no pollution or waste, although they cause initial damage to ecosystems.

However, pumped hydroelectric storage systems have drawbacks, too. Probably the most fundamental is their dependence on unique geological formations. There must be two large-volume reservoirs and sufficient head for building to be feasible. This is uncommon and often forces location in remote places, such as in the mountains, where construction is difficult and the power grid is not present or too distant.

13.12   CONSTANT FREQUENCY, CONSTANT SPEED, AND CONSTANT POWER

There are common applications where design or operation of power generation needs to consider constant frequency, constant speed, and/or constant power flow. Constant frequency is necessary in those cases where the operation of the supplied load is strictly dependent on the line frequency.[61,62] Examples of this are synchronized machine operations in the assembly industry and clock-dependent mechanisms. Constant-speed operations are those cases where a synchronous generator is driven by an induction motor at a particular speed such as for testing in the industry. Constant power flow is the case where the energy is commercially exchanged or bought in bulk between distinct energy companies. A fixed amount of energy is stipulated per day for exchange at certain convenient periods of the year, as, for example, in the rainy and dry seasons, for the electricity production industry.

Some electromechanical variables are used in accordance with the overall network strategy, such as active and reactive power levels or shaft rotation. Such a system is usually a multiple-input/multiple-output control system where the output voltage is associated with the reactive power, and frequency is associated with the mechanical output power. These variables are continuously monitored and fed back to follow reference signals.[1]

When designing the control of an alternative source of energy for electricity generation, it is necessary to consider whether the source is of the rotating or static type. Typical rotating types are related to wind and hydroelectric energy, whereas static types are related to photovoltaic and fuel cells. For the rotating type, three options of generators can be used: dc generator, synchronous generator, or asynchronous or induction generator. In principle, all generators operate internally as ac, but the constructional features of the generator will make the output become dc or ac.

For any ac generator, two variables are most important to control: the output voltage level and the power frequency. When variations occur in the energy levels of the primary source (water or wind), there will be similar active power variations. These variations may be responsible for oscillations. Additionally, renewable energy–based systems are usually small, with reduced shaft inertia easily subject to such oscillations, except for static sources.

One point in common between induction generators and other asynchronous sources of energy, such as photovoltaic and fuel cells, is their asynchronism with respect to the utility grid voltage. Therefore, a similar control philosophy to that used in rotary systems may be extended to them. In the case of induction generators, the control variables are flux, torque, and impressed frequency voltage. For photovoltaic and fuel cells, a dc link is a loop control associated with machine speed, and all the other output variables (torque, flux, and frequency) correspond.

Induction generators may control the input power by mechanical or electromechanical means, or electrically by load matching of the electrical terminals. Mechanical controls are frequently applied in controlling the input shaft power of induction generators; for example, centrifugal weights are frequently used to close or open the admission of water for hydro turbines or for exerting pitch blade control in WTs. Some more sophisticated mechanical controls, including flywheels (to store transient energy), anemometers, or flow meters (to determine the wind intensity or water flow, respectively), in addition to wind tails and wind vanes (for detection of the wind direction), are used to control and limit the turbine rotor speed. Despite being robust, they are slow, bulky, and rough. However, they are still used in practice, as they are considered proven, reliable, and efficient solutions.[63]

Electromechanical controls are more accurate and relatively lighter and faster than mechanical controls. They use governors, actuators, servomechanisms, electric motors, solenoids, and relays to control the primary energy. Some modern self-excited induction generators (SEIG) use internal compensatory windings to keep the output voltage almost constant. Electro-electronic controls use electronic sensors and power converters to regularize energy, speed, and power. They can be quite accurate in adjusting voltage, speed, and frequency; they are lightweight and are easily suitable for remote control and telemetry.

13.13   LINEAR INDUCTION GENERATOR

A linear induction generator (LIG) is similar to a rotating squirrel cage induction motor that has been opened out flat (Figure 13.10). Instead of offering rotary torque to a cylindrical machine, it offers linear force to a flat one as a matter of design. The original rotor bars are smoothed out into a conductor sheet. These features give it advantages such as robustness, no moving parts, natural protection against long-lasting short circuits, silent operation, reduced maintenance, compact size, and ease of control and installation. It is well known that linear induction motors can be self-excited to efficiently produce electrical energy from a linear source.

Thrusts upon an LIG may vary from just a few to thousands of newtons, depending mainly on the size and rating. Linear speeds vary from zero to many meters per second and are determined by design and the desired output frequency and voltage.

Induction machines as generators behave like passive machines during stopping, starting, and reversing procedures. Figure 13.10 illustrates a short stator type machine, where the length of the stator, the translator being as large as the anticipated amplitude, dictates the active length. This is likely to be the preferable configuration in most applications, as the oversized translator is cheaper to construct

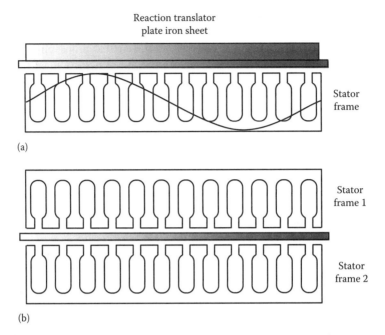

FIGURE 13.10　Linear induction generator. (a) Single stator frame and (b) double stator frame.

than the energized stator, and it is very simple and of light construction. Such construction consists of either conductors embedded in slots of a back iron structure or possibly just a single conducting sheet. Because the rotor diameter of the LIG can be considered a large one, the air gap in the discrete stator sector will be large. Because of that, the generator has a great slip, 10%–15%, and the efficiency will not exceed 80%–85%. A 150 kW prototype machine has been made, and its efficiency is over 65%. Larger machines are still in the development stage.[64–66]

LIGs are more efficient with linear prime movers since there is a significant loss of energy when linear motion is converted into rotary motion. Applications are many and varied, ranging from simple sliding transportation systems to full control of large wind power plants.[67–70]

The principles of using LIGs to convert energy from slow-period oscillations were first developed for ocean wave energy conversion in 1978.[69,70] This machine is a double-sided axial-flux generator. The two-stator sides form a segment of the circumference, and the stator is fixed to the turbine tower. The rotor is a disc that is directly coupled or parallel to the turbine rotor. The construction of the machine is relatively simple and light compared with the conventional design.

Many researchers have demonstrated interest in the LIG features, like Gripnau, Kursten, and Deleroi, who have presented an LIG for direct grid connection. Recent work on the Archimedes Wave Swing prototype have been installed off the coast of Portugal to stimulate additional research on low-frequency linear induction conversion systems as at Delft University of Technology in the Netherlands and the University of Durham in the United Kingdom.[65–69]

LIGs are key components that can convert motions of an oscillating airfoil wind energy absorber (OAWEA) into electrical energy. It works as a self-contained inductive power module configured as a canister that houses a permanent magnet *proof mass* suspended by one or more springs that enable it to move relative to a fixed stator coil. One of the challenges inherent in such an inductive generator is the relatively long period and associated slow speed of the exciting force. This oscillation is expected to be in the frequency range of 0.05–0.5 Hz, extremely low relative to the high frequencies needed to generate significant voltages (typically 2–20 Hz, depending on magnet and coil dimensions).[70,71]

13.14   STAND-ALONE OPERATION

Stand-alone or isolated electricity supply systems are generally used in areas without access to the main grid or in low-power applications. In such situations, these systems are able to provide cost-effective energy services and potentially provide remote communities without electrical energy services. Mains issues here are the quality assurance schemes for improving the reliability of systems and reducing life cycle costs such as maintenance and repair. Many of these systems involve power electronic devices and sophisticated digital electronics, fortunately in industrially modulated products.[72,73]

Power inverters are widely used in association with stand-alone hydro and wind power systems. Inverters are developed and must be chosen for a given application. They are able to convert direct current to alternating current and ensure, among other things, a more suitable behavior for the speed and voltage induction generator controls. Forced commutated inverters used in isolated electrical supply systems convert dc power stored in batteries to ac power that can be used as needed. Naturally, commutated inverters are used for power transmission or back-to-back applications. High-quality stand-alone forced-commutated inverters are available in sizes from 100 W, for powering notebook computers and fax machines, to 1000 kW or more, for powering commercial or industrial installations. Naturally, commutated inverters may go to extremely high powers as in multimegawatt schemes.

The size of an inverter is measured by its maximum continuous output in watts. This rating must be larger than the total power of all the ac loads. The size of the inverter can be minimized if the number and size of the ac loads is kept under control. If forced inverters are expected to run induction motors, like the ones found in automatic washers, dryers, dishwashers, and large power tools, they must be designed to withstand high inrush current during the motor's starting. In many cases, inverters are used to sustain the generator speed by the modulation of its output load through control of the active and reactive power, as discussed in Section 13.2.

The induction generator needs a reasonable amount of reactive power to operate either in a stand-alone or connected scheme to the grid, as discussed in Chapter 4. This reactive power must be fed externally to establish the magnetic field necessary to convert the mechanical power from its shaft into electrical power. In interconnected applications, the synchronous network supplies such reactive power. In stand-alone applications, the load itself, by a bank of capacitors connected across its terminals or by an electronic inverter, must supply the reactive power. For this reason, the external reactive source must remain permanently connected to the stator windings responsible for the output voltage control. When capacitors are connected to induction generators, the system is usually called an SEIG. Economically, self-excitation of the induction generator is usually recommended only for small power plants.[74,75]

When an SEIG is working in stand-alone mode, its output voltage may collapse because of high load or short-circuit current through its terminals, resulting in complete loss of residual magnetism. Although it is an intrinsic protection feature, it may affect the restarting of the overall system as detailed in Chapter 4.

13.15   IG FOR WIND TURBINE MAGNUS

The Magnus effect was established by the German physicist Heinrich Magnus in 1853. However, Isaac Newton had already advocated this effect 180 years before as he watched tennis players at the University of Cambridge. The structure of the turbine based on the Magnus effect has a differential with respect to conventional WTs since it has rotating cylinders around its own axis in place of blades as depicted in Figure 13.11.[76,77] The knowledge about this type of turbine has advanced very slowly along the years, but it usually considers some characteristics of the rotating cylinders related to geometric, kinematics, and energetic parameters. The improvement in performance will depend intrinsically on the type, number, surface pattern, and

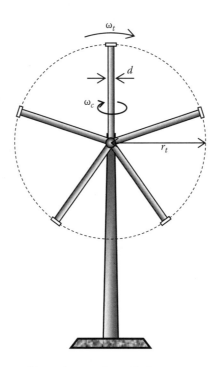

FIGURE 13.11   Magnus turbine using rotating cylinders.

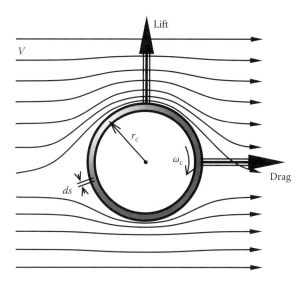

FIGURE 13.12   Lift and drag Magnus effects on cylinders.

dimensions of the cylinders and extrinsically on the optimal control of the turbine rotation concurrently with the cylinders.

The Magnus effect may be described as that which occurs on rotating cylinders or spheres within a fluid. The movement of these solid forms creates a boundary layer of air on their opposite surfaces relative to the fluid direction of movement. This effect is roughly, what happens with the airplane lifting as explained by the Bernoulli principle. The speed of the solid surface at reverse direction of the fluid movement will be higher, and hence, with lower pressure, and conversely, on the side where the velocity of the solid surface is in the same direction of the fluid, the pressure will be higher. This pressure gradient results in a net force upon the body and thereby a turning force vector perpendicular to the tangential velocity of the body relative to the fluid flow direction, as shown in Figure 13.12.

On the other hand, if the rotating body moves through the fluid with an angular velocity ω, the fluid velocity near the body will be higher than ω on one side and lower than ω on the other side. This phenomenon is due to the velocity of the boundary layer surrounding the rotating body, which is added in one side and subtracted in the other side.[78–80] The resulting pressure gradient is a net force on the body in the direction perpendicular to the body velocity vector relative to the fluid flow as shown in Figure 13.12.

13.16   MATHEMATICAL MODEL OF TURBINE MAGNUS

Mechanical stresses on the turbine structure must take into account the wind effect. As suggested earlier, the turbine driving causes a Magnus composition of two forces, called, respectively, lift l, and drag D forces (see Figure 13.13), generated by the

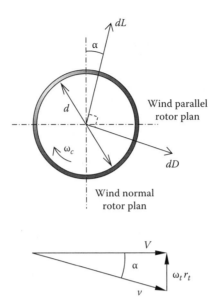

FIGURE 13.13 Differential forces of lift and drag on the cylinder surface.

rotation of a cylinder immersed in the flow of a fluid as described by the Kutta-Joukowski theorem as follows[81]:

$$L = \rho \cdot (\Gamma \times V) \tag{13.10}$$

where
    ρ is the density of the air
    Γ is the circulation vector
    V is the wind speed

The movement produced by the rotation of the cylinder is given by[82]

$$\Gamma = \int V ds = 2\pi \omega_c r_c^2 \tag{13.11}$$

where
    ds is the differential circumference length of the cylindrical surface in m
    V is the angular velocity of the cylinder in rad/s
    r_c is the cylinder radius in m

Thus, the Magnus turbine performance can be evaluated from the torque forces per unit length generated by the lift and drag forces of each cylinder incorporated in a single expression as follows[83]:

$$L = 2\pi \rho r_c^2 \omega_c V \tag{13.12}$$

and the composite mechanical action of the wind on the turbine is

$$P_{mec} = n_c(T_L - T_D)\omega_t \tag{13.13}$$

where
  n_c is the number of cylinders
  T_L and T_D are, respectively, the lift and drag torques in N m

$$T_L = \int_0^{M_L} dT_L = \frac{1}{2}\rho Vd \int_{R_0}^{R} C_L\sqrt{V^2 + \omega_t^2 r_t^2} \cdot r_t dr$$

$$T_D = \int_0^{M_D} dT_D = \frac{1}{2}\rho\omega_t d \int_{R_0}^{R} C_D\sqrt{V^2 + \omega_t^2 r_t^2} \cdot r_t^2 dr$$

where
  ω_t is the angular velocity of the turbine
  C_L and C_D are, respectively, the lift and drag coefficients usually obtained from experimental data
  α is the variable angular position of the cylinder relative to the wind direction

The differential torques exerted by the wind on the turbine shaft in the rotation direction is given by

$$dT_L = r_t\cos\alpha\, dL = r_t \frac{V}{v} dL = \frac{1}{2}\rho VdC_L v r_t dr_t \tag{13.14}$$

$$dT_D = r_t\sin\alpha\, dD = dL \frac{\omega_t r_t^2}{v} = \frac{1}{2}\rho d\omega_t C_D v r_t^2 dr_t \tag{13.15}$$

where v is the wind speed with respect to the movement speed of the cylinder in the air expressed by the following[84,85]:

$$v = \sqrt{V^2 + \omega_t^2 r_t^2} \tag{13.16}$$

With Equations 13.14 and 13.15 integrated and Equation 13.16 taken into account, the length of the radius from the turbine hub surface to the tip of the cylinder results in the lift and drag torques given by

$$T_L = \frac{1}{2}\rho Vd \int_{R_0}^{R} C_L\sqrt{V^2 + \omega_t^2 r_t^2} \cdot r_t dr_t \tag{13.17}$$

$$T_D = \frac{1}{2}\rho\omega_t d \int_{R_o}^{R} C_D \sqrt{V^2 + \omega_t^2 r_t^2} \cdot r_t^2 dr_t \tag{13.18}$$

Because of the torques expressed by Equations 13.17 and 13.18, the mechanical drag and lift actions of the wind on the Magnus turbine is

$$P_{mec} = \frac{1}{2}\rho A V^3 C_{pM} \tag{13.19}$$

where C_{pM} is the Magnus power coefficient.

Therefore, from Equations 13.13 and 13.19, Equation 13.20 can be obtained:

$$C_{pM} = \frac{2n_c(T_L - T_D) \cdot \omega_t}{\rho \cdot \pi \cdot r_t^2 \cdot V^3} \tag{13.20}$$

The terms T_L and T_D of the torques are usually very complex, and a function of the coefficients λ_t and λ_c should be experimentally obtained by curve fitting. These tip speed ratios of the Magnus turbine are related to the angular speed of the turbine ω_t, the radius of the area swept by the cylinder r_t, and the wind speed V as follows:

$$\lambda_t = \frac{\omega_t r_t}{V} \tag{13.21}$$

The turbine Magnus also uses a similar concept, as for conventional turbines, of tip speed ratio using the cylinder angular velocity λ_c relative to the wind expressed by

$$\lambda_c = \frac{\omega_c r_c}{V} \tag{13.22}$$

However, from the mechanical power expressed by Equation 13.13 must be discounted the cylinder friction losses during its rotation in a laminar flow and the motor losses in the cylinder drivers:

$$P_t = P_{mec} - P_{losses} \tag{13.23}$$

The power loss due to friction in the air is expressed by[85,86]

$$P_{losses} = T\omega_c = 1.328 \frac{\rho\pi d^4 \omega_c^3 (r_t - r_o)}{\sqrt{Re}} \tag{13.24}$$

where
  r_t is the radius of the circular area swept by the cylinders
  r_o is the radius of the rotor hub
  Re is the Reynolds number

In general, the physical meaning of the Reynolds number Re is a dimensionless ratio between the inertial forces $v\rho$ and viscosity forces μ/ℓ along the sliding fluid path expressed as follows:

$$Re = \frac{\rho v \ell}{\mu} \tag{13.25}$$

where
  v is the average fluid velocity in m/s
  ℓ is the characteristic length of the wind flow in a tube in m
  μ is the absolute dynamic viscosity of the fluid in Ns/m^2
  ρ is the density of the fluid kg/m^3

In the case of friction along the circular path ℓ described by the cylinder surface the Reynolds number can be taken as follows:

$$Re_d = \frac{\rho(\omega_c r_c)(\pi d)}{\mu} = \frac{\rho \omega_c \pi d^2}{2\mu} \tag{13.26}$$

So, the resulting expression to account for the losses with the number of cylinders has an optimal ratio size of the cylinder $(r_t - r_o)/16$, and the performance of the electric drive motors η_{elet} then is given by the following[85,86]:

$$P_{losses} = T\omega_c = 1.328 \frac{n_c}{16\eta_{elet}} \cdot \frac{\rho \pi d^4 \omega_c^3 (r_t - r_o)}{\sqrt{Re_d}} \tag{13.27}$$

where η_{elet} is the performance of the motorized drive rollers.

These equations are quite involving and full of nuances, and one has to consider experimental results or other forms of approximation to establish its coefficients by curve fitting.

A representation of Equation 13.20 is illustrated in Figure 13.14, where the different effects on the power coefficient are related to the rotation of the cylinder around its shaft and about the shaft of the turbine rotor.

Magnus WT can be explored in a wider range of wind velocities, varying from 2 to 35 m/s with respect to the traditional blade turbines, which are limited to about 3–25 m/s. The reduced rotation of the Magnus cylinder–rotor is about two to three times lower compared with the blades, thus ensuring its high operational safety. This is the main advantage with respect to other types of low efficiency turbines for the most common wind speeds due to their small lift coefficient blades under such conditions. The power coefficient of ordinary WTs drops rapidly to zero at about $V = 4$ m/s. In addition, another advantage of the Magnus WT is its cylinder–rotor rotation aerodynamics with speed regulation preventing from excessive spin up and collapse due to excessive centrifugal forces. In particular, at wind velocities higher than about 35 m/s, the regulation results in diminution of the Magnus force with a cylinder–rotor self-braking.[87–89]

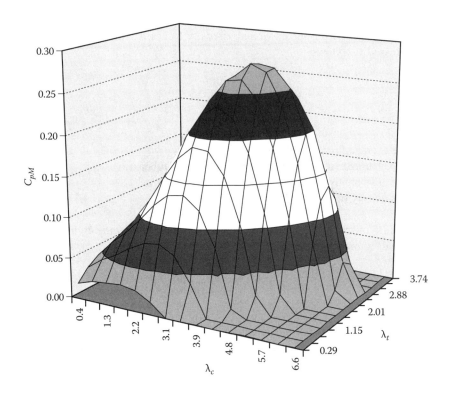

FIGURE 13.14   The Magnus turbine power coefficient as a function of λ_c and λ_t.

As seen from this description, a reasonable modeling of the Magnus turbine can be quite complex because it is difficult to take into account all variations in which the turbine may pass through. A forward chaining controller for the cylinder and turbine speeds that avoids the need of an analytical model for the Magnus WT coupled to an induction generator can overcome all this difficulty. Thus, the three-dimensional graph of Figure 13.14 clearly illustrates the opportunities for the HCC as described in Chapter 10 in order to establish the point of maximum power on a 3D surface of tip speed ratio for either cylinders and rotor. This means that we can search the maximum power point according the wind and generator load variations by adjusting alternately the angular velocity of the cylinders and turbine speeds to the point of maximum power generated.

13.17 DISTRIBUTED GENERATION

Any electric power source connected directly to the distribution network or to the customer through a meter may be called distributed generation (DG). The most common DG primary sources are hydro run-of-river, hydro with a small water dam, wind, photovoltaic, geothermal, biomass, and diesel. These sources and many others can be cogeneration systems with microturbine, fuel cell generation

TABLE 13.2
Large-Scale Renewable DG Energy Systems

Technology	Typical Size per Module
Traditional combustion	
Combined cycle gas turbines	35–400 MW
Internal combustion engines	5 kW–10 MW
Combustion turbines	1–250 MW
Microturbines	35 kW–1 MW
Small-scale combustion	
Stirling engine	2–10 kW
Renewable	
Small hydro	1–100 MW
Microhydro	25 kW–1 MW
WTs	200 W–3 MW
Photovoltaic arrays	20 W–100 kW
Solar thermal, central receiver	1–10 MW
Solar thermal, Lutz system	10–80 MW
Biomass, e.g., based on gasification	100 kW–20 MW
Geothermal	5–100 MW
Ocean energy	100 kW–1 MW
Electricity storage—fuel cells and batteries	
Phosphoric acid fuel cells	200 kW–2 MW
Molten carbonate fuel cells	250 kW–2 MW
Proton exchange fuel cells	1–250 kW
Solid oxide fuel cells	250 kW–5 MW
Battery storage	500 kW–5 MW

Source: Extracted from Gertmar, L., Power electronics and wind power, *Proceedings of the 10th European Conference on Power Electronics and Applications—EPE 2003*, Toulouse, France, 2003.

system, or any kind of alternative energy sources (Table 13.2). A study by the Electric Power Research Institute (EPRI) indicates that in 2010, 25% of new generation systems will be distributed.[22–24,89]

Advantages of DG integration into the public distribution network can be (1) technical (including backup, emergency, and peak shaving systems), environmental (by replacing fossil fuel–powered units), and economic benefits (job opportunities, local primary transformation of goods, rural comforts, and energy for remote areas); and (2) opportunity for distribution utilities to improve the performance of networks by reducing voltage drops along the transmission and distribution lines and its power and energy losses. The current problems related to DG can be grouped into technical (interface to the grid, energy quality, reliability, islanding, safety for the maintenance personnel); commercial (complicated deals with the traditional power suppliers, availability of parts, lack of general standards); and legal problems (permissions, licenses, operating areas, financial capability for withstanding the costs of legal demands). Problems that are more general are related to the operation and control of the DG and perennial system planning and design.

Small and medium wind power and hydropower generating systems are mostly using induction generators since they are relatively easy to connect in parallel with the mains, they do not have an exciter, and they cannot normally sustain a stable island on their own. If they must provide power on a stand-alone basis, they are used only for small generator plants (<15 kW).

As discussed in Chapters 5 and 7, induction generators operate with quite distinct characteristics related to synchronous generators. They do not require precise alignment of frequency and phase angle for as long to avoid voltage dips due to an inrush of magnetizing current. However, induction generators, unless corrected with capacitors or special power converters, operate at relatively low power due to the reactive excitation current drawn from the electricity company's power system. Regardless of generator size, the use of power factor correction capacitors or reactive compensation must be approved by the utility to ensure that issues related to self-excitation and ferroresonance are addressed properly, as discussed in the following.[75,88,89]

In order to assess voltage flicker, the expected number of starts per hour and maximum starting kVA draw data will need to be delivered to the utility company to verify that the voltage dip due to starting is within the acceptable flicker limits according to IEEE 519-1992 and, wherever applicable, the local company regulations, for example, the Con Edison flicker curve requirements used by the Consolidated Edison Company of New York.[89]

Starting or rapid load fluctuations on induction generators can adversely affect the company's distribution system voltage and cause noticeable voltage quality problems for customers on the circuit. Corrective steps include the use of power converters associated to energy sources or capacitor banks, switched capacitors, or other techniques that may be needed to mitigate the voltage flicker and regulation issues that arise. In turn, switched capacitors and sources can cause ferroresonance, which is a serious form of overvoltage that can stress or even damage equipment and loads connected on the system. If the customer's design includes additional capacitors installed on the customer side of the point of common connection, the company will review these measures and may require the customer to install additional equipment to reduce the risk of ferroresonance.

It is important to remember that real loads very often change their output impedance and thus their resonance frequency. For this reason, customers who provide capacitor banks to minimize the voltage drop on the bus at the starting of the generator should provide a way to disconnect them automatically from the generator terminals after the start-up. The customer should perform and submit studies to demonstrate the impact of the capacitors on the system and all precautionary measures on the line to avoid them.

In the event of an electric fault and/or abnormal voltage/frequency condition, most electricity companies consider the independent producer as responsible for tripping their generator intertie breaker and/or contactor and isolating their generator

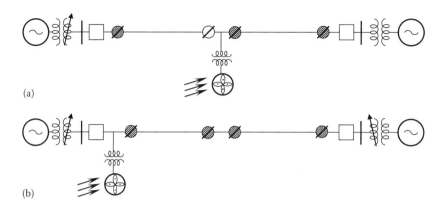

(a)

(b)

FIGURE 13.15 DG Applied for contingency capacity support. (a) Support for both feeders and (b) support for only one feeder.

from the company's distribution system. The protective relaying requirements for a particular facility will depend on the type and size of the facility, voltage level of the interconnection, location on the distribution circuit, disconnecting equipment available, fault levels, and many other factors, as exemplified in Figure 13.15. IEEE Standard 1547-2003 has specific tables that apply to contingency capacity support with recommended default values for the trip settings of distributed generators.

The absolute minimum protective relays that any company should require for any size induction generator should never be less than the relays listed in the following, and on a case-by-case basis, it may be necessary for the utility to require additional protection.[77]

1. Utility grade undervoltage relays (device 27) shall be connected phase to ground on each phase. These relays disconnect the customer from the company's distribution system during faults or when the company feeder is out of service. The default trip settings should conform to IEEE Standard 1547-2003 (Table 13.1). However, for generation greater than 30 kW, the company may require different settings on a case-by-case basis as needed.
2. Utility grade overvoltage relays (device 59) shall be connected phase to ground on each phase. The default trip settings should conform to IEEE Standard 1547-2003 (Table 13.1). However, for generation greater than 30 kW, the company may require different settings on a case-by-case basis as needed.
3. Utility grade over- and under-frequency protection (devices 81/O and 81/U) are used to trip the generator or intertie breaker upon detecting a frequency deviation outside of reasonable operating conditions. The default trip settings should conform to IEEE Standard 1547-2003 (Table 13.2). However, for generation greater than 30 kW, the company may require different settings on a case-by-case basis as needed.

Additional protection could take the form of phase and ground fault overcurrent relays, ground fault overvoltage relays, directional power and/or overcurrent relays, transfer trips, speed matching controls, lockout functions, etc. Most electricity companies have their own specific regulations to cover particular and/or regional aspects.[90,91]

Relays used with induction generators may be single-function or multifunction packages, and they can be mechanical, solid-state, or microprocessor-based types as long as they satisfy the utility grade or type test specifications. Modern microprocessor multifunction relays designed for generator protection that satisfy the required utility grade specifications have recently become much more cost-effective and broadly reliable compared with products of a decade ago and are available from a variety of equipment vendors.

The four main concerns the electricity companies really consider regarding the type of induction generator grounding to utilize are ground fault overvoltages, ferroresonance, harmonics, and ground fault current contribution/detection issues. There is a widely held misconception in the industry that because the induction generator does not normally self-excite, the effects of ground fault overvoltages can be ignored since they would disappear quickly (transient decay). However, partial excitation can still exist on some phases during ground faults, and because an induction generator might self-excite due to capacitors, and even without self-excitation, the transient decay period of its output can cause damage in just a few cycles. The grounding of induction generators and its potential impact must still be treated almost similar to that of a synchronous generator. This means the company may need to specify effective or solid grounding for an induction generator whenever there is concern about ground fault overvoltages on a four-wire multigrounded neutral distribution system.

When interconnecting an IG to ungrounded or unigrounded distribution systems, an ungrounded or impedance-grounded interface to the company distribution system at the point of common connection will usually have to be specified. The final determination as to which ground configuration is most appropriate will be done on a case-by-case basis. It is important to recognize that the type of grounding referred to in this section is the grounding with respect to the utility distribution system, which is a function of not just the generator grounding itself but also the configuration of the interface transformer winding and its ground connection. Therefore, careful electricity distribution system planning has to be established well in advance aiming at an adequate DG sitting for steady-state operation of distribution systems to minimize the electrical network losses and to maintain acceptable reliability levels and voltage profile. As such planning involves very complex systems based on several criteria, a multiobjective analysis defining some quantitative and qualitative parameters has been used, such as the Bellman-Zadeh methodology. Weights (importance levels) are applied to each parameter, which may be classified by the electricity company.[89–93]

The most important advantages of the correct DG sitting using induction generators are related to the significant beneficial impact over the power loss levels and improvement of the voltage levels, mostly of the energy for *peak load shaving*. Furthermore, DG systems using induction generators are not as restricted as the centralized high-power stations, since they can be placed closer to the end-user customers.

This structural rethinking of the electric utility helps to diversify and deploy the electrical energy sources throughout many generation spots, thus providing increased planning flexibilities.

On the other hand, it is important to recognize that the power injected by DG using induction generator units at inappropriate places, or an excess of power generation without voltage regulation through tap changing, may result in increased system losses and undesired voltage levels beyond the acceptable electricity company limits. As a result, there are increasing network costs, sometimes even involving heavy fines to the energy supplying companies, which is a highly undesired outcome. Moreover, DG on these bases may introduce reversed power flows and interfere with the generation protective system. For this reason, sitting DG is extremely important, not only because certain renewable DG technology units require certain geographic conditions but also because they can degrade reliability and power quality for other customers connected to the same feeder.[81,82]

Many of the design requirements that the customer must satisfy are common to all the generator types (that is, SPC, induction, and synchronous generator types; see Chapters 3 and 5). The common requirements include the disconnect switch, certification standards, power quality standards, and IEEE 1547 voltage response tables.[94–97]

13.18   SUGGESTED PROBLEMS

13.1 Research and build a data file including 24 h of electrical demand data for 30 residences, 15 shops, and 5 factories, along with some wind speed data. These data might be reasonably fabricated for your studies, but it should be representative of a small downtown district; that is, each hour is associated to the respective load in kWh.

13.2 Taking into account some wind speed data from Problem 13.1, a second data sheet should be constructed using the generator cost data from USDOE EIA. This data table should have the minimum information needed to calculate the annualized fixed cost values. Write a report with a spreadsheet in order to discuss the capital costs and required capacity factor. Include the operating and maintenance (O&M) and find the annual fixed cost amount for such economics based model. You should ignore the electrical power transmission investment costs for this study because they are usually built with governmental support and contracts.

13.3 Using the data of Problems 13.1 and 13.2, do the following:
1. *Chronological load data*:
   a. From these data, identify the coincident peak and briefly state the significance of the coincident peak. What is the significance of this piece of information to a system planner?
2. *Load duration curve (LDC)*:
   a. Construct the LDC (which, of course, will only be for 24 h, and not 8760 h)
   b. Identify the numerical values of the two *basic pieces of information* provided on your LDC—the system peak and the total energy consumed.

    c. The very important things these convey to you are as follows: you need to build ____MW of capacity, and you need to purchase some calculable (by others) amount of fuel in order to supply ____ MWh of electrical energy.

    d. Given generating technologies fixed and variable costs data, how much capacity of each type would you build in order to have a system with the least cost in terms of production cost?

    e. Assume you can build plants in 50 kW increments. Select only a few technology types to build.

    f. Also plot on the same graph an LDC that includes a reserve margin—some percentage capacity above the peak demand—for reliability purposes. New England historically has a 14% reserve margin. New York has 18%. Other countries often have lower reserve margins.

13.4 Considering nondispatchable technologies for the previous problem:
Now, assume a wind farm has been installed with hourly generation for the same 24 h period given in the posted data file. Make, and state, any assumptions needed to answer the following questions.

    a. Plot the new (net) LDC on the same graph as the preceding one.

    b. What is the energy captured from this wind farm?

    c. What is the capacity factor from this wind farm?

13.5 Design a stand-alone (not connected to any main grid) hybrid system for a small downtown district. The energy sources will be a mix of solar, wind, diesel generator, and battery storage. Some data files are required for this project (you will have to build them yourself):

- 24 h of electrical demand data for 30 residences, 15 shops, and 5 factories, since these data represent a small downtown district.
- 8760 h of wind energy data.
- 8760 h of solar energy data.

    a. Suppose that the daily load demand for this district repeats every day, during 365 days (whole year). Of course, a real project should have data for 8760 h of load, but this is a simplified case study.

    b. If the system could capture all the solar energy during the day and all the wind energy during the night, with a battery to just level out these fluctuations, a simple energy balance equation would be

$$\int_0^{8760} \left(\text{solar radiation*PV total conversion efficiency}\right) dt$$

$$+ \int_0^{8760} \left(\text{wind energy*turbine system total efficiency}\right) dt$$

$$= \int_0^{8760} \left(\text{loads} + \text{losses} + \text{battery} + \text{shunt\_load}\right) dt$$

This equation can have the integration done over different periods, for example, 1 day, 1 month, or one season. The shunt load in this equation may represent an actual load to dump excess energy when needed, or some sort of heating water, or pumping water. Based on the solar energy density (solar radiation), panel efficiency and wind energy density and efficiency, in kW/ m^2, specify the area for the PV array and the number of WTs.

c. Develop the equation further in order to include a diesel generator for critical cases, assuming that the battery size is initially 10% of the daily energy. The diesel generator can be assumed as maintaining a constant power. Size your system for a whole year operation, and carefully describe any possible constraints that you may foresee for emergency needs. Develop a spreadsheet for this case that can be modified for other scenarios.

d. Using the data given earlier, run some scenarios in order to minimize (or zero, if possible) the diesel generator at the expense of increased battery storage. Include fuel price for the diesel generator, and maintenance for the battery, in order to support your cases. Propose a new solution based on such simple optimization study.

e. For this optimized new system, describe all the electrical and electronic devices, and even other auxiliary or mechanical systems that should be purchased, with their specifications and ratings.

You bet! The design of such a stand-alone hybrid system is a very challenging project!

REFERENCES

1. Portolann, C.A., Farret, F.A., and Machado, R.Q., Electronic regulation by the load of the electrical variables and speed of asynchronous micro turbo-generators, *IEEE International Conference on Power Electronics-CIEP96*, Cuernavaca, Mexico, October 1996, pp. 125–130.
2. Bonert, R. and Hoops, G., Standalone induction generator with terminal impedance controller and no turbine controls, *IEEE Proc. B.*, 5, 28–31, 1990.
3. Farret, F.A. and Simões, M.G., *Integration of Alternative Sources of Energy*, Wiley Interscience, Hoboken, NJ, 2006.
4. Patel, M.R., *Wind and Solar Power Systems*, CRC Press, Boca Raton, FL, 1999.
5. Farret, F.A., Gomes, J.R., and Rodrigues, C.R., Sensorless speed measurement associated to electronic control by the load for induction turbogenerators, *Proceedings of the SOBRAEP V Brazilian Power Electronics Conference*, Foz de Iguaçu, Parana, September 1999, pp. 88–93.
6. KWI Architects Engineers Consultants, Status report on variable speed operation in small hydropower, European Commission, Directorate-General for Energy and Transport, Energie, New Solutions in Energy, St. Pölten, Austria, 2000.
7. Durstewitz, M., Hoppe-Kilpper, M., Schmid, J., Stump, N., Windheim, R., Experiences with 3000 MW wind power in Germany, *Proceedings of the European Wind Energy Conference: EWEC '99*, Nice, France, March 1999, pp. 1–5.
8. Wittholz, H. and Pan, D., Étude des capacités de la chaîne d'approvisionnement de l'industrie canadienne de l'énergie éolienne, SYNOVA International Business Development, Ottawa, Ontario, Canada, November 2004.

9. Okpanefe, P.E. and Owolabi, S., Small hydropower in Nigeria, *TCDC Training Workshop on SHP*, Hangzhou, China, 2002.
10. EnBW, Energie Baden-Württemberg AG, Fossil energy, The conventional power stations of EnBW, Karlsruhe, Germany, May 2006.
11. Scottish Environment Protection Agency (SEPA), An economic analysis of water use in the Scotland River Basin District, Summary report, Stirling, Scotland, 2003.
12. Canadian Hydropower Association, Current and planned hydro development in Canada, *Hydropower & Dams*, Issue Two, Ottawa, Ontario, Canada, 2003.
13. RWE, Rheinisch-Westfälisches Elektrizitätswerk AG, RWE's programme for climate protection, Impetus Investment Innovations, Essen/Cologne, Germany, 2006.
14. Commission de Régulation de L'Énergie, Activity report, ISSN 1771-3196, Paris, France, June 2003.
15. Clean Energy Partnership. http://www.cleanenergypartnership.org/.
16. Enmax Corp. http://www.enmax.com/.
17. Hydromax Energy. http://www.hydromaxenergy.com/.
18. Voith Hydro Wavegen. Inverness, United Kingdom, http://voith.com/en/t3364e_Wave_power.pdf, last accessed: August 10, 2014.
19. Blue Energy, Blue Energy Canada Inc., Vancouver, British Columbia, Canada (formerly known as Nova Energy Ltd., Nova Scotia). www.bluenergy.com, last accessed: August 09, 2014.
20. Craig, J., The status and UK resource potential of wave power and tidal stream energy, Irish Sea Forum, Joint Seminar on Irish Sea Renewable Energy Resource, Wales, U.K., October 1999.
21. Jeon, B.-H., Hong, S.-W., and Lim, Y.-K., Autonomous control system of an induction generator for wave energy device, Ocean Development System Lab., Korea Research Institute of Ships and Ocean Engineering, KORDI, Daejeon, Korea, *Proceedings of the Fifth ISOPE Pacific/Asia Offshore Mechanics Symposium*, Daejeon, Korea, November 2002, pp. 17–20.
22. Bedard, R., Mirko, P., Siddiqui, O., Hagerman, G., and Robinson, M., EPRI final survey and characterization, Appendix C, EPRI North American Tidal in Stream Energy Conversion (TISEC), Feasibility Demonstration Project Devices, Palo Alto, CA, November 2005.
23. Previsic, M., Bedard R., and Hagerman, G., E2I EPRI assessment offshore wave energy conversion devices, E2I EPRI WP-004-US-Rev 1, Electricity Innovation Institute, Palo Alto, CA, June 2004.
24. Singh, D., Renewable energy technologies in Australia and New Zealand, United Nations Educational, Scientific and Cultural Organization (UNESCO), Research and Development, Manufacturing and Field Projects (Project REF: SC/EST/99/4596), Perth, Western Australia, Australia, September 1999.
25. Hammarlund, A., European thematic network on wave energy, Project funded through the energy, environment and sustainable development program, Partners: Chalmers University, CRES, Ecole Centrale de Nantes, Edinburgh University, EMU Consult, ESB International, Future Energy Solutions from AEA Technology, Instituto Nacional de Engenharia e Technologia Industrial (INETI), Instituto Superior Tecnico Ponte di Archimede nello Stretto di Messina SpA, Rambøll, Teamwork Technology, University College Cork, European Community, ERK5-CT-1999-20001, Edinburgh, Scotland, 2000/2003.
26. Ocean Energy Technology Overview, Prepared for the U.S. Department of Energy, Office of Energy Efficiency and Renewable Energy, Federal Energy Management Program, DOE/GO-102009-2823, July 2009.
27. Bonneville Power Administration, Technology Innovation Office, Renewable energy technology roadmap, Portland, OR, September 2006.
28. Baker, A.C., *Tidal Power*, Peter Peregrinu, London, U.K., 1991.

29. Boud, R., Status and research and development priorities 2003, wave and marine current energy, UK Department of Trade and Industry of Future Energy Solutions, part of AEA Technology plc, DTI Report Number FES-R-132, AEAT Report Number AEAT/ENV/1054, Aberdeen, U.K., 2002.

30. Camp Dresser and McKee International Inc. in association with GHK (HK) Ltd., Study on the potential applications of renewable energy in Hong Kong, Stage 1 study report, Electrical & Mechanical Services Department, Government of the Hong Kong Special Administrative Region, Hong Kong, China, December 2002.

31. Salequzzaman, M.D. and Newman, P., Environmental impacts of renewable energy: case study of tidal power, *Proceedings of the Australasian Universities Power Engineering Conference (AUPEC-2001)*, Millennium Power Vision, Perth, Western Australia, Australia, September 2001.

32. Derby Tidal Power Project, Tidal Energy Australia, Report and recommendations of the Minister for the Environment and Heritage under Section 43 of the Environmental Protection Authority Act 1986, Bulletin 1071, Perth, Western Australia, Australia, October 2002.

33. Renewable Energy Research Development & Demonstration, Programme Overview, Sustainable Energy Ireland (SEI), brochure published and funded by the Irish Government under the National Development Plan 2000–2006 with programs part financed by the European Union, Dublin, Ireland, 2006.

34. Tidal Impact Co., Tidal impact, feasibility study of tidal power in UK, Memorandum submitted to UK Parliament, London, U.K., 2001.

35. Solomon, D.J., Fish passage through tidal energy barrages, Energy Technology Support Unit, Harwell, Contractor's Report No. ETSU TID4056, Johnson City, TN, 1988, p. 76.

36. Turnpenny, A.W.H., Memorandum submitted to UK Parliament, Fawley Aquatic Research Laboratories, London, U.K., January 2001.

37. Turnpenny, A.W.H., Mechanisms of fish damage in low-head turbines: An experimental appraisal. In: *Fish Migration and Fish Bypasses*, edited by M. Jungwirth, S. Schmutz, and S. Weiss, Blackwell, Oxford, U.K., 1998, pp. 300–314.

38. Public Utilities Commission of the State of California, Energy Division, Resolution E-3957, October 2005.

39. WhisperGen UK Limited. Binley Industrial Estate Binley, Coventry, United Kingdom, http://christchurch.yalwa.co.nz/ID_102193003/WhisperGen-Limited.html, last accessed: August 10, 2014.

40. Van Arsdell, B.H., *The Stirling Engine Reference, Guide and Catalog*, American Stirling Company, San Diego, CA, 2002.

41. Häggmark, S., Neimane, V., Axelsson, U., Holmberg, P., Karlsson, G., Kauhaniemi, K., Olsson, M., and Liljegren, C., Aspects of different distributed generation technologies, Connection of distributed energy generation units in the distribution network and grid, CODGUNet Work Package WP 3, Nordic Countries, March 2003.

42. Scott, S.J., Holcomb, F.H., and Josefik, N.M., Distributed electrical power generation, Summary of alternative available technologies, ERDC/CERL SR-03-18, US Corps of Engineers, Engineering Research and Development Center, September 2003.

43. Sigma Elektroteknisk AS, Micro CHP technology & economic review, Oslo, Norway, June 2003.

44. US DOE, An introduction to distributed generation interconnection, US DOE in collaboration with the State of Wisconsin, Division of Energy, Alliant Energy and Unison Solutions, March 2004.

45. Morton, A., Maximizing the penetration of intermittent generation in the SWIS, Econnect Project No. 1465, South West Interconnected System (SWIS), Prepared for Office of Energy, Perth, Western Australia, Australia, July 2005.

46. Seman, S., Niiranen, J., and Arkkio, A., Ride-through analysis of doubly fed induction wind-power generator under unsymmetrical network disturbance, *IEEE Trans. Power Syst.*, April 2006, pp. 1782–1789.
47. AWEA—Electrical guide to utility scale wind turbines, AWEA Grid Code White Paper, Policy Department, Washington, DC, March 2005.
48. Petersson, A., Analysis, modeling and control of doubly-fed induction generators for wind turbines, Thesis for the Degree of Doctor of Philosophy, Chalmers University of Technology, Göteborg, Sweden, 2005.
49. Kretschmann, J., Wrede, H., Mueller-Engelhardt, S., and Erlich, I., Enhanced reduced order model of wind turbines with DFIG for power system stability studies, *First International Power and Energy Conference, PECon 2006*, Putrajaya, Malaysia, November 2006.
50. Marko, S., Darul'a, I., and Vlcek, S., Development of wind farm, models for power system studies, *Comm. J. Electr. Eng.*, 56(5–6), 165–168, 2005.
51. Boldea, I., *The Electric Generators Handbook: Variable Speed Generators*, The Electric Power Engineering Series, series editor L.L. Grigsby, CRC/Taylor & Francis Group, Boca Raton, FL, 2006.
52. Natural Resource Canada. Collection of publications, http://www.canren.gc.ca/, last accessed: August 10, 2014.
53. Caldwell, J., Goodwin, M., Haller, M., Miller, N., Sims, R., Scher, M., and Zavadil, B., *Electrical Guide to Utility Scale Wind Turbines*, for the American Wind Energy Association (AWEA), March 2005.
54. Matthew, G., Final report: Small hydro survey for IEUA facilities, IEUA Headquarters, Moccasin, Canada, 2005.
55. Hydropower Micro-Systems, A Buyer's guide, prepared by the Hydraulic Energy Program, Renewable Energy Technology Program, CANMET Energy Technology Centre (CETC), Natural Resources, Ottawa, Ontario, Canada, 2004.
56. Sigma Engineering Ltd., Green energy study for British Columbia, Phase 2: Mainland, Small Hydro, BC Hydro, Vancouver, British Columbia, Canada, October 2002.
57. Jayaprakash, P., Varma, G., Prasad, K.P., Aravind, P.V., and Sunil, R., Field testing of an induction generator load controller for micro hydel station using pump as turbine and induction motor as generator, Closure report, submitted by Integrated Rural Technology Centre, Mundur, Palakkad, sponsored by KRPLLD of the Centre for Development Studies, Thiruvananthapuram, Kerala, March 2004.
58. EWI, The economic value of storage in renewable power systems—the case of thermal energy storage in concentrating solar plants, http://www.ewi.uni-koeln.de/content/research_and_consulting/ewi_models/gems/storage/, last accessed: August 19, 2014.
59. Seitz, B., Harwood, G., Salire, C., Peterson, J., and Hess, H., Small remote hydroelectric generating station, *IEEE Industry Applications Society Annual Conference*, Seattle, WA, October 2004.
60. Bocquel, A., *300 MW Variable Speed Drives for Pump-Storage Plant Application*, published by Alstom & Company, Goldisthal, Germany, May 2004.
61. Schreiber, D. and Venac, I., *Power Electronics for Wind Turbines*, Semikron Elektronik GmbH & Co KG, Nürnberg, Germany, June 2006.
62. Soens, J., Thong, V.V., Driesen, J., and Belmans, R., Modeling wind turbine generators for power system simulations, *European Wind Energy Conference EWEC*, Madrid, Spain, June 2003.
63. Gertmar, L., Power electronics and wind power, *Proceedings of the 10th European Conference on Power Electronics and Applications—EPE 2003*, Toulouse, France, 2003.

64. Mueller, M.A., Baker, N.J., and Brooking, P.R.M., Electrical power conversion in direct-drive wave energy converters, *Fifth European Wave Energy Conference*, Cork, Ireland, September 2003.

65. Baker, N.J., Linear generators for direct drive marine renewable energy converters, PhD thesis, School of Engineering, University of Durham, Durham, U.K., 2003.

66. Force. Why use linear motors, http://www.force.co.uk/, last accessed: August 19, 2014.

67. Spooner, E. and Mueller, M.A., Comparative study of linear generators and hydraulic systems for wave energy conversion, DTI/Pub URN 01/1506 Publication Code RPS003, Contractor, Durham Centre for Renewable Energy, School of Engineering, University of Durham, Durham, U.K., Report: ETSU V/06/00189/00/00, 2001.

68. Sustainable Energy Programmes, Wave energy: Linear versus hydraulic generators for energy conversion, flexible coupling translator sea surface electrical output sea bed or dynamic reference weight stator buoy, Project Summary 003, Leeds, U.K., January 2002.

69. Boud, R., Status and research and development priorities, wave and marine current energy, dti report no. FES-R-132 and AEAT report no. AEAT/ENV/1054 prepared for the UK Department of Trade and Industry by of Future Energy Solutions, part of AEA Technology plc, 2003.

70. Woolsey, C., Hagerman, G., and Morrow, M., A self-sustaining, boundary-layer-adapted system for terrain exploration and environmental sampling—Phase I, Final report, NIAC-NASA Institute for Advanced Concepts, May 2005.

71. Omholt, T., A wave-activated electric generator, *OCEANS'78*, Washington, DC, December 1978, pp. 585–589.

72. Costa, L., Kariniotakis, G., Kamarinopoulos, A., and Hartziargyriou, N., WP2—Investigation of low-cost, innovative RES and wastewater treatment solutions, Report on State of the Art on Low-Cost Innovative RES Technologies, Contract INCO-CT-2004-509161, Paris, France, July 2005.

73. Wright, A.J. and Formby, J.R., *Overcoming Barriers to Scheduling Embedded Generation to Support Distribution Networks*, Department of Trade and Industry, EA Technology, New and Renewable Energy Technology, London, U.K., 2000.

74. Distributed Generation Technical and Policy Committee, Alberta distributed generation interconnection guide, Parts 1 and 2, Alberta, Canada, July 2002.

75. Distributed Utility Associates and Endecon Engineering, Distributed generation interconnection manual, Public Utility Commission of Texas, US Department of Energy, Office of Energy Efficiency and Renewable Energy, Texas, May 2002.

76. Goňo, R., Rusek, S., and Hrabčík, M., Wind turbine cylinders with spiral fins, Czech Science Foundation, Ostrava, Czech Republic, 2009.

77. Bychkov, N.M., Dovgal, A.V., and Sorokin, A.M., Parametric optimization of the magnus wind turbine, *16th International Conference on Methods of Aerophysical Research (ICMAR)*, Kazan, Russia, 2008.

78. Corrêa, L.C., Lenz, J.M., Ribeiro, C.G., Trapp, J.G., and Farret, F.A., Maximum power point tracking for magnus wind turbines, *39th Annual Conference of the IEEE Industrial Electronics Society*, Vienna, Austria, November 2013, pp. 1716–1720.

79. Zhao, J., Hou, Q., Jin, H., Zhu, Y., and Li, G., CFD analysis of ducted-fan UAV based on magnus effect, *IEEE International Conference on Mechatronics and Automation*, Chengdu, China, August 2012, pp. 1722–1726.

80. Luo, D., Huang, D., and Wu, G., Analytical solution on magnus wind turbine power performance based on the blade element momentum theory, *J. Renew. Sustainable Energy*, 3(3), 033104, 2011.

81. Burton, T., Jenkins, N., Sharpe, D., and Bossanyi, E., *Wind Energy Handbook*, 2nd edn., John Wiley & Sons Ltd., Chichester, U.K., 2011.

82. Luo, D., Huang, D., and Wu, G., Analytical solution of magnus wind turbine power performance based on blade element momentum theory, *J. Renew. Sustainable Energy*, 3, 033104, June 2011, doi:10.1063/1.3588039, American Institute of Physics, pp. 033104-1–033104-13.

83. Barbero, A., García-Matos, J.A., Cantizano, A., and Arenas, A., Numerical tool for the optimization of wind turbines based on Magnus effect, *Ninth World Wind Energy Conference and Exhibition (WWEC 2010)*, Istambul, Turquia, 2010.

84. Padrino, J.C. and Joseph, D.D., Numerical study of the uniform steady-state flow past a rotating cylinder, *J. Fluid Mech.*, 557, 191–223, 2006, Cambridge University Press, doi: 10.1017/S0022112006009682.

85. Rouse, H., *Elementary Mechanics of Fluids*, Dover Publications Inc., New York, 1946, pp. 275, 376.

86. Bychkov, N.M., Dovgal, A.V., and Kozlov, V.V., Magnus wind turbines to an alternate to the blade ones, *J. Phys.: Conf. Series*, 75(1), 012004, 2007.

87. Emerging Small Wind Technology, Working paper no. 7, The Applied Research Institute for Prospective Technologies, Vilnius, Lithuania, 2008.

88. Sengupta, T.K. and Talla, S.B., Robins—Magnus effect: The continuing saga, *Curr. Sci.*, 86(7), 1033–1036, April 10, 2004.

89. Morton, A., Maximizing the penetration of intermittent generation in the SWIS, Econnect Project No. 1465, South West Interconnected System (SWIS), Prepared for the Office of Energy, Perth, Western Australia, Australia, December 2005.

90. Consolidated Edison Company of New York (Con Edison). http://www.coned.com/, last accessed: August 19, 2014.

91. UK Government, Wind Power in the UK. A guide to the key issues surrounding onshore wind power development in the UK, Sustainable Development Commission, London, U.K., May 2005.

92. Barin, A., Pozzatti, L.F., Canha, L.N., Machado, R.Q., Abaide, A.R., and Farret, F.A., Analysis of multi-objective methods applied to distributed generation systems, *Proceedings IEEE International Conference on Power Engineering, Energy and Electrical Drives (POWERENG)*, Setúbal, Portugal, April 2007.

93. El-Khattam, W. and Salama, M.M.A., Distributed generation technologies, definitions and benefits, Department of Electrical and Computer Engineering, University of Waterloo, Waterloo, Ontario, Canada, January 2004.

94. Western Power Corporation, Transmission and distribution annual planning report prepared by Networks Business Unit, Perth, Western Australia, Australia, 2006.

95. PB Power, A study into the economic renewable energy resource in Northern Ireland and the ability of the electricity network to accommodate renewable generation up to 2010, Report prepared by PB Power, Auckland, New Zealand, June 2003.

96. Dyko, A., Burt, G., and Bugda, R., Novel protection methods for active distribution networks with high penetrations of distributed generation, Year II report, DTI Centre for Distributed Generation and Sustainable Electrical Energy, London, U.K., June 2006.

97. Katiraei, F., Abbey, C., and Bahry, R., Analysis of voltage regulation problem for a 25 kV distribution network with distributed generation, *Proceedings of the IEEE-PES General Meeting*, Montreal, Quebec, Canada, 2006, pp. 1–8.

14 Economics of Induction Generator–Based Renewable Systems

14.1 SCOPE OF THIS CHAPTER

The vision of the interaction of values, principles, and resource dynamics, of policies and social effects, constitutes the paradigm of economics, the environment, and distributed generation. Such a paradigm brings together the views of environmentalists, economists, and end users. Environmentalists are interested in maintaining the equilibrium, renewal, and regeneration of ecosystems; their main perspective is embodied in the sustainability of natural systems. Economists and market decision makers are interested in how natural resources or natural services can be used for human benefit. From this perspective, only things useful to human beings have any value. Standard economic analysis so far does not consider the ecosystem to have any intrinsic value, whereas the ecological view suggests that natural systems must be protected independently of their use or economical value to humans. The end user is interested in converting the available resources into products and services. If electric energy is good for home and industrial needs, the end user is interested in how electrical energy will be priced and how reliable its availability will be.

Public utilities sell power to users at tariffs based on historic load demand and investments made to bring electricity to the end user. Public utilities also regulate how end users are tied into the distribution system and monitor power quality, reliability, and harmonic pollution affecting the end user. If the end user is interested in investing in distributed generation because of reliability, harmonic pollution, or power quality concerns, he or she may not be concerned about the ecological impact. On the other hand, if the end user is investing in distributed generation of community or small-scale power, he or she may be interested in keeping up with the environmental issues involved in the broader impact of such installation. This chapter covers the fundamental economic ideas that support decision-making criteria for renewable energy systems, particularly those that need induction generators to convert the energy source into electricity.

14.2 OPTIMAL AND MARKET PRICE OF ENERGY IN A REGULATORY ENVIRONMENT

A comprehensive understanding of the interaction of the environment and electric energy users through an economic analysis sometimes binds incompatible perspectives. Electric energy is not found in nature directly (except in lightning and statics).

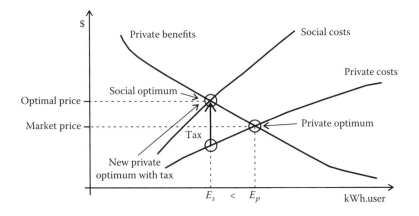

FIGURE 14.1 Optimal and market price of energy in terms of private benefits and social costs.

Electric energy has to be converted from fossil fuels (oil, coal, and natural gas), nuclear power, hydropower (large-scale renewable), and small-scale renewable sources like wind and solar. Nonrenewable sources have a very high-energy density at the expense of generating wastes and polluting by-products. The costs associated with pollution and resource depletion must be taken into consideration when balancing the social costs of energy production with the social benefits.

Figure 14.1 shows the private benefit of a given electric power infrastructure: demand for a kWh of energy consumed by users. The private cost of supply is indicated in the figure with a private optimum at energy E_p. However, there are costs not included under utilities' perspectives: some losses may occur that affect people. Perhaps the landscape will be transformed to make room for a coal thermal plant, or pollution will transform the perceived impact of the environment on people. Accumulated carbon dioxide may increase due to loss of land area, including beaches and wetlands, thinning of species and forest area with heat waves, droughts, and disruption of water supplies. Anything related to climate change is hard to quantify as a social cost but surely represents a curve to be considered to find an optimal social cost for an energy investment. If all private benefits are captured by social benefits, the new intersection E_s represents a new equilibrium point representing a compromise between the need for energy and the desire to keep a resilient ecosystem.

One way to have an ecosystem-friendly market-oriented policy is to raise the price of energy with a pollution tax, as indicated by the arrow in Figure 14.1. A pollution tax looks like an easy way to make consumers buy less of the *pollution source* in compliance with tight regulatory government guidelines. On the other hand, a subsidy could also be implemented to put projects forward that create external benefits. Figure 14.2 shows such a situation where green energy would be fostered by encouraging private investors in small-scale DGs—for example, smaller, more efficient hydropower plants, combining heat-power cycles (cogeneration), and combining agricultural processing machines with local energy production. Provisions for

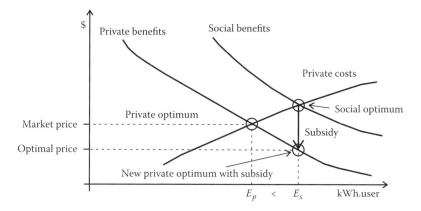

FIGURE 14.2   Effects of subsidies on social optimal market price.

hybrid systems incorporating hydro, solar, and wind installations will make each technology capable of complementing or circumventing the gaps in power generation, physical limitations, or economic efficiencies of the other sources. The arrow in Figure 14.2 indicates a subsidy that lowers the costs through a tax rebate or offers payments for green power investments and the purchase of on-site generated power.

With environmental policies, the government acts to modify market outcomes through taxes, subsidies, and regulations that will consider major economic benefits. For example, hydroelectric power can bring a stable water supply for irrigation and flood control at the cost of flooding existing farmlands and creating distress for wildlife and local communities. Economic theory is generally in favor of removing subsidies. However, there are perceived social benefits from shifting to a renewable fuel economy and subsidizing the development of new renewable and energy efficient technologies.

Policy conditions for a shift to renewable energy include a mix of free market competition and regulation with environmental taxes correcting distortions, temporary subsidies to support the market entry of these renewable energies, and the removal of hidden subsidies to conventional sources.

14.3   WORLD CLIMATE CHANGE RELATED TO POWER GENERATION

The atmosphere is a global public good into which individuals and companies release gases and particulates. Despite two conferences dealing with this issue, at Rio de Janeiro, Brazil, in 1992 and at Kyoto, Japan, in 1997, progress on combating global climate change has been slow.

World consumption of energy since the Industrial Revolution has affected the concentration of greenhouse gases. In addition to increased burning of fossil fuels like coal, oil, and natural gas, emissions of man-made chemicals like chlorofluorocarbons (CFCs), methane, and nitrous oxide from agriculture and industry contribute to the deterioration of the atmosphere.

Primitive man used to burn 200 kcal/day. Energy consumption grew in one million years to almost 250,000 kcal/day per person.[1] This enormous growth of per-capita energy consumption is due to

- Increased use of coal as a source of heat and power in the nineteenth century
- Use of internal combustion engines with massive use of petroleum and its derivatives
- Electricity generation from thermoelectric plants
- Degree of comfort demanded by the modern society

In 1990, world average yearly per-capita energy consumption was about 1.5 TOE (tons of oil equivalent); that is, the average energy consumption was 30,000 kcal/day. With a current population of almost six billion people, this could lead to an overall consumption of approximately 8×10^9 TOE. There is a huge gap between the per-capita energy consumption of the industrialized countries where 25% of the world's people live and the less developed countries where the remaining 75% live. The United States alone, with 6% of the world's population, consumes 35% of the world's energy, contributing proportionally to carbon (CO_2) emitted per unit of energy.

Most economists agree that action is necessary, though the options considered differ drastically depending on government interests. More than 2500 economists, including eight Nobel laureates, endorsed a report for the Intergovernmental Panel on Climate Change in favor of a global approach to climate change, recommending unified environmental, economic, social, and geopolitical decisions. Economic studies have found that there are many potential policies that would reduce greenhouse-gas emissions for which the total benefits outweigh the total costs.

For the United States in particular, sound economic analysis shows that there are policy options that would slow climate change without harming American living standards, and these measures may in fact improve US productivity in the long run. The most efficient approach to slowing climate change is through market-based policies. In order for the world to achieve its climatic objectives at minimum cost, a cooperative approach among nations is required—for example, an international emission trading agreement. The United States and other nations can most efficiently implement their climate policies through market mechanisms, such as carbon taxes or auction of emissions permits. The revenues generated from such policies can effectively be used to reduce the deficit or to lower existing taxes.

Carbon emissions C can be related to energy consumption E and economy measured as gross domestic product GDP and population as shown by Kaya[2]:

$$E = \left(\frac{C}{E}\right)\left(\frac{E}{GDP}\right)\left(\frac{GDP}{P}\right)(P) \qquad (14.1)$$

where
  C/E is the index of carbonization, measured in tons of carbon per TOE
  E/GDP is the energy intensity, the energy needed to produce one unit of GDP, measured in TOE per GDP
  GDP/P is the per capita gross domestic product
  P is the population

The rate of increase of carbon emissions $\Delta C/C$ is given as the sum of four factors:

$$\frac{\Delta C}{C} = \frac{\Delta \dfrac{C}{E}}{\dfrac{C}{E}} + \frac{\Delta \dfrac{E}{GDP}}{\dfrac{E}{GDP}} + \frac{\Delta \dfrac{GDP}{P}}{\dfrac{GDP}{P}} + \frac{\Delta P}{P} \qquad (14.2)$$

14.3.1 Economy of Renewable Sources and Hydrogen

Stabilization of CO_2 can be evaluated by identities (Equations 14.1 and 14.2) suggesting that $\Delta C/C = 0$ could be reached by (1) reducing population growth, (2) reducing per capita income, and (3) reducing the rate of energy intensity. However, the only long-range policy is to actually increase the rate of decarbonization by massive investments in clean energy solutions, renewable or fusion technologies.

Hydrogen can facilitate the transition from fossil fuels to renewable energy sources, playing an essential role in the decarbonization of the global energy system, ultimately improving the outlook with regard to climate change. Hydrogen can provide a major hedge by enabling decentralized micropower plants, vehicles based on efficient fuel cells, integration with solar and wind technologies, and the use of fuel-cell-powered vehicles in homes and rural residences.

Figure 14.3 shows historical data of global share[3] of and estimated evolution toward a hydrogen-based economy. The first step toward a hydrogen-based economy would probably be the use of natural gas as the main fuel that will be reformed to supply hydrogen for fuel cell power plants. By avoiding burning natural gas at its maximum efficiency of 35% through the thermodynamic cycle but converting the

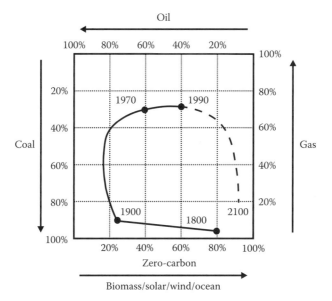

FIGURE 14.3 Evolution toward a hydrogen-based economy.

energy through the electrochemistry cycle at an efficiency of 70%, the reserves of fossil fuels would be extended with a corresponding decrease in carbon emissions.

The second step would be complete generation through renewable primary sources like hydro, wind, solar, and tidal. Although solar is primarily reliant on static converters, induction generators are the major devices for the other sources and could act as flywheel-based storage devices.

A third effort could be concentrated on the direct conversion of electricity from carbon, without burning and tied to biomass systems. Fusion technology, depending on international agreements and regulatory issues, is still a promise that would require massive investments. It has been noted that valuation and cost–benefit analysis can be applied to an environmental economic analysis, providing insights for policymaking. Cline[4] suggested that limiting global warming would require aggressive action, which would appear to have higher costs than benefits for several decades, but a long-term effect with a very low discount rate would sway from *what is better for the economy* to *what is better for the environment*. This means that climate stabilization should be the goal, rather than economic optimization of costs and benefits.

14.3.2 ENERGY VERSUS ENVIRONMENT ECONOMY

Balancing the valuation of human life, intrinsic value of ecosystems, and ethical considerations under the pressures of population and economic growth requires a system perspective.[5] Therefore, energy resources and power conversion systems play an important role in reducing greenhouse emissions. Cogeneration of heat and power improves overall efficiency, and fuel cells can implement a sustainable paradigm integrating economic needs and environmental constraints through distributed generation.

Hydrogen can currently be produced with a diversified mix of technologies: steam reforming of natural gas, gasification of biomass, and efforts for dry reforming using ultra-high-temperature solar reactors, but those technologies are still in the research and development phase. Electrolysis from renewable sources (wind and solar) is already competitive today, and reversible fuel cells capable of combining electrolysis (electrical energy to H_2 and O_2), H_2 fuel cell mode (power from stored gases), and propane/air fuel cell mode (power from propane and air) are in the initial stages of research.

There are two types of measures to address climate change: preventive measures, which tend to lower or mitigate the greenhouse effect, and reactive measures dealing with the consequences of the greenhouse effect and trying to minimize their impact. Preventive measures include the following:

- Reducing emissions of greenhouse gases, either by reducing the level of economic activities responsible for it or by shifting to more energy-efficient technologies that allow the same level of economic activity at lower levels of CO_2 emissions.
- Enhancing greenhouse gas sinks. Since forests recycle CO_2 into oxygen, maintaining forested areas intact and implementing significant campaigns of reforestation would reduce the concentration of CO_2 in the atmosphere.

An economic approach suggests that we should apply cost-effectiveness analysis in considering such policies. This differs from cost–benefit analysis in having a more modest goal. Rather than attempting to decide which policy should be implemented, cost-effectiveness analysis asks what is the most efficient way to reach a policy goal. In general, economists favor approaches that work through market mechanisms to achieve their goals. Market-oriented approaches are considered cost-effective. Rather than attempting to control market actors directly, they shift incentives so that individuals and firms change their behavior to take into account external costs and benefits. Pollution taxes and transferable permits are potentially useful tools for greenhouse gas reduction. Other relevant economic policies include measures to create incentives for the adoption of renewable energy sources and energy-efficient technology.

14.4   APPRAISAL OF INVESTMENT

As a basis for financing a renewable energy technology project, a feasibility analysis is required, particularly due to the need for bank assistance and financial commitments. A scenario must be constructed for this appraisal including the following topics.

- Does a renewable energy system make sense where the grid is present?
- Is there a demand for the project?
- Is there an administrative capacity to carry out the project?
- Is there information about the market?
- Is there a historical study about environmental conservation?
- What are the capital, operating, maintenance, and replacement costs of the renewable energy options in comparison to conventional energy sources?
- What are the advantages of the natural resources to be harnessed?
- Is it possible to obtain sufficient technical data to define the equipment needs?
- What are the technical options and what can they do?
- Does the country (if public sector) or the company (if private sector) have the financial capacity or access to sufficient funds to carry out the project?
- If private sector, how would this investment contribute to overall profit?
- If public sector, how would this project contribute to the economic growth of the country?
- Who benefits and by how much? Who pays and how much? Are there other social objectives to be achieved (stabilization, diversification), and what disadvantages do renewable energy technologies have relative to fossil fuel-based options?

The feasibility analysis can initially be viewed as an internal project appraisal effort. Given a particular load profile and range of available energy sources, the final choice of energy system will be influenced by the initial capital costs, operating costs, and maintenance costs. Technical alternatives should be compared using life-cycle costs, not just initial capital costs. The following factors impact the cash flow and the payback from an induction-generator-based project:

- Electricity demand
- Rate of interest and structure of the loan
- Utilization factor of the generator site
- Incentives for external benefits
- Economic and financial conditions
- Sustainability
- Population growth
- Financing
- Environment (local, regional, global)
- Socioeconomic factors (e.g., employment)
- Security and diversity of fuel supply (import dependency)
- Integration (including decentralized versus centralized supply)
- Purchase price for the generated power
- Tax liability and savings from depreciation

Several factors will determine the interaction of the investor with the utilities. The utility will be avoiding fuel costs for contracted power. Buyback rates reflect the relative costs of providing power by the utility, with power injected into the grid during peak-load periods earning higher rates than power generated in off-peak periods. A meter to record exported and imported energy in the various periods enables accurate billing to occur.

Another approach used by some utilities for small-scale domestic buyback rates is that of *net metering*. In this scheme, the utility pays the same rate for power generated as it charges for power consumed by the consumer. In effect, the meter runs backwards when the system is exporting power to the grid. This is a simple approach, as a new meter is not necessarily required. However, a utility may still install an additional meter to get an indication of the amount of energy exported.

For larger systems, the price paid for exported power is generally set by negotiation. The electricity tariffs charged to consumers are quite complex. These can be broadly split into time-of-use tariffs and general tariffs. The general tariffs base the cost on energy consumption only. A consumer on this tariff may find it advantageous to install a renewable energy source to reduce total energy consumption but still maintain the reliability of the grid system.

Wind energy uses the force of moving air to power generators connected to the shaft of turbines. Electricity is produced from the rotational energy in units consisting of a tower, rotor (with blades, hub, and shaft), generator, control equipment, and power conditioning and protection equipment. A gearbox is usually used to match the generator speed with the natural impressed velocity of the wind. Some sort of braking is used to protect the system against high winds and severe weather conditions. Combining wind power with other sources of generation like photovoltaic and sometimes diesel generators form more reliable hybrid systems.

In the past few years, wind energy has become competitive on a large scale on a cost-per-kilowatt-hour basis. The typical lifetime of a wind generator is assumed

20–25 years for a modern turbine design, during which time capital costs will be amortized. Economies of scale apply to capital costs, though the price increases as the turbine size increases. A rule of thumb for costs is US$1000/kW capacity on average, with the marginal tower cost at US$1600/m. Installation costs include foundations, transportation, road construction, utilities, telephones and other communications, substation, transformer, controls, and cabling. Depending on soil conditions and distance to power lines, a safety margin must be applied. Operation and maintenance costs generally increase throughout the life of a wind power project.

Electricity costs may range from about US$0.008 to US$0.014/kWh generated (although some authors claim even lower numbers), and increase at a rate of about 2.5% per year. The high end (US$0.014/kWh) would include a major overhaul of equipment such as rotor blades and gearboxes, which are subject to a higher rate of wear and tear. Other operating costs include plant monitoring and semiannual inspections.

The potential income from wind energy is based on annual energy production, which in turn is dependent on average annual wind speed. Wind speed varies globally, regionally, and locally following seasonal patterns. The duration and force of the wind is critical to having a lucrative operation of wind turbines to pay the investments on the equipment, maintenance, and operational costs. Large-scale wind turbines require at least an annual average wind speed of 5.8 m/s (13 mph) at 10 m height, while small-scale wind turbines require 4 m/s (9 mph).

Energy production is also a function of turbine reliability and availability. Even though the induction generator is a small fraction of all the required investments, indeed the unit converts mechanical energy from the shaft to electricity. Therefore, even tiny improvements in the efficiency of the induction generator pay off. Better electrical and magnetic design associated with insulation, thermal, and mechanical design will make the energy converter unit extract power that is more reliable. Controllers capable of putting the induction generator in the best operating condition, programming the optimum magnetic flux level, and producing required lagging/leading unit power factor energy contribute to faster amortization and a productive investment.

Hydropower converts the potential and kinetic energy of water conveyed through a pipeline or canal to the turbine into electrical energy. The energy in the water enters at high pressure, rotates the generator shaft, and leaves at lower pressure. Power generation can be amplified by increases in water elevation, river or stream flow, or the size of the watershed. Small hydropower generators are usually run-of-the-river systems that use a dam or weir for water diversion but not for reservoir storage. The amount of pressure at the turbine is determined by the actual flow of the stream or river and the head or height of the water above the turbine.

Controlling the flow of a river can have uses other than power generation including irrigation, flood control, municipal and industrial water supply, and recreation. Those parameters affect the financial return from such projects and sometimes have long-term benefits that may interest local officials. Hydropower is characterized by extremely high up-front costs, low operating costs, and long life cycles. This makes projects sensitive to financial variables, construction timing, bank interests, and discount rates.

Hydro projects usually have long lead times, typically taking 10 years from analysis to deployment. Such projects have acerbic requirements on resource assessment and environmental and social considerations such as water use, displacement of homes, and distance to transmission lines. Small power plants on the order of 10–200 kW for rural applications in distributed energy systems do not have the same requirements as larger projects (greater than 1 MW). However, those small-scale applications for a target cost of $2000/kW will need a 35% investment in civil engineering, 40% for plant installations and commission, 7% for overall design and management, 8% for electrical engineering, and contingencies on the order of 10%. The induction generator may need a peak power tracking control, a dummy load, or any type of storage system and pumping-back schemes are sometimes used in conjunction of water flow needs.

A cost–benefit analysis can help in the construction of a list of economic, environmental, and social indicators. Economic figures are naturally quantified, but social and environmental indicators are generally hidden impacts and may be viewed either as external costs to the environment, local or global, or as external benefits, for example, job creation. An appraisal must be undertaken in order to provide a basis for selection or rejection of projects by ranking them in order of profitability or social and environmental benefits and ensuring that investments are not made in projects that earn less than the cost of capital, generally expressed as a minimum rate of return.

Cost comparisons in industry often assume that the annual cost of capital is a fixed percentage of the investment. However, a long-term renewable energy project needs a cash flow analysis to assess the difference between generated revenue and ongoing expenses. Some of the analytical tools that can be used are as follows:

- Benefit–cost ratio (BCR)
- Net present value (or discounted cash flow)
- Internal rate of return
- Payback period
- Least cost analysis
- Sensitivity analysis

14.4.1 BENEFIT–COST RATIO

The BCR is the ratio of discounted total benefits to total costs. For a project to be acceptable, the ratio must have a value of 1 or greater. Among mutually exclusive projects, the rule is to choose the project with the highest BCR. The disadvantages of the BCR are that it is typically sensitive to the choice of discount rate and can provide incorrect analysis if differences in size or scale of the various projects being compared are great.

14.4.2 NET PRESENT VALUE (OR DISCOUNTED CASH FLOW)

The net present value (NPV) approach (also referred to as the discounted cash flow approach) measures the present value of money exclusive of inflation. NPV uses the time value of money to convert a stream of annual cash flow generated by a project to a single value at a chosen discount rate. This approach also allows one to incorporate income tax implications and other cash flows that may vary from year to year. The discounted cash flow or net present value method takes a spread of cash flows over a period and discounts the cash flows to yield the cumulative present value. When comparing alternative investment opportunities, the project with the highest cumulative NPV is the most attractive one. The only serious limitation with this approach is that it should not be used to compare projects with unequal time spans.

14.4.3 INTERNAL RATE OF RETURN

The internal rate of return (IRR) and net present value approaches are very similar. As outlined, the NPV determines today's values of a future cash flow at a given discount rate. On the other hand, the IRR approach seeks to determine the discount rate (or interest rate) at which the cumulative net present value of the project is equal to zero. This means that the cumulative NPV of all project costs would exactly equal the cumulative NPV of all project benefits if both were discounted at the internal rate of return. In the private sector, this computed financial internal rate of return (FIRR) is compared with the company's actual cost of capital. If the FIRR exceeds the company's cost of capital, the project is considered financially attractive. The higher the IRR compared with the cost of capital, the more attractive the project. On the other hand, if the IRR is less than the company's cost of capital, then the project is not considered financially attractive.

For projects financed in whole or in part by the public sector, the discounted cash flow may need to be adjusted to account for social benefits or economic distortions such as taxes and subsidies, economic premiums for foreign exchange earnings that accrue from the project, or employment benefits. The resulting statistic would be the economic internal rate of return (EIRR) and would be compared with the country's social opportunity cost of capital. If the EIRR exceeds the social opportunity cost of capital, the project would provide economic benefits to the society.

14.4.4 PAYBACK PERIOD

Payback period is the easiest and most basic measure of financial attractiveness of a project. The payback period reflects the length of time required for a project's cumulative revenues to return its investment through the annual (nondiscounted) cash flow. A more attractive investment is one with a shorter payback period. In development settings, however, there is little reason to assume that projects with short payback periods are superior investments. In addition, the criterion has a bias against long gestation projects such as renewable energy.

14.4.5 LEAST-COST ANALYSIS

The least-cost analysis method is used to determine the most efficient (least costly) way to perform a given task, reach a specified objective, or obtain a set of benefits measured in terms other than money. For example, the objective might be to supply a fixed quantity of potable drinking water to a village. The alternatives might be wind pumping, run-of-river off-take, or impoundment. One would calculate all costs, capital and recurrent, to achieve the objective, apply economic adjustments, and discount the resulting stream of costs for each alternative examined. The one with the lowest NPV would be the most efficient one.

14.4.6 Sensitivity Analysis

Sensitivity analysis refers to the testing of key variables in the cash flow pro forma balance sheet to determine the sensitivity of the project's NPV to changes in these variables. For example, in a renewable energy project proposal, one may increase fuel costs or fuel transport costs, remove import restrictions on solar panels, lower labor costs, and increase land acquisition at different rates to determine the corresponding impact on the NPV. It is useful to test in the cash flow pro forma a variable that appears to offer significant risk or probability of occurring. The analysis becomes another useful tool when combined with others to improve the decision-making process.

While each index provides data needed to make efficient decisions, multicriteria analysis (combining one or more tools with other project data and benefits) can be helpful in evaluating future financial performance. For example, other criteria might include the distribution of benefits, ease and speed of implementation, and replicability that might be combined with one of the quantifiable indices illustrated earlier.

14.5 CONCEPT SELECTION AND OPTIMIZATION OF INVESTMENT

In induction-generator-based renewable energy systems, the energy source is typically a low-speed prime mover: hydropower or wind power. The investments required for such power plants are basically in three areas: (1) a power generation section, consisting of the turbine, gear box, generator, and all structural elements to support them; (2) a protective and control operational center capable of withstanding environmental and operational hazards; and (3) a utility power interface section, tailoring the power output to the requirements of the electrical grid.

One important economic concept that is used to support capital investment analysis is the capacity factor (CF). The CF measures the operational hours, seasonal constraints, and source and demand fluctuations that limit the full utilization of the installed power plant by dividing the actual energy produced during a specified period by the amount of the energy that would have been produced under full power:

$$CF = \frac{\text{Actual energy}}{\text{Energy at full use}} \qquad (14.3)$$

A wind turbine rated at 100 kW would produce 876.00 kWh of electricity per year if operating 100% of the time. However, wind velocity fluctuation, maintenance down time, and even demand matching will constrain the annual output power to a lower value found by multiplying CF by the energy at 100% use. For average production, the average input must be computed, and the given average shaft velocity would determine a gearbox that optimizes the power transference. However, the operational generator speed would oscillate about this optimum point, and the electronic system must compensate for such deviation. Therefore, there is a typical cost of energy generated by the system that must be taken into consideration for the amortization of the capital, operation, and maintenance. Since hydropower and wind power systems do not require fuel for operation, the following equation suggested by Ramakumar and Hughes[6] expresses the cost of energy, neglecting taxes, surcharges, and insurance:

$$C = \left[\frac{r(1+r)^n}{(1+r)^n - 1} + m \right] \frac{P}{87.6(CF)} \tag{14.4}$$

where
C is the generation cost in US cents per kilowatt hour
CF is the capacity factor
m is the fraction of the capital costs needed per year for operation and maintenance of the unit
n is the amortization period in years
P is the capital cost in US dollars per kilowatt
r is the annual interest rate per unit (0.01 times the annual percentage rate)

If the induction generator operates with a hybrid system, for example, diesel, synthetic fuels from biomass, gas, or fuel cell, there is another term to be added to Equation 14.4, shown in Equation 14.5, where F is the fuel input per volume and D is the fuel cost in US cents per volume:

$$C = \left[\frac{r(1+r)^n}{(1+r)^n - 1} + m \right] \frac{P}{87.6(CF)} + FD \tag{14.5}$$

Multiple connections of renewable energy sources will have an average cost given by Equation 14.6:

$$C_{avg} = \frac{\sum_i \left[\frac{r(1+r)^n}{(1+r)^n - 1} + m \right] P_i R_i}{(87.6) \sum_i (R_i CF_i)} \tag{14.6}$$

where
C_{avg} is the average generation cost in US cents per kilowatt hour
i is the index related to the device
CF_i is the capacity factor for the ith device
m_i is the operation and maintenance of the ith device
n_i is the amortization period in years for the ith device
P_i is the capital cost in US dollars per kilowatt for the ith device
r is the annual interest rate per unit (0.01 times the annual percentage rate)
R_i is the rating in kilowatts of the ith device

If the amortization period and the operation and maintenance charge rates are taken to be the same for all the devices, then Equation 14.7 applies:

$$C_{avg} = \frac{T_{cr}I}{(87.6)R_{eq}} \qquad (14.7)$$

where

I is the total investment in US dollars $= \sum_i P_i R_i$

R_{eq} is the equivalent continuous rating in kilowatts $= \sum_i R_i CF_i$

T_{cr} is the total charge rate $= \dfrac{r(1+r)^n}{(1+r)^n - 1} + m$

If the assumption of the same amortization period, operation and maintenance charge rates for all is not valid, a conservative approach is to use the smallest n_i and the largest m_i for both n and m in the aforementioned equations.

14.6   FUTURE DIRECTIONS

With the present worldwide situation as volatile as it is, the need to allocate capital efficiently has become more apparent for companies to both preserve good credit ratings and invest in growth opportunities at the same time. There have been some educational efforts related to energy decision making.[7] Risk-adjusted return on capital approaches[8] and chance-constrained programming[9] have been introduced recently where the investor might wish to obtain the highest possible expected return but also protect against too risky projects. The accepted risk level is based on the decision maker's risk aversion. The extent to which violations are permitted to happen is a matter of policy—for example, how far will the government allow companies to go above the environmental tax limitation? Such recent approaches formulate an optimization problem with probabilistic constraints and seem suitable for economical evaluation of renewable energy systems.

14.7   SOLVED PROBLEMS

1. Create a flowchart to be used by a decision maker to determine his or her investment in a renewable energy project.

   Solution

   Figure 14.4 shows a flowchart with the decision making methodology.

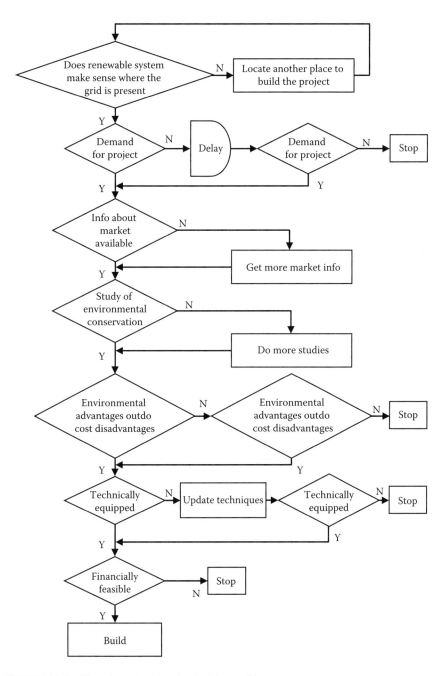

FIGURE 14.4 Flowchart showing the decision-making process.

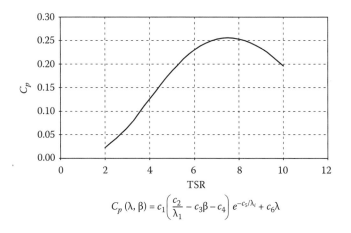

$$C_p\,(\lambda,\,\beta) = c_1\!\left(\frac{c_2}{\lambda_1} - c_3\beta - c_4\right) e^{-c_5/\lambda_i} + c_6\lambda$$

FIGURE 14.5   Turbine aerodynamic power coefficient.

2. If a site has an average wind speed of 10 m/s and turbines with a diameter of 12 m are to be installed, what is the capacity factor for 10 turbines rated for 14 m/s? Plot a seasonal power generation graph and discuss ways to compensate for fluctuations.

Solution

Figure 14.5 shows the turbine aerodynamic power coefficient which is used to calculate the following wind power equation:

$$P_{wind} = \frac{\rho\pi R^2 V^3 C_p}{2} = 0.5\rho A V^3 C_p$$

where
   ρ is the mass density of the air
   A is the area swept by the rotor blades
   V is the wind speed
   C_p is the nondimensional turbine power coefficient

$$\text{Capacity factor} = \frac{0.5\cdot 10\cdot\rho\cdot A\cdot(10)^3\cdot(C_p)_1}{0.5\cdot 10\cdot\rho\cdot A\cdot(14)^3\cdot(C_p)_2} = 0.36\cdot\left[\frac{(C_p)_1}{(C_p)_2}\right]$$

where
   $(C_p)_1$ is the turbine power coefficient for wind speed at 10 m/s
   $(C_p)_2$ is the turbine power coefficient for wind speed at 14 m/s

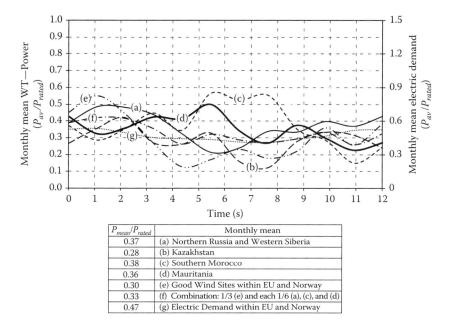

P_{mean}/P_{rated}	Monthly mean
0.37	(a) Northern Russia and Western Siberia
0.28	(b) Kazakhstan
0.38	(c) Southern Morocco
0.36	(d) Mauritania
0.30	(e) Good Wind Sites within EU and Norway
0.33	(f) Combination: 1/3 (e) and each 1/6 (a), (c), and (d)
0.47	(g) Electric Demand within EU and Norway

FIGURE 14.6 Seasonal wind power variation for some different regions. (Figure taken from Czisch, G. and Ernst, B., High wind power penetration by the systematic use of smoothing effects within huge catchment areas shown in a European example, Windpower 2001, American Wind Energy Association (AWEA), Washington, DC, 2001.)

Seasonal Variation

In general the wind power is greater than the power demand but there are seasons where such scenario is reversed; Figure 14.6 shows seasonal wind power variation for some different regions. A simple way to compensate this is to use a combined system of wind-hydro storage. In such system, the excess wind power is used to pump water to the storage at high place. When demand is higher than the wind power, this stored water can be utilized to run a turbine to generate power (Figure 14.6).

3. Neglecting taxes, surcharges, and insurance, what would be the capital cost of energy over 20 years of a wind power system whose generation cost is $0.01/kWh, at an annual interest rate of 0.075, annual maintenance and operation costs of 2.5% of the capital cost needed, and a capacity factor of 0.5? Extend this cost to five identical units under similar overall conditions.

Solution

(a) From Equation 14.4:

$$C = \left[\frac{r(1+r)^n}{(1+r)^n - 1} + m \right] \frac{P}{87.6(CF)}$$

where

C is the generation cost in US cents per kilowatt hour

CF is the capacity factor

m is the fraction of the capital costs needed per year for operation and maintenance of the unit

n is the amortization period in years

P is the capital cost in US dollars per kilowatt

r is the annual interest rate per unit, (0.01 times the annual percentage rate)

Thus, the capital cost P can be calculated as

$$n = 20$$

$$C = 1$$

$$r = 0.075$$

$$m = 0.025$$

$$CF = 0.5$$

$$P = \frac{1*87.6*0.5}{\left[\dfrac{0.075*(1.075)^{20}}{(1.075)^{20}-1} + 0.025 \right]} = 355.8\$/kW$$

(b) From Equation 14.7, one has

$$C_{avg} = \frac{T_{cr}I}{87.6*R_{eq}}$$

where

$$T_{cr} = \frac{r*(1+r)^{n}}{(1+r)^{n}-1} + m = 0.123$$

$$R_{eq} = \sum_{i=1}^{5} R_i * CF_i = 0.5 \sum_{i=1}^{5} R_i$$

$$I = \sum_{i=1}^{5} R_i * P_i = P \sum_{i=1}^{5} R_i$$

Therefore, for such a case, P will remain same, that is, $P = 355.8\$/kW$.

14.8 SUGGESTED PROBLEMS

14.1 Compare the decision to install a small power (up to 20 kW) synchronous generator with an induction generator. In this analysis, you should include initial purchase price, cost of repairs and maintenance, operational costs, overall efficiency, product availability, and unit replacement.

14.2 Implement the cost equation (Equation 14.4) with a spreadsheet program. Make WHAT-IF simulations for several economic scenarios.

14.3 For a three-phase induction machine on the range of 10–50 hp, find the typical power factor and calculate the required self-excitation capacitances in order to build a self-excited induction generator isolated system. Prepare a scenario for studying cost of capacitance versus cost of machine for this range. Extend the machine rated power further to 100 hp or beyond 200 hp and study which capacitances are required and their required cost. Propose a break even decision that makes such self-excitation capacitance still possible. Now, consider the use of a three-phase inverter with a battery on the dc-link to provide VARs for an isolated induction generator and compare to the previous studies; suggest for which application range it is recommended the use of excitation capacitances versus using a three-phase inverter for providing excitation for the machine.

REFERENCES

1. Goldemberg, J., *Energy, Environment and Development*, Earthscan Publications, London, U.K., 1996.
2. Kaya, Y., *Impact of Carbon Dioxide Emission Control on GNP Growth: Interpretation of Proposed Scenarios*, IPCC Energy and Industry Subgroup, Hampshire, U.K., 1990.
3. Nakicenovic, N., Grübler, A., and McDonald, A., *Global Energy Perspectives*, Cambridge University Press, Cambridge, U.K., 1998.
4. Cline, W.R., *The Economics of Global Warming*, Institute for International Economics, NW, Washington, DC, 1992.
5. Harris, J.M. and Codur, A.M., Sustainable production in agriculture through soil care management: an enigma of environmental and ecological paradigm, *Int. J. Sust. Dev. Green Eco.*, 1, 1, 2012.
6. Ramakumar, R. and Hughes, W.L., Renewable energy sources and rural development in developing countries, *IEEE Trans. Educ.*, E-24(3), 242–251, August 1981.
7. Shaten, R.J., *The Energy Resources and Economics Workbook*, Society for Energy Education, University of Virginia School of Medicine, VA, June 2003.
8. Kaminker, Ch. and Stewart, F. The role of institutional investors in financing clean energy, OECD Working Papers on Finance, Insurance and Private Pensions, No.23, OECD Publishing, 2012.
9. Gurgur, C. and Luxhoj, J.T., Application of chance-constrained programming to capital rationing problems with asymmetrically distributed cash flows and available budget, *Eng. Econ.*, 48(3), 241, 2003.
10. Czisch, G. and Ernst, B., High wind power penetration by the systematic use of smoothing effects within huge catchment areas shown in a European example, Windpower 2001, American Wind Energy Association (AWEA), Washington, DC, 2001.

Appendix A: Introduction to Fuzzy Logic

Fuzzy logic has been making a name for itself as the most successful application of artificial intelligence. It has advanced significantly in recent years and has found widespread applications. It has also caught the attention of the public in the past few years. Now, people ask what it is, mathematicians study how it works, engineers work on what it does, and marketers wonder what it can do. Engineering applications generally use fuzzy logic for controllers in noisy, nonlinear, and time-variant systems, tuning problems, and estimation of multiple input/output relations. However, the range of applications is amazing. Fuzzy systems can be integrated with neural networks to extract hidden rules out of numerical data. Fuzzy reasoning can be used to model events in politics, military planning, and appraisal processes. There are fuzzy computers being constructed. The topic is so vast that this book can only introduce some concepts.

The theoretical foundations of fuzzy logic were laid in 1965 when Prof. LotfiZadeh published the paper "Fuzzy Sets," in the journal *Information and Control*; fuzzy set theory was born, but no one paid much attention to it. After all, the term *fuzzy* had (and still has) a sort of negative connotation in English. Before the publication of this paper, Zadeh was already well known in the area of linear systems theory for his contributions to the analysis of discrete systems. However, he believed that as the complexity of a system increases, the ability to make precise and significant statements about its behavior diminishes. In his words, "The closer one looks at a real-world problem, the fuzzier becomes its solution."

Linear systems theory assumes several properties that are not actually found in real applications. Therefore, conventional methods of control are good for simple systems, while fuzzy systems are suitable for complex problems or in applications that involve human descriptive or intuitive thinking. A fuzzy system can estimate input–output functions and can be trainable for pattern recognition applications and dynamic systems control. Unlike statistical tools and conventional (continuous or binary) logic, it can estimate functions without a mathematical model of the system. Therefore, fuzzy systems are model-free estimators able to *learn from experience* with a linguistic description of the system's operation.

Mamdani reported the first application of fuzzy logic for control of a dynamic process in a steam engine. The problem was to regulate the engine speed and boiler steam pressure by means of the heat applied to the boiler and the throttle setting of the engine. Since the system was nonlinear, noisy, and strongly coupled, and no mathematical model was available, the fuzzy control was designed from the operator's experience and performed much better than expected. Following this work, Mamdani tried unsuccessfully to secure funding for his research from the British granting agencies. Unable to obtain support, he ultimately abandoned this line of investigation. However, his work did not go unnoticed in Japan.

Ten years later, the Sendai Subway Automatic Train Operations Controller captured the attention of control engineers around the world. The Hitachi team designed, developed, and compared fuzzy and conventional PID-type controllers in 300,000 simulations tests, as well as 3,000 riderless subway runs with real hardware. Strategies used by experienced train operators were implemented in fuzzy rules that performed a *predictive fuzzy control*. Speed control during cruising, braking control near station zones, and switching of control were determined by fuzzy rules that processed sensor measurements and considered factors such as the comfort and safety of travelers. In operation since 1986, this controller has reduced the stopping distance by 2.5 times, doubled the comfort index, and reduced power consumption by 10%.

While Western scientists and engineers ignored or attacked fuzzy logic during the 1970s and 1980s, Japanese companies applied fuzzy technology to air conditioners; auto engines, brakes, transmissions, and cruise control; cameras and camcorders; washing machines; dryers; and elevators. By the time the first US fuzzy conference was held in Austin, Texas, in June 1991 (at Microelectronics and Computer Technology Corp.), the Japanese had already held international meetings and established two fuzzy research centers: Laboratory for International Fuzzy Engineering (LIFE) and Fuzzy Logic Systems Institute (FLSI).

A.1 NATURAL VAGUENESS IN EXPERIMENTAL OBSERVATIONS

Most everyday situations and observations are imprecise, carrying a certain degree of fuzziness in the description of their nature. This vagueness may be associated with variables like position, color, texture, shape, or even the language used to describe a phenomenon. Very often, the same concept will have different meanings in different contexts or at different times. A *hot* day in winter is not the same as a *hot* day in summer. A car going *fast* on a highway does not have the same speed as a car going *fast* in a parking lot. Such imprecision or fuzziness associated with descriptions and observations of natural events is common in all fields of study addressed by scientists, philosophers, and business analysts. We are able to formulate ideas, make decisions and recognize concepts that have a high level of vagueness. Consider the following statements as examples:

- The room is hot.
- Mike is quite tall and Mary somewhat short.
- My son said that I am old.
- My grandmother said that I am young.
- That computer's price is around twice the competitor's price.
- You should buy a less expensive tire, which will last as long as a brand name tire.

These propositions are incompatible with traditional computer-based modeling and information systems, which require precise numerical data. A common-sense statement like "When the engine temperature is hot, increase the fan speed" can be readily understood by a fuzzy system, even if there is no exact threshold point where the engine is considered warm or hot. A fuzzy system can evaluate descriptions like this one and make decisions for necessary actions. The power of fuzzy logic is to bridge the gap between machine systems and human nature.

Conventional computers and digital systems follow either–or (Boolean) crisp logic where the answer is always yes or no. Take Figure A.1a for example. Is this a bowl of apples? The answer is yes. How about Figure A.1b? Is it a bowl of apples? The answer in this case is no, since it is a bowl of oranges. This is an example of crisp logic in which the bowl contains either apples or oranges.

Now, take the case of the bowl in Figure A.1c where someone has swapped an apple for one of the oranges; is it a bowl of apples? The answer is "It is a bowl of mostly apples." If we keep switching the apples and oranges and asking the same question, there will be a situation where the bowl will contain half apples and half oranges as shown in Figure A.1d. Such a bowl is neither a bowl of apples nor a bowl of oranges. This is the fundamental difference between crisp logic and fuzzy logic: the violation of the Law of No Contradiction where the set A must not contain the set not-A. Here is a situation where the bowl belongs to both the set of bowls of oranges and the set of bowls of apples.

In conventional logic, number 1 can indicate that an object belongs to a set, while 0 can indicate that the object is not a member of a set. On the other hand, in fuzzy logic, there are degrees of membership that vary from 0 to 1. To see how this definition is used in constructing an actual fuzzy set, let us consider the concept of *adult*.

We usually agree that somebody who is 14 years old is barely if at all an adult, while a person about 25 years old might be already established in some profession, be independent, or be married and could be considered an adult. In classical set theory, we are forced to choose an arbitrary cutoff point, say, 18 years old. Since the boundaries

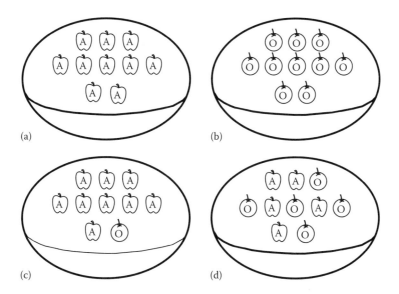

FIGURE A.1   Is each one of these a bowl of apples?

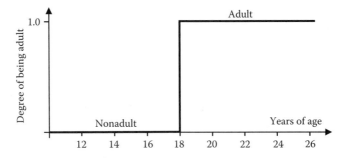

FIGURE A.2   Boundary between adult and no adult definitions.

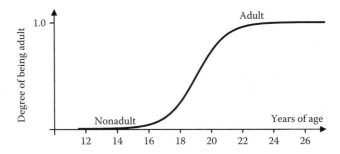

FIGURE A.3   Various degrees of being adult.

between what is in a set and what is outside a set are sharp, we can draw an imaginary line that splits adult from no adult on the scale of years as shown in Figure A.2.

We can draw different lines at different ages, sometimes without a good reason for them. The sale of alcoholic beverages is usually allowed to people 21 and older. Car rental companies might rent cars only to people over 25. The common sense is that the truth of the concept of being an adult grows with age. Therefore, the fuzzy principle views adult as a curve, as depicted in Figure A.3, and not as a sharp line as indicated in Figure A.2. Such a curve that relates how much something belongs to a fuzzy set is called a membership function.

Fuzziness should not be confused with probability. The two can occur together, but they are different concepts. Fuzziness describes the ambiguous or vague case, whereas probability assigns a degree of certainty to a statement. With probabilities, an object either is or is not something; it cannot be both. For example, tomorrow it will either rain or not rain. The uncertainty lies in which of the two possible outcomes will occur. However, the personal perception of the rain is described by their fuzziness: drizzle, light rain, moderate rain, or downpour.

A.2   FUZZY LOGIC OPERATIONS

People intuitively use logical operations every day. A typical thought before leaving the office in the afternoon could be "If I do not get stuck in traffic today and it's not raining, I will play a little tennis." The connective *and* assumes the truth of both statements, *not* stuck and *not* raining, to conclude that a tennis game will be possible. Logic operations like *and*, *or*, and *not* originated in Aristotle's syllogistic logic and came of the age when George Boole published *An Investigation of the Laws of Thought on which are Founded the Mathematical Theories of Logic and Probabilities* in 1854. Boole believed that the processes of human reasoning could be reduced to an algebra or other formal systems, modeled after two valued (true/false, 1/0) logic. Although his vision was never verified, he furnished the conceptual basis for a formal system in which logical propositions could be represented and manipulated like algebraic symbols, setting the stage for the invention of computers a century later.

Of course, the missing link was that human reasoning is never restricted to yes/ no, true/false, and black/white answers. There are degrees of truth and shades of gray that are not handled by crisp logic. Therefore, Boolean logic operations must be extended to fuzzy logic to manage the notion of partial truth, truth-values between *completely true* and *completely false*.

The fuzzy nature of a statement like "X is low" was presented earlier. Suppose now that there are two fuzzy sets, low and high, where a typical logical operation could be given as

X is low *and* Y is high

What is the truth-value of the *and* operation? The logic operations with fuzzy sets are performed with the membership functions. Although some researchers have explored several interpretations for fuzzy logic operations, the following definitions are the most convenient:

truth(X *and* Y) = Min(truth(X), truth(Y))
truth(X *or* Y) = Max(truth(X), truth(Y))
truth(*not* X) = 1 − truth(X)

The *and* operation between two fuzzy sets takes the minimum between the membership functions of the two fuzzy operands; *or* takes the maximum between the membership functions of the two fuzzy operands, and *not* is performed by subtracting the membership function from 1. Note that if we plug the values 0 and 1 into these definitions, we get the same truth table as expected from conventional Boolean logic. This is known as the extension principle, which states that the classical results of Boolean logic are recovered from fuzzy logic when all fuzzy membership grades are restricted to the traditional set (0, 1). This effectively establishes fuzzy sets logic as a true generalization of classical set theory logic, so there is no conflict between fuzzy and crisp methods.

To illustrate how the fuzzy logic operations work, let us assume an air-conditioner vent. Suppose the vent has blades that control the openings and can be moved so that the blades tilt downward or upward. This blade angle sends the airflow toward the floor or the ceiling. Figure A.4a and b show the two fuzzy sets (downward and upward) that describe the position of the vent blades. If the blades are totally rotated to −45° with respect to the horizontal, then they are downward. On the other hand, if the blades are totally rotated to +45°, they are completely upward.

The range of the input variable (angle) is called the universe of discourse, which could be conveniently scaled to per-unit values (values between −1 and +1)

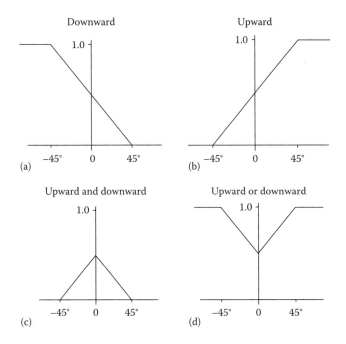

FIGURE A.4   (a) Fuzzy set downward, (b) fuzzy set upward, (c) fuzzy set upward *and* downward, (d) fuzzy set upward *or* downward.

by dividing the input angle by the maximum value of the range. Any position in between has degrees of being downward or upward. The *and* operation of the two fuzzy sets (upward *and* downward) is shown in Figure A.4c. The *and* operation is the minimum between the two membership functions. Figure A.4d shows the *or* operation between the two fuzzy sets (upward *or* downward). It is easily seen that upward = *not* downward, since the membership function of upward is equal to the subtraction of the membership function of downward from the value 1.

A.3   RULES OF INFERENCE

Statements assert facts or states of affairs; they give descriptions that can be organized in rules of reasoning. Several philosophers, mainly Descartes, Newton, Galileo, and Bacon, developed the application of mathematical descriptions and the use of logical rules for formulating hypotheses. They were influenced by the earlier syllogistic logic, in which premises were manipulated to produce true conclusions. A typical syllogistic rule of inference that goes by the Latin name *modus ponens* (affirmative mode) can be given as

If A is true
And A implies B
Then B is true

where the connectives *and*, *or*, and *not* are essential to derive the truth of these rules.

In Boolean logic, the laws of thought are expressed by the rules of operations based on symbols, which are based on the algebra of the numbers 0 and 1. However, there are real-world situations where the statements given by rules are only partially true; they are *fuzzy*. A fuzzy rule applies common sense and helps to get the knowledge out of the system's operation. As an example, let us assume two cars traveling at the same speed along a straight road. The distance between the cars becomes one of the factors that lead the second driver to brake his or her car to avoid collision. The second driver might use the following rules.

Rule 1: *If* the distance between two cars is medium *and* the speed of the car is medium,
Then brake medium for speed reduction.
Rule 2: *If* the distance between two cars is small *and* the speed of the car is medium,
Then brake hard for speed reduction.

Judging how short the distance is and how fast is the speed may involve factors such as weather conditions, day or night, and the driver's fatigue. Therefore, rules like these have a degree of truth with respect to other related statements. A set of statements is known as a rule base, and the process that combines all the statements of a rule base to produce a meaningful conclusion is called inference. Although concepts such as height, speed, age, or temperature are fuzzy in nature, an actual number indicates how many feet, miles/hour, years, or degrees are related to the variable that is being measured or controlled. Thus, the input information for a fuzzy system needs to be initially converted into the correspondent fuzzy sets. This is done by the *fuzzification* process that is the computation of the degree of membership of the input variables to the given fuzzy sets.

Figure A.5 shows how the two rules mentioned earlier are processed into a fuzzy system. The top of the figure shows the evaluation of Rule 1, where the distance D between the two cars is medium, the speed S is medium, and the braking action should be medium. The *and* operation between the distance and the speed is performed by selecting the minimum truth-value (μ_1). The output fuzzy set of Rule 1 is cut off at the strength indicated by the rule firing (height μ_1), turning the triangle into a trapezoid shape. The bottom of the figure shows the evaluation of Rule 2, where the distance D is small, the speed of the car is medium, and the braking action should be hard. Again, the lower membership function (μ_2) is selected to perform the *and* operation for Rule 2 giving an output fuzzy set for Rule 2 with height μ_2, where the area above the height μ_2 is lopped off.

Each rule in the rule base is fired in accordance with the truth-value for the fuzzy logic operation of the rule. Since any rule can be fired, there is an *or* configuration for such input information. This *or* operation is performed at each output fuzzy subset by taking the maximum of the membership functions. The bottom part of Figure A.5 shows the construction of the total output fuzzy set. The final fuzzy set output has to be converted to a crisp value in order to serve an output to an external interface. In the figure, it is done by the centroid method, which takes the center of the gravity of the geometric figure as the final crisp value.

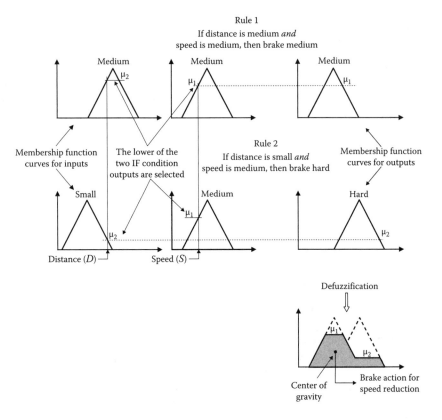

FIGURE A.5 Fuzzy inference for two rules that relate speed, distance, and brake action for a vehicle in a highway.

A.4 FUZZY SYSTEMS

Most commercial fuzzy products are rule-based systems that receive current information in a feedback loop from a device as it operates and controls the operation of a mechanical or other device. A fuzzy logic system has four blocks as shown in Figure A.6. Crisp input information from the device is converted into fuzzy values for each input fuzzy set with the fuzzification block. The universe of discourse of the input variables determines the required scaling for correct per-unit operation.

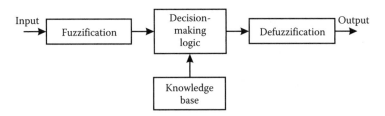

FIGURE A.6 Block diagram of a fuzzy system.

Scaling is very important because the fuzzy system can be retrofitted with other devices or ranges of operation by just changing the scaling of the input and output. The decision-making logic determines how the fuzzy logic operations are performed (Sup-Min inference) and, together with the knowledge base, determine the outputs of each fuzzy *if–then* rule. Those are combined and converted to crisp values with the defuzzification block. The output crisp value can be calculated by the center of gravity as follows:

$$I = \frac{\int U \cdot \mu(U) dU}{\int \mu(U) dU} \tag{A.1}$$

or the weighted average (height method of defuzzification) can be used:

$$I = \frac{\sum_{i=1}^{n} U_i \cdot \mu(U_i)}{\sum_{i=1}^{n} \mu(U_i)} \tag{A.2}$$

The following six steps are involved in the creation of a rule-based fuzzy system:

1. Identify the inputs and their ranges and name them.
2. Identify the outputs and their ranges and name them.
3. Create the degree of fuzzy membership function for each input and output.
4. Construct the rule base under which the system will operate.
5. Decide how the action will be executed by assigning strengths to the rules.
6. Combine the rules and defuzzify the output.

A.5 EXAMPLE OF FUZZY LOGIC CONTROL FOR AIR-CONDITIONING SYSTEMS

An ordinary air conditioner has a thermostat that takes a temperature setting like 20°C and turns the cooling system on or off to keep the temperature within a range, say, 17°C–23°C. The currents rushing in and the time delay in bringing down the temperature after the system turns on make such a rigid system dramatically ineffi-cient. Mitsubishi Electric Corp. has developed a room air-conditioning system where the controller allows degrees of operation from on to off for optimal operation. The company reports that this improves the room's temperature variation, is three times as stable as a crisp system, and provides a 24% power savings. The temperature sen-sor can detect whether the room is occupied and can direct the air upward if people are present or downward if the room is empty. Mitsubishi's fuzzy air conditioner can also learn the room's characteristics and fine-tune its own operation.

The following discussion is a simplified implementation of an air-conditioning system with a temperature sensor. The temperature is acquired by a microprocessor that has a fuzzy algorithm to process an output to continuously control the speed of

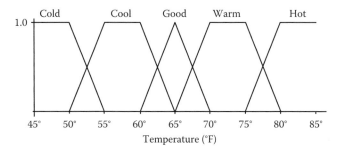

FIGURE A.7   Fuzzy sets for temperature.

a motor that keeps the room at a *good* temperature. It also directs the vent upward or downward as necessary. Figure A.7 illustrates the process of fuzzification of the air temperature. There are five fuzzy sets for temperature: cold, cool, good, warm, and hot.

The membership function for fuzzy sets cool and warm are trapezoidal, the membership function for good is triangular, and the membership functions for cold and hot are half-triangular with shoulders indicating the physical limits of the process. (Staying in a room with a temperature lower than 45°F or above 85°F would be quite uncomfortable.) The way to design such fuzzy sets is a matter of degree and depends solely on the designer's experience and intuition. Most probably, an Eskimo and an Equatorian would draw very different membership functions for such fuzzy sets!

To build the rules that will control the motor, we could watch how a human expert would adjust the settings to speed up and slow down the motor in response to the temperature, obtaining the rules empirically. If the room temperature is good, keep the motor speed medium. If it is warm, turn the knob of the speed to fast and blast the speed. On the other hand, if the temperature is cool, slow down the speed. Stop the motor if it is cold. This is the beauty of fuzzy logic: to turn common sense, linguistic descriptions into a computer-controlled system.

Figure A.8 shows how the five fuzzy sets of temperature are related to the five fuzzy sets of motor speed. The input–output pairs are in the regions defined by the intersections, which are defined by the fuzzy rules. The clusters are patches in the geometric plan of temperature–speed. They relate the given information (temperature) with the required action (machine speed set point). More rules give more clusters, leading to more fuzzy sets and more accuracy.

A fuzzy system can have two or more inputs that are related with *and, or,* and *not* operations. It can also have more than one output. This air-conditioning system has an extra output that controls the angle of the blade of a vent. It is known that as the temperature gets hotter, the vent should be directed upward, because the blown cold air would be better distributed in an occupied room. As the temperature gets colder or the room less occupied, the blown air is somewhat warmer. Since the natural convection makes the air go up, the vent should be direct downward. Consequently, Table A.1 gives a complete rule base for such a system.

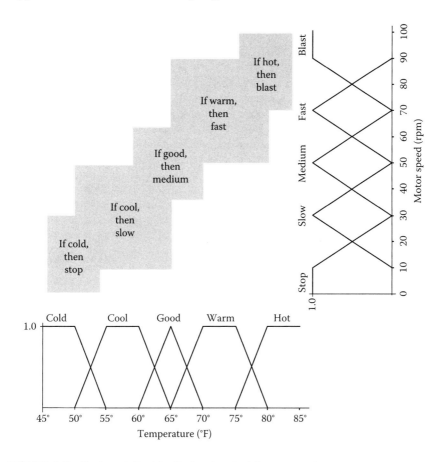

FIGURE A.8 Regions defined by the input–output fuzzy reasoning.

TABLE A.1

Rules for Air-Conditioning Control

Rule 1 If temperature is cold, then stop motor—vent is downward.

Rule 2 If temperature is cool, then motor speed is slow—vent is downward.

Rule 3 If temperature is comfortable, then motor speed is medium—vent is horizontal.

Rule 4 If temperature is warm, then motor speed is fast—vent is upward.

Rule 5 If temperature is hot, then blast motor speed—vent is upward.

These five rules will be fired in accordance with the membership functions of the room temperature. Such membership functions give the truth-value for each rule and will weigh the average of the outputs, controlling continuously the motor speed from 0 to 100 rpm and the blade angle from −45° to +45°. Figure A.9 shows an example where the input temperature of 76.5°F fires rules 4 and 5, leading to the two outputs indicated in the figure.

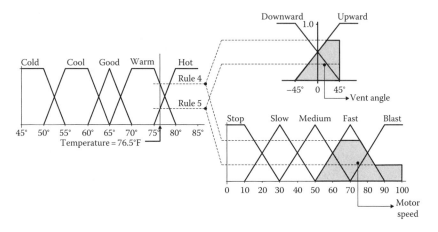

FIGURE A.9 Fuzzy inference where the input temperature fires two rules that relate the motor speed and vent angle for the air-conditioning system.

A.6 SUMMARY

Fuzzy systems have a bright future in consumer products, industrial and commercial systems, and decision-support systems. The term *fuzzy* refers to the ability to deal with imprecise or vague inputs. Instead of using complex mathematical equations, fuzzy logic uses linguistic descriptions to define the relationship between the input information and the output action. In engineering systems, fuzzy logic provides a convenient and user-friendly front end to develop control programs, helping designers to concentrate on the functional objectives, not on the mathematics. This appendix introduced a brief history of fuzzy logic, discussed the nature of fuzziness in the real-world problems, and showed how fuzzy operations are performed and how fuzzy rules can incorporate the underlying knowledge. The required steps to construct a fuzzy system were demonstrated in a case study of an air-conditioning system. Fuzzy logic is a very powerful tool that is pervading every field with successful implementations.

GLOSSARY

Defuzzification: The process of determining the best numerical value to represent a given fuzzy set.
Degree of membership (μ): A number between 0 and 1 that expresses the confidence that a given element belongs to a fuzzy set.
Fuzzification: The process of converting nonfuzzy input variables to fuzzy variables.
Fuzzy composition: A method of deriving fuzzy control output from given fuzzy control inputs.
Fuzzy implication: Same as fuzzy rule.
Fuzzy inference: Same as fuzzy composition.
Fuzzy model: Fuzzy rules and membership functions that describe the fuzzy relationship between the output and input of a controlled system.

Fuzzy rule: If–then rules relating input (conditions) fuzzy variables to output (actions) fuzzy variables.

Fuzzy rule base: A set of rules that define a fuzzy control or fuzzy model.

Fuzzy set (or fuzzy subset): A set consisting of elements having degrees of membership varying between 0 (nonmember) and 1 (full member). It is usually associated with linguistic values, such as small, medium, large, etc.

Height defuzzification method: A method of calculating a crisp output from a composed fuzzy value, by performing a weighted average of individual fuzzy subsets.

Membership function: A function that defines a fuzzy subset, by associating every element in the set with a number between 0 and 1.

Sup-min or max-min composition method: A composition (or inference) method for constructing the output membership function by using maximum and minimum principle.

Universe of discourse: The range of values associated with a fuzzy variable.

BIBLIOGRAPHY

Boole, G., *An Investigation of the Laws of Thought on Which are Founded the Mathematical Theories of Logic and Probabilities*, (1854), reprinted by Dover Publications, Inc., New York, 1958.

Brubaker, D.I., Fuzzy-logic basics: Intuitive rules replace complex math, EDN, 111–115, June 18, 1992.

Cox, E., The *Fuzzy Systems Handbook,* AP Professional, Boston, MA, 1994.

Kosko, B., *Fuzzy Thinking: The New Science of Fuzzy Logic*, Hyperion, New York, 1993.

Mamdani, E.H., Application of fuzzy algorithms for control of a simple dynamic plant, *Proc. Inst. Elect. Eng.*, United Kingdom, 121, 1585–1588, 1974.

McNeill, F.M. and Thro, E., *Fuzzy Logic: A Practical Approach*, AP Professional, Boston, MA, 1994.

Munakata, T. and Jani, Y., Fuzzy systems: An overview, *Commun. ACM*, 37, 68–76, March 1994.

Schwartz, D.G., Japanese advances in fuzzy systems and case-based reasoning, Technical report No. 91-111, Department of Computer Science, Florida State University, Tallahassee, FL, November 1991.

Simões, M.G. and Bose, B.K., Application of fuzzy logic in the estimation of power electronic waveforms, *IEEE/IAS Annual Meeting Conference Record*, Toronto, Canada, 1993, pp. 853–861.

Simões, M.G. and Bose, B.K., Fuzzy neural network based estimation of power electronic waveforms, *III Congresso Brasileiro de Eletrônica de Potência (COBEP'95)*, São Paulo, Brazil, December 1995, pp. 211–216.

Simões, M.G., Fuzzy logic and neural network based advanced control and estimation techniques in power electronics and AC drives, PhD dissertation, The University of Tennessee, Knoxville, TN, December 1995.

Sousa, G.C.D., Bose, B.*K., and Simões, M.G.,* A simulation-implementation methodology of a fuzzy logic based control system, *III Congresso Brasileiro de Eletrônica de Potência (COBEP'95)*, São Paulo, Brazil, December 1995, pp. 223–228.

Sousa, G.C.D., Bose, B.K., and Spiegel, R., Fuzzy logic based intelligent control of a variable speed cage machine wind generation system, *IEEE Power Electronics Specialists Conference,* Atlanta, GA, June 1995, pp. 389–395.

Zadeh, L.A., Fuzzy sets, *Inform. Contr.*, 8, 338–353, 1965.

Appendix B: C Statements for the Simulation of a Self-Excited Induction Generator

```c
/* The following program simulates the stand alone operation
of a self-excited induction generator (SEIG) */
/* wr1 is the speed of the prime mover. It is assumed to be
constant here but any function of time can */
/* be assigned to it based on load variations */

#include<stdio.h>
#include<math.h>
#include<conio.h>
#include<stdlib.h>


/* Generator1 parameters */

#define c1 180e-6              /* Self exciting capacitance, F */
#define rs1 0.262              /* Stator resistance, Ohms */
#define xs1 0.633              /* Stator reactance, Ohms */
#define rr1 0.447              /* Rotor resistance, Ohms */
#define xr1 1.47               /* Rotor reactance, Ohms */

#define tsim 3                 /* Simulation time, secs */
#define deltat 0.0001

double itrmax = 1+tsim/deltat;   /* Maximum number of
                                    iterations */
double pi = 22.0/7.0;
double wr1 = 377.1428;           /* Rotor speed, rad/sec */

void main()
{
    /* Initialization of variables */
    FILE *fp1;                 /* To save the required data */
    int ii;
    double ids1 = 0.0,iqs1 = 0.0,idr1 = 0.0,iqr1 = 0.0,vld1 =
    0.0,vlq1 = 0.0,ild1 = 0.0,ilq1 = 0.0;

    /* Initializing residual magnetism */
    double vds1 = 10.0,vqs1 = 10.0,vdr1 = 0.0,vqr1 = 0.0;
```

```
/* h-Runge-Kutta parameter, r1 and l1 are load parameters
initialized to approximate an open circuit */
    double im1,ls1,lr1,k1 = 0.0,m1 = 0.25,xm1,h = 0.0001,r1 =
    1000.0,l1 = 100.0,ti;

/* Runge-Kutta parameters */
    double a1,b1,c11,d1,e1,f1,g1,h1,a2,b2,c2,d2,e2,f2,g2,h2,a3,
    b3,c3,d3,e3,f3,g3,h3,a4,b4,c4,d4,e4,f4,g4,h4;

    /* Initializing functions for Runge-Kutta procedure */
    double fids(double,double,double,double,double,double,double,
    double,double,double);

    double fiqs(double,double,double,double,double,double,double,
    double,double,double);
    double fidr(double,double,double,double,double,double,double,
    double,double,double,double);

    double fiqr(double,double,double,double,double,double,double,
    double,double,double,double);

    double fvld(double,double);

    double fvlq(double,double);

    double fild(double,double,double,double);

    double filq(double,double,double,double);

    /* Instantaneous voltage and current values are stored in
    data.dat */
     fp1 = fopen("data.dat,""w");

    fflush(stdin);
    printf("Simulating an isolated induction generator\n");

    /* Running the induction generator for tsim seconds */

    for(ii = 0;ii<itrmax;ii = ii+1)
    {
        fflush(stdin);
          ti = ii*deltat;

          /* Making the residual magnetism zero after one
          iteration */
          if(ii>0)
          {
                  vds1 = 0.0;
                  vqs1 = 0.0;
```

```
              vdr1 = 0.0;
              vqr1 = 0.0;
          }
     xm1 = wr1*m1;                /* Mutual reactance, Ohms */
     ls1 = (xs1+xm1)/wr1;    /* Stator inductance, H */
     lr1 = (xr1+xm1)/wr1;    /* Rotor inductance, H0 */
     k1 = 1/((m1*m1)-(lr1*ls1));

     /* Using Runge-Kutta fourth order method to solve
     for D-Q parameters */

     a1 = fids(m1,ids1,iqs1,idr1,iqr1,vld1,vds1,vdr1,k1,
     lr1);
     b1 = fiqs(m1,ids1,iqs1,idr1,iqr1,vlq1,vqs1,vqr1,k1,
     lr1);
     c11 = fidr(m1,ids1,iqs1,idr1,iqr1,vld1,vds1,vdr1,k1,
     ls1,lr1);
     d1 = fiqr(m1,ids1,iqs1,idr1,iqr1,vlq1,vqs1,vqr1,k1,
     ls1,lr1);
     e1 = fvld(ids1,ild1);
     f1 = fvlq(iqs1,ilq1);
     g1 = fild(vld1,ild1,r1,l1);
     h1 = filq(vlq1,ilq1,r1,l1);

   a2 = fids(m1,(ids1+h*0.5*a1),(iqs1+h*0.5*b1),(idr1+
   h*0.5*c11), (iqr1+h*0.5*d1),(vld1+h*0.5*e1),vds1,vdr1,k1,
   lr1);

   b2 = fiqs(m1,(ids1+h*0.5*a1),(iqs1+h*0.5*b1),(idr1+
   h*0.5*c11), (iqr1+h*0.5*d1),(vlq1+h*0.5*f1),vqs1,vqr1,
   k1,lr1);

   c2 = fidr(m1,(ids1+h*0.5*a1),(iqs1+h*0.5*b1),(idr1+
   h*0.5*c11), (iqr1+h*0.5*d1),(vld1+h*0.5*e1),vds1,vdr1,
   k1,ls1,lr1);

   d2 = fiqr(m1,(ids1+h*0.5*a1),(iqs1+h*0.5*b1),(idr1+
   h*0.5*c11), (iqr1+h*0.5*d1),(vlq1+h*0.5*f1),vqs1,vqr1,
   k1,ls1,lr1);
     e2 = fvld((ids1+h*0.5*a1),(ild1+h*0.5*g1));
     f2 = fvlq((iqs1+h*0.5*b1),(ilq1+h*0.5*h1));
  g2 = fild((vld1+h*0.5*e1),(ild1+h*0.5*g1),r1,l1);
  h2 = filq((vlq1+h*0.5*f1),(ilq1+h*0.5*h1),r1,l1);

   a3 = fids(m1,(ids1+h*0.5*a2),(iqs1+h*0.5*b2),(idr1+
   h*0.5*c2), (iqr1+h*0.5*d2),(vld1+h*0.5*e2),vds1,vdr1,
   k1,lr1);

b3 = fiqs(m1,(ids1+h*0.5*a2),(iqs1+h*0.5*b2),(idr1+h*0.5*c2),
(iqr1+h*0.5*d2),(vlq1+h*0.5*f2),vqs1,vqr1,k1,lr1);
```

```
      c3 = fidr(m1,(ids1+h*0.5*a2),(iqs1+h*0.5*b2),(idr1+
      h*0.5*c2),(iqr1+h*0.5*d2),(vld1+h*0.5*e2),vds1,vdr1,k1,
      ls1,lr1);
      d3 = fiqr(m1,(ids1+h*0.5*a2),(iqs1+h*0.5*b2),(idr1+h*0.5
      *c2), (iqr1+h*0.5*d2),(vlq1+h*0.5*f2),vqs1,vqr1,k1,ls1,
      lr1);
         e3 = fvld((ids1+h*0.5*a2),(ild1+h*0.5*g2));
         f3 = fvlq((iqs1+h*0.5*b2),(ilq1+h*0.5*h2));
         g3 = fild((vld1+h*0.5*e2),(ild1+h*0.5*g2),r1,l1);
         h3 = filq((vlq1+h*0.5*f2),(ilq1+h*0.5*h2),r1,l1);

      a4 = fids(m1,(ids1+h*a3),(iqs1+h*b3),(idr1+h*c3),
      (iqr1+h*d3),(vld1+h*e3),vds1,vdr1,k1,lr1);

b4 = fiqs(m1,(ids1+h*a3),(iqs1+h*b3),(idr1+h*c3),(iqr1+h*d3),
(vlq1+h*f3),vqs1,vqr1,k1,lr1);

      c4 = fidr(m1,(ids1+h*a3),(iqs1+h*b3),(idr1+h*c3),(iqr1+
      h*d3),(vld1+h*e3),vds1,vdr1,k1,ls1,lr1);

      d4 = fiqr(m1,(ids1+h*a3),(iqs1+h*b3),(idr1+h*c3),(iqr1+
      h*d3),(vlq1+h*f3),vqs1,vqr1,k1,ls1,lr1);
         e4 = fvld((ids1+h*a3),(ild1+h*g3));
         f4 = fvlq((iqs1+h*b3),(ilq1+h*h3));
         g4 = fild((vld1+h*e3),(ild1+h*g3),r1,l1);
         h4 = filq((vlq1+h*f3),(ilq1+h*h3),r1,l1);

      /* Instantaneous D-Q parameters obtained after solving
      the differential equations */
      ids1 = ids1+(a1+2*a2+2*a3+a4)*(h/6.0);
      iqs1 = iqs1+(b1+2*b2+2*b3+b4)*(h/6.0);
      idr1 = idr1+(c11+2*c2+2*c3+c4)*(h/6.0);
      iqr1 = iqr1+(d1+2*d2+2*d3+d4)*(h/6.0);
      vld1 = vld1+(e1+2*e2+2*e3+e4)*(h/6.0);
      vlq1 = vlq1+(f1+2*f2+2*f3+f4)*(h/6.0);
      ild1 = ild1+(g1+2*g2+2*g3+g4)*(h/6.0);
      ilq1 = ilq1+(h1+2*h2+2*h3+h4)*(h/6.0);

      /* Calculating the magnetization current */
      im1 = sqrt((((ids1+idr1)*(ids1+idr1))+((iqs1+iqr1)*(iqs1+
      iqr1))));

      /* Calculating the mutual inductance
      /* This relation between the mutual inductance and
      magnetization current
      /* is obtained from experimental data */
      m1 = (0.0423*exp(-0.0035*im1*im1))+0.0236;
```

```
        /* Switching the load at ti seconds */
        if(ti>2)
    {
            r1 = 20;
            l1 = 20e-3;
        }

        /* Storing terminal voltage and current in a file */
        fprintf(fp1,"\n%5.4lf\t%5.4lf,"vlq1,iqs1);
    }
    printf("simulation complete");
        fclose(fp1);
    fflush(stdin);
    getch();
}


/* Functions to evaluate expressions for D-Q parameters */

double fids(double m1,double ids1,double iqs1,double
idr1,double iqr1,double vld1,double vds1,double vdr1,double
k1,double lr1)
{
    double dids;
    dids = k1*((rs1*lr1*ids1)-(wr1*m1*m1*iqs1)-(rr1*m1*idr1)-
    (wr1*m1*lr1*iqr1)+(lr1*vld1)-(lr1*vds1)+(m1*vdr1));
    return dids;
}


double fiqs(double m1,double ids1,double iqs1,double idr1,double
iqr1,double vlq1,double vqs1,double vqr1,double k1,double lr1)
{
    double diqs;
    diqs = k1*((wr1*m1*m1*ids1)+(rs1*lr1*iqs1)+(wr1*m1*lr1*
    idr1)-(m1*rr1*iqr1)+(lr1*vlq1)-(lr1*vqs1)+(m1*vqr1));
    return diqs;
}

double fidr(double m1,double ids1,double iqs1,double idr1,double
iqr1,double vld1,double vds1,double vdr1,double k1,double
ls1,double lr1)
{
    double didr;
    didr = k1*(-(rs1*m1*ids1)+(wr1*m1*ls1*iqs1)+(rr1*ls1*idr1)+
    (wr1*ls1*lr1*iqr1)-(m1*vld1)+(m1*vds1)-(ls1*vdr1));
    return didr;
}
```

```c
double fiqr(double m1,double ids1,double iqs1,double
idr1,double iqr1,double vlq1,double vqs1,double vqr1,double
k1,double ls1,double lr1)
{
    double diqr;
    diqr = k1*(-(wr1*m1*ls1*ids1)-(rs1*m1*iqs1)-(wr1*ls1*lr1*
    idr1)+(rr1*ls1*iqr1)-(m1*vlq1)+(m1*vqs1)-(ls1*vqr1));
    return diqr;
}

double fvld(double ids1,double ild1)
{
    double dvld;
    dvld = (ids1/c1)-(ild1/c1);
    return dvld;
}

double fvlq(double iqs1,double ilq1)
{
    double dvlq;
    dvlq = (iqs1/c1)-(ilq1/c1);
    return dvlq;
}

double fild(double vld1,double ild1,double r1,double l1)
{
    double dild;
    dild = (vld1/l1)-r1*(ild1/l1);
    return dild;
}

double filq(double vlq1,double ilq1,double r1,double l1)
{
    double dilq;
    dilq = (vlq1/l1)-r1*(ilq1/l1);
    return dilq;
}
```

Appendix C: Pascal Statements for the Simulation of a Self-Excited Induction Generator

Pascal program for simulation of the self-excited induction generator;
{Post-Graduation Program in Electrical Engineering - Federal University of Santa Maria
Solution for the state equations of the transient model for the self-excited induction generator using a 4th order Runge Kutta method.}

```
USES CRT,PRINTER;

type
register1 = record
    diD,diQ,diqr,didr,dilD,dilQ :real
end;
register2 = record
    dVcD,dVcQ,va : real;
end;

var
    arq1 : text ;
    arq2 : text ;
    reg1 : register1;
    reg2 : register2;
    X0,Y0,h,x, teste : real ;
    k1,k2,k3,k4 : real ;
    q1,q2,q3,q4 : real ;
    p1,p2,p3,p4 : real ;
    r1,r2,r3,r4 : real ;
    s1,s2,s3,s4 : real ;
    t1,t2,t3,t4 : real ;
    u1,u2,u3,u4 : real ;
    z1,z2,z3,z4 : real ;
    VD,Vdr,VQ,Vqr : real ;
    Pm,Rs,Rr,Ls,Lr,Xs,Xr,M,w,ws,Xm,Im,R,L,C : real ;
    dVcD,iD0,iQ0,iqr0,idr0,VcD0,VcQ0,ilD0,ilQ0 : real ;
```

```pascal
A1,A2,A3,t0,tf,t,difiD,dif,ma : real ;
difference,y,npoints,TQ,theta,modulev : real ;
k,j : integer ;
line : integer ;
step : real ;
a : char ;

FUNCTION f1 (VD,Rs,diD,dVcD,Vdr,w,M,diQ,Lr,diqr,Rr,didr,A1,A2 :
real) : real ;
  begin
  f1: = 0;
  f1 : = A1*(VD-Rs*diD-dVcD)+A2*(Vdr-w*M*diQ-w*Lr*diqr-Rr*didr) ;
  end ;
FUNCTION f2 (VQ,Rs,diQ,dVcQ,Vqr,w,M,diD,Rr,diqr,Lr,didr,A1,A2 :
real) : real ;
  begin
  f2: = 0;
  f2 : = A1*(VQ-Rs*diQ-dVcQ)+A2*(Vqr+w*M*diD-Rr*diqr+w*Lr*didr) ;
  end ;
FUNCTION f3 (VQ,Rs,diQ,dVcQ,Vqr,w,M,diD,Rr,diqr,Lr,didr,A2,A3 :
real) : real ;
  begin
  f3: = 0;
  f3 : = A2*(VQ-Rs*diQ-dVcQ)+A3*(Vqr+w*M*diD-Rr*diqr+w*Lr*didr) ;
  end ;
FUNCTION f4 (VD,Rs,diD,dVcD,Vdr,w,M,diQ,Lr,diqr,Rr,didr,A2,A3 :
real) : real ;
  begin
  f4: = 0;
  f4 : = A2*(VD-Rs*diD-dVcD)+A3*(Vdr-w*M*diQ-w*Lr*diqr-Rr*didr) ;
  end ;
FUNCTION f5 (diD,dilD,C : real) : real ;
  begin
  f5: = 0;
  f5 : = (diD/C)-(dilD/C) ;
  end ;
FUNCTION f6 (diQ,dilQ,C : real) : real ;
  begin
  f6: = 0;
  f6 : = (diQ/C)-(dilQ/C) ;
  end ;
FUNCTION f7 (dVcD,R,dilD,L : real) : real ;
  begin
  f7: = 0;
  f7 : = (dVcD/L)-(R*dilD/L) ;
  end ;
FUNCTION f8 (dVcQ,R,dilQ,L : real) : real ;
  begin
  f8: = 0;
  f8 : = (dVcQ/L)-(R*dilQ/L) ;
  end ;
```

```
begin
  assign(arq1,'irleber.m');
  settextbuf(arq1,reg1.diD);
  settextbuf(arq1,reg1.diQ);
  settextbuf(arq1,reg1.diqr);
  settextbuf(arq1,reg1.didr);
  settextbuf(arq2,reg2.dVcD);
  settextbuf(arq2,reg2.dVcQ);
  settextbuf(arq1,reg1.dilD);
  settextbuf(arq1,reg1.dilQ);
  rewrite(arq1);
  clrscr ;
  writeln('Input the initial values:');
  writeln;
  write('Enter with the calculation step (h):');
  readln(h) ;
  write('Enter with the initial time:');
  readln(t0) ;
  write('Enter with the final time:');
  readln(tf) ;

  R: = 1000;
  L: = 100;
  C: = 160e-6;
  Rs: = 0.575;
  Xs: = 1.3074;
  Rr: = 0.0404;
  Xr: = 1.3074;
  Pm: = 4;
  w: = 2*pi*60;
  id0: = 0;
  iQ0: = 0;
  idr0: = 0;
  iqr0: = 0;
  VcD0: = 0;
  VcQ0: = 0;
  ilQ0: = 0;
  k: = 0;
  t : = t0 ;
  dif: = 1;
  Ma: = 0.0659;
  M: = 0.0659;
  im: = 0;
  reg1.diD : = iD0 ;
  reg1.diQ : = iQ0 ;
  reg1.diqr : = iqr0 ;
  reg1.didr : = idr0 ;
  reg2.dVcD : = VcD0 ;
  reg2.dVcQ : = 0.00001 ;
  reg1.dilD : = ilD0 ;
  reg1.dilQ : = ilQ0 ;
```

```
while t< = tf do
   begin
   k: = k+1;
   if k = 1 then
   begin
   modulev: = sqrt(sqr(reg2.dVcD)+sqr(reg2.dVcQ));
   theta: = arctan(abs(reg2.dVcD/reg2.dVcQ));
   reg2.va: = (modulev/sqrt(2))*sin((w*t+theta*pi/180));
   writeln(arq1,reg1.did/sqrt(2));
   flush(arq1);
   end;
   if k = 20 then
   k: = 0;
   clrscr;
   gotoxy(10,10);
   writeln('tempo',t);


   k1: = 0 ;
   k2: = 0 ;
   k3: = 0 ;
   k4: = 0 ;
   q1: = 0 ;
   q2: = 0 ;
   q3: = 0 ;
   q4: = 0 ;
   p1: = 0 ;
   p2: = 0 ;
   p3: = 0 ;
   p4: = 0 ;
   r1: = 0 ;
   r2: = 0 ;
   r3: = 0 ;
   r4: = 0 ;
   s1: = 0 ;
   s2: = 0 ;
   s3: = 0 ;
   s4: = 0 ;
   t1: = 0 ;
   t2: = 0 ;
   t3: = 0 ;
   t4: = 0 ;
   u1: = 0 ;
   u2: = 0 ;
   u3: = 0 ;
   u4: = 0 ;
   z1: = 0 ;
   z2: = 0 ;
   z3: = 0 ;
   z4: = 0 ;
```

```
if t = t0 then
   VD: = 5
else
   VD: = 0;
if t = t0 then
   VQ: = 5
else
   VQ: = 0;
if t = t0 then
   Vdr: = 0
else
   Vdr: = 0 ;
if t = t0 then
   Vqr: = 0
else
   Vqr: = 0 ;

im: = (sqrt(sqr(reg1.diD+reg1.didr)+sqr(reg1.diQ+reg1.diqr))) ;
writeln('dVcD',reg2.dVcD) ;
writeln('dvcQ',reg2.dVcQ) ;
M: = (0.0423*exp(-0.0035*(sqr (Im)))+0.0236);
TQ: = -1.5*Pm/2*(reg1.diQ*reg1.didr-reg1.diD*reg1.diqr)*M ;
Xm: = M*w;
Ls: = (Xs+Xm)/w;
Lr: = (Xr+Xm)/w;
A1: = -Lr/(sqr(M)-Lr*Ls);
A2: = M/(sqr(M)-Lr*Ls) ;
A3: = -Ls/(sqr(M)-Lr*Ls);
dif : = abs(Ma-M);
Ma: = M;

k1: = h*f1(VD,Rs,reg1.diD,reg2.dVcD,Vdr,w,M,reg1.diQ,Lr,reg1.
diqr,Rr,
reg1.didr,A1,A2);
q1: = h*f2(VQ,Rs,reg1.diQ,reg2.dVcQ,Vqr,w,M,reg1.diD,Rr,reg1.
diqr,Lr,
reg1.didr,A1,A2);
p1: = h*f3(VQ,Rs,reg1.diQ,reg2.dVcQ,Vqr,w,M,reg1.diD,Rr,reg1.
diqr,Lr,
reg1.didr,A2,A3);
r1: = h*f4(VD,Rs,reg1.diD,reg2.dVcD,Vdr,w,M,reg1.diQ,Lr,reg1.
diqr,Rr,
reg1.didr,A2,A3);
s1: = h*f5(reg1.diD,reg1.dilD,C);
t1: = h*f6(reg1.diQ,reg1.dilQ,C);
u1: = h*f7(reg2.dVcD,R,reg1.dilD,L);
z1: = h*f8(reg2.dVcQ,R,reg1.dilQ,L);

k2: = h*f1(VD,Rs,reg1.diD+(k1/2),reg2.dVcD+(s1/2),Vdr,w,M,reg1.
diQ+(q1/2),Lr,reg1.diqr+(p1/2),Rr,reg1.didr+(r1/2),A1,A2);
```

```
q2: = h*f2(VQ,Rs,reg1.diQ+(q1/2),reg2.dVcQ+(t1/2),Vqr,w,M,reg1.
diD+(k1/2),Rr,reg1.diqr+(p1/2),Lr,reg1.didr+(r1/2),A1,A2);
p2: = h*f3(VQ,Rs,reg1.diQ+(q1/2),reg2.dVcQ+(t1/2),Vqr,w,M,reg1.
diD+(k1/2),Rr,reg1.diqr+(p1/2),Lr,reg1.didr+(r1/2),A2,A3);
r2: = h*f4(VD,Rs,reg1.diD+(k1/2),reg2.dVcD+(s1/2),Vdr,w,M,reg1.
diQ+(q1/2),Lr,reg1.diqr+(p1/2),Rr,reg1.didr+(r1/2),A2,A3);
s2: = h*f5(reg1.diD+(k1/2),reg1.dilD+(u1/2),C);
t2: = h*f6(reg1.diQ+(q1/2),reg1.dilQ+(z1/2),C);
u2: = h*f7(reg2.dVcD+(s1/2),R,reg1.dilD+(u1/2),L);
z2: = h*f8(reg2.dVcQ+(t1/2),R,reg1.dilQ+(z1/2),L);


k3: = h*f1(VD,Rs,reg1.diD+(k2/2),reg2.dVcD+(s2/2),Vdr,w,M,reg1.
diQ+(q2/2),Lr,reg1.diqr+(p2/2),Rr,reg1.didr+(r2/2),A1,A2);
q3: = h*f2(VQ,Rs,reg1.diQ+(q2/2),reg2.dVcQ+(t2/2),Vqr,w,M,reg1.
diD+(k2/2),Rr,reg1.diqr+(p2/2),Lr,reg1.didr+(r2/2),A1,A2);
p3: = h*f3(VQ,Rs,reg1.diQ+(q2/2),reg2.dVcQ+(t2/2),Vqr,w,M,reg1.
diD+(k2/2),Rr,reg1.diqr+(p2/2),Lr,reg1.didr+(r2/2),A2,A3);
r3: = h*f4(VD,Rs,reg1.diD+(k2/2),reg2.dVcD+(s2/2),Vdr,w,M,reg1.
diQ+(q2/2),Lr,reg1.diqr+(p2/2),Rr,reg1.didr+(r2/2),A2,A3);
s3: = h*f5(reg1.diD+(k2/2),reg1.dilD+(u2/2),C);
t3: = h*f6(reg1.diQ+(q2/2),reg1.dilQ+(z2/2),C);
u3: = h*f7(reg2.dVcD+(s2/2),R,reg1.dilD+(u2/2),L);
z3: = h*f8(reg2.dVcQ+(t2/2),R,reg1.dilQ+(z2/2),L);


k4: = h*f1(VD,Rs,reg1.diD+k3,reg2.dVcD+s3,Vdr,w,M,reg1.diQ+q3,Lr,
reg1.diqr+p3,Rr,reg1.didr+r3,A1,A2);
q4: = h*f2(VQ,Rs,reg1.diQ+q3,reg2.dVcQ+t3,Vqr,w,M,reg1.diD+k3,Rr,
reg1.diqr+p3,Lr,reg1.didr+r3,A1,A2);
p4: = h*f3(VQ,Rs,reg1.diQ+q3,reg2.dVcQ+t3,Vqr,w,M,reg1.diD+k3,Rr,
reg1.diqr+p3,Lr,reg1.didr+r3,A2,A3);
r4: = h*f4(VD,Rs,reg1.diD+k3,reg2.dVcD+s3,Vdr,w,M,reg1.diQ+q3,Lr,
reg1.diqr+p3,Rr,reg1.didr+r3,A2,A3);
s4: = h*f5(reg1.diD+k3,reg1.dilD+u3,C);
t4: = h*f6(reg1.diQ+q3,reg1.dilQ+z3,C);
u4: = h*f7(reg2.dVcD+s3,R,reg1.dilD+u3,L);
z4: = h*f8(reg2.dVcQ+t3,R,reg1.dilQ+z3,L);


   reg1.diD: = reg1.diD+((k1+2*k2+2*k3+k4)/6) ;
   reg1.diQ: = reg1.diQ+((q1+2*q2+2*q3+q4)/6) ;
   reg1.diqr: = reg1.diqr+((p1+2*p2+2*p3+p4)/6) ;
   reg1.didr: = reg1.didr+((r1+2*r2+2*r3+r4)/6) ;
   reg2.dVcD: = reg2.dVcD+((s1+2*s2+2*s3+s4)/6) ;
   reg2.dVcQ: = reg2.dVcQ+((t1+2*t2+2*t3+t4)/6) ;
   reg1.dIlD: = reg1.dIlD+((u1+2*u2+2*u3+u4)/6) ;
   reg1.dIlQ: = reg1.dIlQ+((z1+2*z2+2*z3+z4)/6) ;
   t: = t+h ;
   dif: = 1 ;
   if t>15 then
     begin
     R: = 10;
```

```
   L: = 0.009;
   end;
 end;
 close(arq1);
 clrscr ;
 step : = t0 ;
 writeln ;
 gotoxy(25,2) ;
 writeln('                     => RESULTS <=') ;
 writeln ;
 writeln('INTERVAL : ',h) ;

 line : = 0 ;
 for k : = 0 to (1000) do
 begin
 gotoxy(5,7+line) ;
 writeln('INTERVAL : ',step) ;
 gotoxy(40,7+line) ;
 writeln('dVcD: ',reg2.dVcD) ;
 step : = step+h ;
 line : = line+1 ;
 if line = 14 then
    begin
    gotoxy(5,23) ;
    write('Press <ENTER> to continue…') ;
    a : = readkey ;
    clrscr ;
    line : = 0 ;
    end ;
   end ;
END.
```

Appendix D: Power Tracking Curve-Based Algorithm for Wind Energy Systems

Dimitri Torregrossa, Abdellatif Miraoui,
Maurizio Cirrincione, and Marcello Pucci

There exist many optimization control algorithms for tracking the angular speed of induction generators in order to extract the maximum (peak) power from the wind. It is possible to improve the overall efficiency of energy extraction by 4%–5% compared with a nonoptimized fixed speed control system. Since most induction generators for these applications are larger than 50 kW, the gain in power will be very significant.

Several techniques are available for controlling the speed of induction generators for optimized power extraction:

- Mathematical modeling of wind turbines
- Measurement or estimation of wind speed
- Hill climbing or fuzzy search algorithm
- Calculation of the optimal coupling reference speed

In addition to maximizing the turbine energy capture, a tracking algorithm allows the turbine and the generator to be consistently sized. In this appendix, the simulated system has a double converter of the type described in Figure 7.34, where a converter connected to the induction generator runs a direct vector control (DVC) algorithm and another converter injects power into the grid. The key to this optimization technique is the computation of the optimal induction generator torque, for many turbine speeds, to maximize the captured energy, as explained in the following.[1–4]

Equations D.1 and D.2 can give the turbine power P_W:

$$P_W = \frac{1}{2}\rho C_p(\lambda)\pi R^2 v^3 \tag{D.1}$$

where
    R is the turbine radius
    v is the wind velocity
    ρ is the air density
    $C_p(\lambda)$ is the power coefficient

λ is the tip speed ratio (TSR) defined as

$$\lambda = \frac{\omega_T R}{v} \tag{D.2}$$

where ω_T is the turbine angular speed.

The turbine torque is therefore

$$T_T = \frac{P_T}{\omega_T} \tag{D.3}$$

Solving for ω_T from Equation D.2, then

$$\omega_T = \frac{\lambda R}{v} \tag{D.4}$$

Figure D.1a and b indicates both torque coefficient $C_T(\lambda)$ and power coefficient $C_P(\lambda)$, where

$$C_T(\lambda) = \frac{C_p(\lambda)}{\lambda}$$

and the torque is given by

$$T_T = \frac{1}{2}\rho\pi R^3 C_T(\lambda) v^2 \tag{D.5}$$

obtained by using Equations D.1 and D.4 in Equation D.3. It should be noted that the optimal TSR λ_{opt} for which the extracted power is maximized does not give the maximum torque. This maximum power torque is the reference torque, which will be given for the DVC of the machine.

In order to calculate this reference torque from Equation D.2, the wind velocity is computed by

$$v = \frac{\omega_T R}{\lambda} \tag{D.6}$$

Replacing this value into Equation D.5 yields

$$T_T = \frac{1}{2}\rho\pi R^3 C_T(\lambda)\frac{\omega_T^2}{\lambda^2} \tag{D.7}$$

If the turbine is forced to work at λ_{opt} for which $C_{Topt} = C_T(\lambda_{opt})$, therefore

$$C_{Topt} = \frac{C_{p\max}}{\lambda_{opt}} \tag{D.8}$$

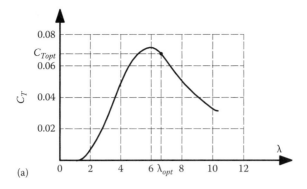

(a)

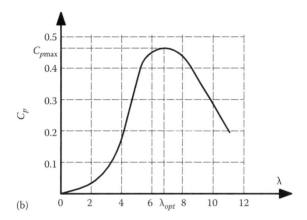

(b)

FIGURE D.1   (a) Torque coefficient $C_T(\lambda)$ and (b) power coefficient $C_p(\lambda)$.

and

$$T_{Topt}\left(\omega_T\right) = \frac{1}{2}\rho\pi R^5 C_{ToptT}\frac{\omega_T^2}{\lambda_{opt}^2} \tag{D.9}$$

and the reference generator torque is given by

$$T_G' = T_{ToptT}\left(\omega_T\right) \tag{D.10}$$

Figure D.2 shows in dashed line the graph of Equation D.10 and the way the optimization technique works. In this figure, consider also the graphs of Equation D.5 of the turbine torque for different speed velocities. This figure also shows how the angular speed of the turbine converges to the point of maximum power extraction following the changes in wind velocity if the generator torque follows the graph of Equation D.10. Suppose that v_1 is the initial wind velocity and that ω_A is the initial value of the angular speed, then the turbine torque is greater than that of the generator and the system accelerates up to the point ω_B where the turbine torque equals the generator torque in Equation D.11. The reverse is true if the initial angular speed ω_C and wind velocity v_1

correspond to a turbine torque less than that of the generator; the generator decelerates toward the speed ω_B of the maximum extracted power where the two torques are equal. If the wind velocity changes, for example, it becomes v_2 with a resulting turbine torque higher than that of the generator, and the system reacts by accelerating until reaching the new angular speed ω_D and equal torque. Of course, if a lower power is required, then the reference generator torque is increased so that the working point is pushed to the left with a resulting smaller angular speed and, obviously, a decreased efficiency.

Figure D.3 shows the block diagram of a DVC scheme; there is a gearbox for the torque/speed matching. The mechanical equation is given by

$$\frac{d}{dt}\omega_T = \frac{T_T - T_G'}{J_T + J_G'} \tag{D.11}$$

where

T_G' is the reflected generator torque corresponding to the turbine angular speed ω_T
J_T is the turbine inertia
J_G' is the reflected generator inertia corresponding to the turbine angular speed ω_T,
    which are based on the following relationships:

$$\omega_G = \omega_T n \tag{D.12}$$

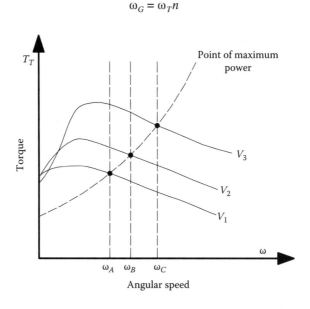

FIGURE D.2 Locus of power operating point in respect to the wind speed.

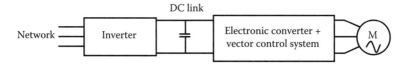

FIGURE D.3 General scheme of the induction generator for wind power.

$$T_G' = T_G n \tag{D.13}$$

$$J_G' = J_G n^2 \tag{D.14}$$

where
  ω_G is the generator angular speed
  n is the gearbox multiplication ratio

Figure D.3 shows the general scheme of the induction generation system driven by the wind turbine and based on a back-to-back power converter structure.

Figure D.4 shows the block diagram of the wind generation system, as implemented in MATLAB®/Simulink for the numerical simulations. It is composed of four main blocks:

1. The DVC subsystem to implement the field-oriented control of the induction generator using the data presented in Table D.1
2. The H-Bridge 1 block to simulate the three-phase voltage source inverter supplying the induction generator, including its pulse width modulation (PWM) driving technique
3. The induction machine block to implement the classic space-vector dynamic model of the induction machine
4. The H-Bridge 2 block to simulate the three-phase voltage source inverter connecting the dc link to the grid, including its control and PWM driving technique
5. The turbine torque block to compute the torque of the turbine using the data presented in Table D.2
6. The power injected into the grid block computing the generated active power
7. The power turbine and λ block to compute the characteristic values of the turbine
8. The torque reference block to compute the reference torque for the DCV to work at the optimal point of maximum extracted power

Figure D.5 shows the expansion of the DVC block. The phase currents of the induction machine are acquired and low-pass filtered by Butterworth filters before sampling and acquisition. The three-phase currents are then transformed into the two-phase system in the stationary reference frame (block 2ph-3ph). Afterward, the stator currents are transformed into the rotor flux-oriented reference frame (block sin&cos) based on the knowledge of the rotor flux angle. The DVC rotor controller block implements the torque and current control of the machine based on the reference and estimated torque of the machine and the stator current components in the flux-oriented reference frame. The outputs of the DVC controller block are the direct and quadrature components of the stator voltage in the flux-oriented reference frame. These voltages are first transformed into the two-phase stationary reference frame (block rotation 1) and then in the three-phase reference frame (block 2ph-3ph).

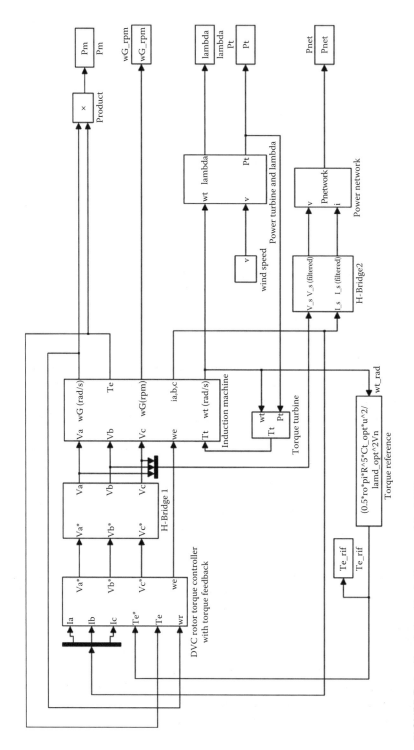

FIGURE D.4 Simulation block diagram of the DVC.

TABLE D.1
Induction Machine Parameters

Stator resistance	0.5814 Ω
Rotor resistance	0.4165 Ω
Stator leakage inductance	4.15 mH
Rotor leakage inductance	3.48 mH
Three-phase magnetizing inductance	82.23 mH
Pole pairs	2
Inertia	0.02 kg m²

TABLE D.2
Turbine Parameters

Inertia	10 kg m²
Turbine radius	2.5 m
Optimal	7
Optimal C_t	0.055
Maximum C_p	0.45
Gear ratio	4.86
Air density	1.26 kg/m³

Figure D.6 shows the implementation of the rotor flux estimator block. The rotor flux amplitude and angle are estimated based on the rotor equations of the induction machine in the stationary reference frame. Finally, the rotor flux speed is estimated based on the rotor flux in direct and quadrature components in the stationary reference frame.

Figure D.7 shows the implementation of a classic carrier-based PWM technique. The output phase voltages of the inverter are computed by comparing the reference values with a triangular carrier. The dc-link voltage is assumed constant here.

To show the effectiveness of this algorithm, the wind speed profile shown in Figure D.8 has been given for three constant wind speeds of 3 m/s, then 5 m/s, and then back to 3 m/s, with a linear variation between them.

The first simulation results have been obtained by using the earlier-described scheme with the optimal power tracking described in Equation D.9 where λ has been set to the optimal value of 7, to which correspond the optimal C_t = 0.055 and the maximum C_p = 0.45.

Figure D.9a shows the time waveform of λ during this test. It can be observed that after an initial transient λ, it converges to the optimal value of 7 independently from the wind speed. Figure D.9b shows the torque generated by the induction machine during the same test. For each wind speed, there exists a steady-state torque corresponding to the maximum power. Figure D.9c shows the corresponding waveform of the generator speed during the same test. By selecting different reference torques for each wind speed, the optimal power-tracking algorithm makes the machine work at different speed values for each wind speed.

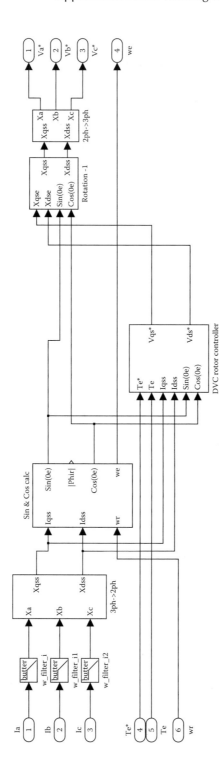

FIGURE D.5   Simulation block diagram of the expanded DVC.

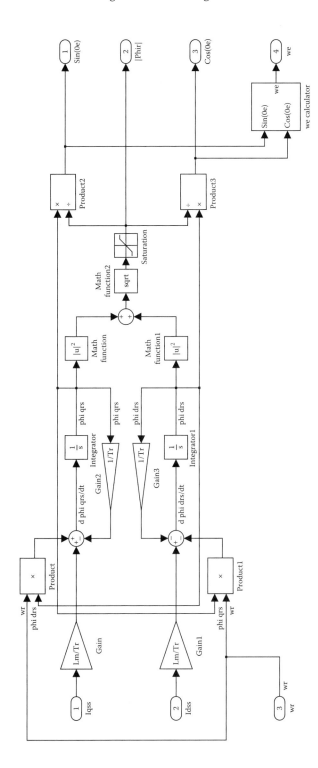

FIGURE D.6 Flux alignment calculation.

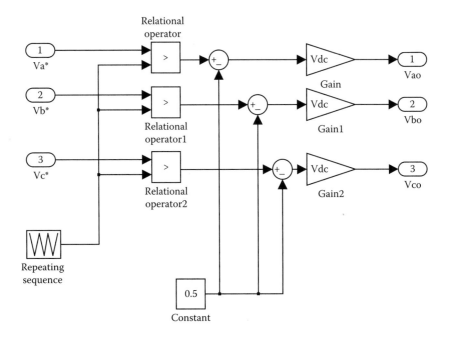

FIGURE D.7   Sinusoidal pulse width modulator.

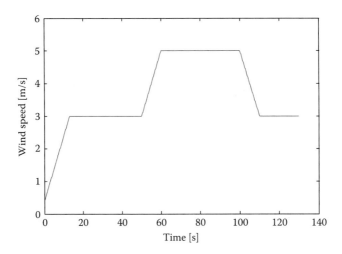

FIGURE D.8   Profile of wind speed.

Figure D.9d shows the power generated by the system and injected into the grid at the different wind speeds. It should be noticed that each wind speed always works in optimal power tracking conditions; obviously, the higher the wind speed, the greater the generated power.

In the following, it will be shown that, under the same test, the generated power diminishes for values of λ different from the optimal one.

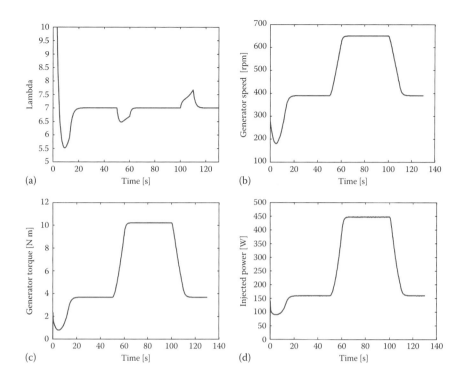

FIGURE D.9   (a) Waveform of λ; (b) generator torque; (c) generator speed; (d) generated power.

Figure D.10 shows the power generated by the system and injected into the grid at the different wind speeds as a result of two simulations made for different values of λ (set to 5 and 8 m/s). It can be seen that for both cases the power injected into the grid is less than the one obtained with the optimal power-tracking algorithm with the optimal value of λ. While for λ = 8 the generated power is at its maximum

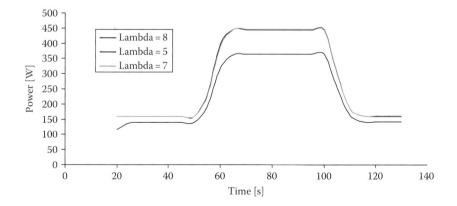

FIGURE D.10   Power injected into the grid for the optimal λ and other values of λ.

TABLE D.3
Power Reduction (%)

Wind Speed (m/s)	= 5	= 8
3	12.20	0.70
5	24.50	0.32

obtained for $\lambda = 7$, with $\lambda = 5$, there is a strong reduction of the generated power. This is explainable by the curvature of C_p around its maximum point.

Table D.3 shows the percentage reduction of power injected into the grid for values of λ equal to 5 and 8 in respect to the maximum power extracted with $\lambda = 7$.

REFERENCES

1. Leidhold, R., Garcia, G., and Valla, M.I., Maximum efficiency control for variable speed wind-driven generators with speed and power limits, *Proceedings of the 28th Annual International Conference of the IEEE Industrial Electronics Society*, Sevilla, Spain, November 2002, pp. 157–162.
2. Torregrossa, D., Cirrincione, M., and Pucci, M., *Vector Control of Eolic Generators with Asynchronous Machine*, Final report for the Electrical Engineering Diploma of the University of Palermo, Palermo, Italy, Chapters 6–7, November 2004.
3. Grantham, C., Steady-state and transient analysis of self-excited induction generators, *IEE Proc.*, 136(2), 61–68, March 1989.
4. Doxey, B.C., Theory and application of the capacitor-excited induction generator, *The Engineer Magazine*, pp. 893–897, November 1963.

Index